Copernicus, Darwin, & Freud

Copernicus, Darwin, & Freud: Revolutions in the History and Philosophy of Science

Friedel Weinert

A John Wiley & Sons, Ltd., Publication

This edition first published 2009
© 2009 by Friedel Weinert

Blackwell Publishing was acquired by John Wiley & Sons in February 2007. Blackwell's publishing program has been merged with Wiley's global Scientific, Technical, and Medical business to form Wiley-Blackwell.

Registered Office
John Wiley & Sons Ltd, The Atrium, Southern Gate, Chichester, West Sussex, PO19 8SQ, United Kingdom

Editorial Offices
350 Main Street, Malden, MA 02148-5020, USA
9600 Garsington Road, Oxford, OX4 2DQ, UK
The Atrium, Southern Gate, Chichester, West Sussex, PO19 8SQ, UK

For details of our global editorial offices, for customer services, and for information about how to apply for permission to reuse the copyright material in this book please see our website at www.wiley.com/wiley-blackwell.

The right of Friedel Weinert to be identified as the author of this work has been asserted in accordance with the UK Copyright, Designs, and Patents Act 1988.

Library of Congress Cataloging-in-Publication Data

Weinert, Friedel, 1950–
Copernicus, Darwin, & Freud : revolutions in the history and philosophy of science / Friedel Weinert.
     p. cm.
  Includes bibliographical references and index.
     ISBN 978-1-4051-8184-6 (hardcover : alk. paper) – ISBN 978-1-4051-8183-9 (pbk. : alk. paper)
  1. Science–Philosophy. 2. Copernicus, Nicolaus, 1473–1543. 3. Darwin, Charles, 1809–1882.
  4. Freud, Sigmund, 1856–1939. I. Title.

  Q175.W49 2009
  501–dc22
                                                2008031751

A catalogue record for this title is available from the British Library.

Set in 10/13pt Galliard
by SPi Publisher Services, Pondicherry, India
Printed and bound in Singapore
by Utopia Press Pte Ltd

1   2009

Contents

Contents

Preface

The present volume originates from seminars the author first gave at Victoria University in Wellington (New Zealand) and later at the University of Bradford (UK). The idea for the seminars and the book goes back to Freud's claim to have completed the Copernican revolution. The seminars and the book were conceived in the spirit of the author's conviction that a tight connection exists between scientific and philosophical ideas.

The completion of the manuscript was facilitated by the award of a Visiting Fellowship in the Department of Philosophy, Logic and Scientific Method at the London School of Economics (June to December 2006). I would like to thank members of the Department for their hospitality. I owe particular thanks to Stephan Hartmann for encouraging me to apply and for acting as my referee for research grant applications. I gratefully acknowledge the financial help from the School of International and Social Studies at the University of Bradford. I also benefited from a Small Research Grant, awarded by the British Academy, which made my LSE visit financially viable. Robert Nola of the University of Auckland (NZ) read the first draft of the manuscript and gave me valuable advice on how to improve it. I hereby express my gratitude to him both for his encouragement and for acting as my referee in research grant applications. I am also grateful to several anonymous referees who provided useful comments on the manuscript. Finally, I owe thanks to Nick Bellorini, Senior Commissioning Editor at Blackwell, for supporting this project. I would also like to thank Annette Abel, my copy-editor, for her careful reading of the manuscript and for stylistic improvements.

I hope the reader will enjoy reading this book as much as I enjoyed writing it.

Friedel Weinert
University of Bradford

Acknowledgments

The author and publisher acknowledge permission to reproduce the following illustrations:

Aristotle (354–322 BC). *Source*: Deutsches Museum (German Museum, Munich). *Photo*: Foto Deutsches Museum.

Claudius Ptolemy (AD 100–75). *Source*: Wikimedia.

Nicolaus Copernicus (1473–1543). *Source*: Deutsches Museum (German Museum, Munich). *Photo*: Foto Deutsches Museum.

Tycho Brahe (1546–1601) *Source*: *Nature* **15** (1876–7), p. 406

Johannes Kepler (1571–1630). *Source*: Deutsches Museum (German Museum, Munich). *Photo*: Foto Deutsches Museum.

Galileo Galilei (1564–1642). *Source*: Wikimedia.

Isaac Newton (1642–1727). *Source*: Deutsches Museum (German Museum, Munich). *Photo*: Foto Deutsches Museum.

Gottfried Wilhelm Leibniz (1646–1716). *Source*: Deutsches Museum (German Museum, Munich). *Photo*: Foto Deutsches Museum.

William Paley (1743–1805). *Source*: Wikimedia.

Robert Boyle (1627–91). *Source*: Deutsches Museum (German Museum, Munich). *Photo*: Foto Deutsches Museum.

Jean Baptiste Lamarck (1744–1829). *Source*: Wikimedia.

Charles Darwin (1809–82). *Source*: *Nature* **10** [1874], Frontispiece

Alfred Wallace (1823–1913). *Source*: Wikimedia

Louis Agassiz (1807–73). *Source*: *Nature* **19** [1879], Frontispiece

Julia Pastrana. *Source*: E. Haeckel, *The Evolution of Man* [London 1906], Vol. I, 161.

Thomas Huxley (1825–95). *Source*: *Nature* 9 [1874], p. 256.

Ernst Haeckel (1834–1919). *Source*: E. Haeckel, *Last Word on Evolution* [London 1910], Frontispiece

Francis Bacon (1561–1626). *Source*: Deutsches Museum (German Museum, Munich). *Photo*: Foto Deutsches Museum.

John C. Adams (1819–32). *Source*: *Nature* 34 [1886], Frontispiece.

Immanuel Kant (1724–1804). *Source*: Deutsches Museum (German Museum, Munich). *Photo*: Foto Deutsches Museum.

Friedrich Nietzsche (1844–1900). *Source*: Wikimedia.

Sigmund Freud (1856–1938). *Source*: Deutsches Museum (German Museum, Munich). *Photo*: Foto Deutsches Museum.

Johann Gustav Droysen (1808–84). *Source*: Wikimedia.

Wilhem Dilthey (1833–1911). *Source*: Wikimedia.

Max Weber (1864–1920). *Source*: Mohr Siebeck Verlag.

Le Livre Noir de la Psych-analyse. *Source*: Les Arènes, Paris.

Introduction

This book deals with issues in the area of intersection between the history and philosophy of science. Using Copernicanism, Darwinism, and Freudianism as scientific traditions, it aims to show that a tight connection exists between science and philosophy. There are many more connections between Copernicus, Darwin, and Freud than their respective contributions to the completion of the Copernican revolution. The study of Copernicanism, Darwinism, and Freudianism shows that scientific approaches to the world naturally and inevitably lead to philosophical consequences.

Freud saw that scientific ideas change the way we think about the world. With his heliocentric view, Copernicus displaced humans from the physical center of the universe (1543). With his evolutionary theory, Darwin inserted humans into the organismic order of nature (1859). According to Freud, both Copernicus and Darwin dealt severe blows to the proud image of humans as masters of the universe. Freud saw himself as completing the cycle of disparagement by destroying the belief that humans were "masters in their own house" (1916).

But the impact of scientific ideas on human self-images is only a small part of the philosophical consequences that scientific theories typically have. The present book is a study of three revolutions in thought and their philosophical consequences. It is an exercise in an integrated approach to the history and philosophy of science. Chapter I is devoted to the transition from geocentrism to heliocentrism. Chapter II focuses on the momentous shift in views on the nature of organic life that will forever be associated with the name of Charles Darwin. Chapter III discusses Freud's shift from Enlightenment rationality to unconscious motivations as the driving force of human behavior. A glance at the table of contents will show the reader that each chapter selects a number of philosophical issues, which derive from the study of the respective traditions in the history of scientific ideas.

The title of the book, *Copernicus, Darwin, and Freud: Revolutions in the History and Philosophy of Science*, conveys, as illustrated in the discussion of Copernicanism, Darwinism, and Freudianism, a threefold message: first, that human views of the surrounding natural and social world constantly adapt to new scientific discoveries;

second, that as science progresses, the achievements of former ages come under further scrutiny; and third, that philosophical consequences are necessarily implicated in these changes. I do not mean to claim that a process of constant uprooting and rebuilding of the edifice of knowledge takes place. Such a discontinuous view of the growth of scientific knowledge is mistaken, as the discussion of scientific revolutions will attempt to show. I mean that established knowledge will be inserted into new problem situations, which will result in modifications on many levels. Popper was right in saying that all scientific knowledge is conjectural knowledge. And as long as humans inhabit the solar system, it is to be expected that new ways of thinking will emerge and will reevaluate the place of human beings in the wider cosmos. Even though today's humans possess more quantitative knowledge, boast more technological prowess than Newton's contemporaries, Newton's abiding image still rings true: we are still like children, playing with pebbles on a beach, while a vast unknown ocean lies before us.

I

Nicolaus Copernicus: The Loss of Centrality

The mathematician who studies the motions of the stars is surely like a blind man who, with only a staff to guide him, must make a great, endless, hazardous journey that winds through innumerable desolate places. [Rheticus, Narratio Prima (1540), 163]

1 Ptolemy and Copernicus

The German playwright Bertold Brecht wrote his play *Life of Galileo* in exile in 1938–9. It was first performed in Zurich in 1943. In Brecht's play two worldviews collide. There is the **geocentric worldview**, which holds that the Earth is at the center of a closed universe. Among its many proponents were Aristotle (384–322 BC), Ptolemy (AD 85–165), and Martin Luther (1483–1546). Opposed to geocentrism is the **heliocentric worldview**. Heliocentrism teaches that the sun occupies the center of an open universe. Among its many proponents were Copernicus (1473–1543), Kepler (1571–1630), Galileo (1564–1642), and Newton (1643–1727).

In Act One the Italian mathematician and physicist Galileo Galilei shows his assistant Andrea a model of the Ptolemaic system. In the middle sits the Earth, surrounded by eight rings. The rings represent the crystal spheres, which carry the planets and the fixed stars. Galileo scowls at this model. "Yes, walls and spheres and immobility," he complains. "For two thousand years people have believed that the sun and all the stars of heaven rotate around mankind." And everybody believed that "they were sitting motionless inside this crystal sphere." The Earth was motionless, everything else rotated around it. "But now we are breaking out of it," Galileo assures his assistant. In the new model stars are no longer "fixed to a crystal vault"; they are allowed to "soar through space without support." [Brecht 1963; Blumenbach 1981, Vol. III, 762–82]

In Act Two learned scholars, a Mathematician and a Philosopher, visit Galileo in his study to look at the Jupiter moons through the newly discovered telescope. Galileo briefly explains the failings of the Ptolemaic system to them. It simply is not consistent with the facts. The planets are not "where in principle they ought to be." And the motions of the Jupiter moons around their planet, Galileo's great discovery, can simply not be explained on the Ptolemaic system. So much for words! Seeing is better than talking. Rather naively, Galileo asks his learned guests whether they would "care to start by observing these satellites of Jupiter." Unfortunately for Galileo both the Mathematician and the Philosopher refuse Galileo's invitation. Rather than observations, they demand "a formal dispute" in the scholastic tradition. "Mr. Galileo," asks the Philosopher, "before turning to your famous tube, I wonder if we might have the pleasure of a disputation? Its subject can be: Can such planets exist?" Galileo simply wants them to "look through the telescope" and convince themselves. "Of course, of course," says the Mathematician, "I take it you are familiar with the opinion of the ancients that there can be no stars, which turn round centers other than the Earth, nor any, which lack support in the sky?" Brecht only dramatized a real event. In a letter to Johannes Kepler (dated August 19, 1610), Galileo laments the steadfast refusal of scholastic professors, like Cesare Cremonini, a humanist at the University of Padua, to view the moon and the planets through the newly invented telescope. [Blumenberg 1955, 637]

2   A Clash of Two Worldviews

In his play, Brecht captures the clash of two worldviews brilliantly as he charts out the dialogue which might have developed between Galileo and his scholarly visitors. The disputation ends to the dissatisfaction of both parties. Soon the visitors leave without ever having glanced through the telescope. Adherence to the geocentric (Earth-centered) worldview makes Galileo's visitors disparage his appeal to observational evidence. Adherence to the heliocentric (sun-centered) worldview makes Galileo distrust the usefulness of learned disputations. In order to understand how the respective supporters of the two opposing worldviews came to clash so violently, as dramatized in Brecht's play, we have to look more closely at their presuppositions. We have to scrutinize the structure of the geocentric and the heliocentric worldviews.

Geocentrism predates heliocentrism by a millennium and a half. Copernicus knew of an ancient precursor: Aristarchos of Samos, who had proposed the conception of a moving Earth. But geocentrism remained the official explanation of the structure of the universe until its slow erosion in the sixteenth and seventeenth centuries. The dialogue between Galileo and his visitors could have taken place in the summer of 1610. The Copernican hypothesis had been known for 67 years. It would take another 77 years, until the publication of Newton's *Principia* (1687), before the geocentric worldview finally conceded defeat. It took 144 years of active debate and research for the Copernican view to establish itself. Can a scientific revolution take that long? What is important about a revolution is not its length but its depth. What makes a change revolutionary is its upheaval in an established structure, a reversal of viewpoints, a replacement of presuppositions. It is a general rearrangement of elements in a network, be it conceptual, political, or social. Some elements in the system are displaced, some replaced, and others remain. To understand a scientific revolution – the tangle of philosophical and scientific elements – we need to understand the system *before* its rearrangement. So to understand the Copernican revolution, we need to understand the geocentric worldview.

2.1   The geocentric worldview

Now the ancients build up one heaven upon another, like layers in a wall, or, to use a closer analogy, like onion skins: the inner supports the outer (...). [Kepler, Epitome (1618–21), Bk. IV, Part I, §3 (21)]

Geocentrism is much more than the view that the Earth resides motionless at the center of the universe. It amounts to a worldview that emerged in two phases. First, Aristotle provided a physical cosmology – the larger architecture of the cosmos. His cosmology included an important theory of motion. Aristotle advanced some unsatisfactory ideas about the motions of the planets. In a second phase Ptolemy furnished the mathematical astronomy – the geometry of the planetary motions. The Greek division of labor between physical cosmology and mathematical astronomy hindered the development of astronomy for centuries. [Dikjsterhuis 1956, §77, 146; Mittelstraß 1962, Ch. 4.4; de Solla Price 1962] For it separates the dynamic question of physical causes – why planets move in particular ways – from the kinematic question of motion – how the motion of these bodies can be described mathematically. In his *Almagest* (published around AD 150), Ptolemy explicitly embraces this distinction. Physics deals with the corruptible bodies on Earth; it amounts to no more than guesswork, which is due to the "unclear nature of matter." Mathematics, however, provides certain knowledge, since it investigates the nature of "divine and heavenly things." [Ptolemy 1984, 36] This separation was to last until Kepler's discovery of planetary laws at the beginning of the seventeenth century. As a worldview, geocentricism claimed to provide a scientific account of what was then regarded as the cosmos. It engaged its adherents in a number of philosophical commitments. It presented to its believers a comprehensive and coherent view of the universe. So did heliocentrism. With so much at stake, Brecht's play rightly depicted the frosty encounter of three scholarly men in 1610 as a clash between worldviews.

2.2   Aristotle's cosmology

Aristotle constructs his cosmology on the basis of a two-sphere universe and a theory of motion. Later Ptolemy provided some mathematical refinements.

1   Aristotle constructs a **two-sphere universe**. It is divided into the *supralunary sphere*, which includes the moon and the region lying beyond it, and the *sublunary sphere*. This is the region between the Earth and the moon. The Earth is a tiny sphere suspended stationary at the geometric center of the much larger

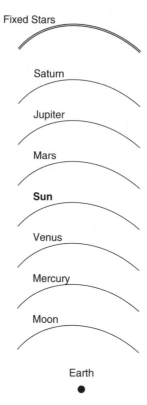

Fixed Stars

Saturn

Jupiter

Mars

Sun

Venus

Mercury

Moon

Earth

Figure 1.1   The circular orbs and order of the planets in Antiquity. The sun is situated among the other planets. The Earth sits motionless at the center

rotating sphere which carries the stars. The stars are markings on the outer sphere. In this picture it is the steady rotation of the outer sphere that produces the daily (diurnal) circles of the stars. Between the outer sphere and the Earth, smaller concentric spheres carry the then known six planets, including the sun. [Figure 1.1]

The supralunary sphere is, according to Aristotle, a region of utmost *perfection*, *symmetry*, and *regularity*. The Greeks ordained the *circle* as a perfect geometric shape. It is therefore in accordance with the perfection of the supralunary sphere that the stars and planets should move in perfect circles. By contrast, the sublunary sphere is the region of *change*, *flux*, and *decay*. The sublunary sphere is filled with four elements: earth, water, fire, and air. If undisturbed, they would settle in concentric shells around the central region of the Earth. But owing to the movement of the sphere of the moon, the elements get mixed throughout the sublunary world.

Aristotle (354–322 BC)

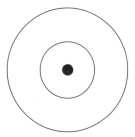

Figure 1.2 A simple homocentric model. The Earth is located centrally. Nesting concentric rings or shells (orbs) envelop it. [See Andersen/Barker/Chen 2006, Ch. 6.4; Barker 2002] These carry the planets. The outer ring carries the distant stars. The model fails because planets move at varying distances from the Earth

The motions of the lunar sphere are therefore responsible for all change and almost all variety observed in the sublunary world. [Kuhn 1957, 82–3]

This cosmology of a "cosmic onion" sounds very obscure to modern ears. To ancient eyes Aristotelianism presented the most comprehensive and convincing theory of the architecture of the cosmos. It seemed to account for some of the naked-eye observations available at the time. The centrality of the Earth, so it seemed, could be inferred from the path of falling objects on Earth and the circular motion of the stars.

Following the Greek philosopher and astronomer Eudoxos (408–355 BC), Aristotle assumed that the planets and the stars moved in concentric shells (or orbs) around the central Earth. [Figure 1.2] On closer inspection, this simple model must fail. It did not even fit Greek observations. For instance, if the sun were carried around the central Earth on a concentric shell, night and day would always retain equal lengths. Yet the Greeks knew from their observations that day and night have variable lengths, depending on the seasons. [See Section 3.2] The Greeks also noticed that planets move at varying distances around the Earth. The model of homocentric spheres had to be dropped. It was in contradiction with elementary observations. It was Ptolemy's achievement to have constructed a geometric model on the basis of the more complicated geometry. It involved the invention of new geometric devices: eccentrics, epicycles, deferents, and equants. [Dijksterhuis 1956, §68, 147; Rosen 1959, Introduction; Copernicus, *Commentarioulus* 1959, 57; Copernicus 1543, Bk. V, §3; Dreyer 1953, 143]

2   Although Aristotle's rudimentary views of planetary orbs were quickly replaced, his **theory of motion** proved to be a much more lasting contribution. Aristotle devised his influential theory of motion to support his cosmology. His model of the cosmic "onion" made the Earth a central, stationary object. How could this centrality be justified? The theory of motion claimed to provide a physical mechanism to account for the trajectory of all objects – earthly and celestial.

According to the Aristotelian theory of motion, objects either remain at rest or move in a straight line. A stone will fall back toward the center of the universe, occupied by the Earth. Smoke will rise upward toward the sky, in search of its natural place. The upward and downward motions constitute the object's *natural* motions. In order to deflect objects from their natural motions an external push or force is needed. To move, objects need a mover, which moves them. There is no motion without a mover: *Omne quod movetur ab alio movetur.* [Aristotle 1952a, Bk. VII, VIII] Of course, Aristotle could observe that projectiles do not behave in this way. A stone hurled through the air or an arrow released from a bow will normally fly in a parabola before returning to Earth. Aristotle could explain the projectile's motion. After the release of the object from the mover, disturbed air became the source of the external push. It prolonged the projectile's motion.[1] Eventually the object would succumb to its inclination to return to the Earth.

The natural motion for heavenly objects is circular. Circular motion is continuous and infinite. Aristotle states that continuous motion – the rotatory locomotion of the planets – is caused by an unmoved mover, a Deity. [Aristotle 1952a, Bk. VIII, §10]

Thus things have natural and unnatural motions. They also occupy natural places in the universe. Aristotle held that the four building blocks of the universe – earth, water, fire, and air – hold natural positions on and near the Earth. If wrestled from their natural positions, the elements *strive* to regain their natural position. When a stone is lifted from the ground and released, it seeks to regain its natural position. When a fire is lit, flames and smoke leap up toward their natural positions at the periphery of the terrestrial region. The natural position of the Earth is at the geometric center of the universe. For something at rest must exist at the center of the revolving heaven. Therefore, Aristotle concludes, the Earth must exist. [Aristotle 1952b, Bk. II, §3] A piece of Earth will always fall to where it naturally belongs, i.e., the geometric center of the universe. From these arguments from terrestrial physics Aristotle derived not just the centrality of the Earth but also its stability and sphericity. [Aristotle 1952b, Bk. II, §§13–4] In lunar eclipses, he points out, the outline of the Earth's shadow on the moon "is always curved"; and as observers travel north and south along a longitude, different stars become visible to them in the sky. Later Ptolemy added some further arguments. The sun, the moon, and the stars are seen to rise earlier for inhabitants of eastern regions of the globe than "for those toward the west." [Ptolemy 1984, §I.4]

Aristotle's physical cosmology and his theory of motion form a logical link. The theory of motion renders the cosmology reasonable. And his cosmology provides the necessary framework for physical phenomena to be arranged into two separate spheres. The Aristotelian laws of motion govern the sublunary sphere. In the sublunary sphere terrestrial physics rules. The laws of motion account for the apparent observations in this region of space: the drop of heavy objects and the

---

[1] Kuhn [1957], 119; Dijksterhuis [1956], I, §§30–9. Contrast this account with the Newtonian explanation. According to Newton's First Law objects are either at rest or perform a constant rectilinear motion (if undisturbed). Rectilinear motion has become a "natural" motion, for which no external force is required.

rise of light objects. In the supralunary sphere celestial physics rules. This is the region of perfection. It admits only of spherical shapes and circular motion. It is a finite region. In Brecht's play Galileo complains about the "walls" and "immobility" of the geocentric universe. The distance to the stars was estimated to be 20,000 Earth radii, which is less than today's Earth–sun distance. [Zeilik 1988, 29–31] The outer boundary is marked by the sphere of fixed stars. Although this sphere rotates in a period of 24 hours once around the motionless Earth, the stars appear fixed because after each rotation they reappear in the same location as in the previous periods. The planets, by contrast, are "wandering stars," because they perform observable, traceable movements across the sky. The Aristotelian cosmos is an energy-deficient universe. Its energy-deficiency is a direct consequence of Aristotle's theory of motion. There is no motion without a mover. Heavenly bodies are moved on their spheres by a mover, residing outside the outer sphere. The Aristotelian universe requires an energy-input from beyond the fixed stars – it is finite. As we shall see, the Copernican universe is also finite but it is no longer energy-deficient.

2.3   Ptolemy's geocentrism

Aristotle gave us a cosmology and a theory of motion. This was the first step in the construction of the geocentric worldview. The second step was completed several

hundred years later. It took a consummate geometer to do it: Claudius Ptolemy. Ptolemy was the first astronomer to design a complete mathematical system of the universe, which accurately predicted planetary motions to within 5° of modern values. His was a *geocentric* model, built by means of *geometric* reasoning. Later Copernicus would construct a *heliocentric* system, also built by means of *geometric* reasoning. Ptolemy uses geometry to describe astronomical observations. He agrees with Aristotle that the celestial spheres, which carry the planets, perform *uniform* motions. He assumes that the Earth is *spherical*, at the *center* of the cosmos, and *stationary*. If the Earth

Claudius Ptolemy
(AD 100–75)

were not central, he argues, the equinox would not occur, and "intervals between summer and winter solstice would not be equal." [Ptolemy 1984, §I.5] Ptolemy offered perfectly good reasons for rejecting as ridiculous any motion of the Earth. Aristarchos of Samos (310–230 BC) is said to have taught the daily and annual rotation of the Earth. [Dreyer 1953, Ch. VI, 136–48; Dijksterhuis 1956, Ch. I, §§78–9; Kuhn 1957, Ch. I; Koestler 1964, 50–2] But if the Earth moved, its inhabitants would feel the effects drastically – objects thrown straight up into the air would not return in a straight line to the spot from where they had been launched; buildings would crumble under the force of the motion; birds would never fly from west to east. [Ptolemy 1984, §I.7; cf. Copernicus 1543, Bk. I, §7] To the Greek mind it was commonsense that the Earth was at rest.

But there was a problem. The Earth-bound observer does not observe the uniformity of planetary motion against the background of the fixed stars. Equipped with his basic presuppositions, Ptolemy, like other astronomers before him, had to explain two main variations in the motions of the planets. *First*, there was their **nonuniform motion** and *second* their **retrograde motion**. The problem arises because the observations do not conform to the Ptolemaic presuppositions about planetary motions. The basic problem of ancient astronomy is to construct geometric models, which satisfy the a priori presuppositions and seemingly account for the apparent motions of the planets. It is the *problem of saving the appearances*, rather than constructing realistic models of the solar system. Like his Greek predecessors, Ptolemy relied on geometric models to solve these problems. Ptolemy tried to fit the observations to his unquestioned presuppositions: the circular motion of celestial objects and the Aristotelian theory of motion. But Ptolemy improved the usefulness of the models. Some of Ptolemy's predecessors, like Hipparchus (190–125 BC), had invented new geometric devices to deal with these problems. *Eccentric motion* was used to solve the first; *epicyclic motion* was used to solve the second. In his *Almagest* Ptolemy made frequent references to Hipparchus's work, usually with the intention of improving it. He introduced a new geometric device (the equant) to achieve a better geometric model of the appearances. He treated each problem separately. For instance, in dealing with the apparent annual motion of the sun around the Earth, he ignored its apparent daily motion. Unlike the Copernican model, the Ptolemaic model does not present a system in which the appearances are due to a common factor – like the motion of the Earth around the sun.

The first problem was the **nonuniform motion** of the planets through the zodiac, irrespective of the effect of retrograde motion. As Kepler later showed, planets do not orbit the sun at uniform speed. The nearer they are to the sun the faster they move, and the further they are away from the sun the more slowly they amble. But the Greeks could not accept such nonuniform motion as real. It had to be apparent.

How can uniform circular motion account for apparent nonuniform motion? The answer is eccentric motion. [Figure 1.3]

The sun is modeled as moving around the Earth on an eccentric circle at uniform speed. The circle is called *eccentric* because the Earth does not occupy its center. While the sun moves around the center of the eccentric, the Earth is slightly displaced from its center. The distance between these two points accounts for the appearance of variation in motion. As seen from the center of the eccentric, the planet moves through equal angles in equal times. But as seen from the Earth, the planet sweeps through different angles in equal times. For the Earth-bound observer, when the planet is closer to the Earth, it *seems* to be moving faster.

The second problem is the apparent anomalous westward motion of a planet with respect to the stars: its **retrograde motion**. It is accompanied by a change in brightness. For outer planets it occurs near the time of opposition, when the planet is opposite the sun in the sky. For inner planets, like Mercury and Venus, it occurs at inferior conjunction, when they are seen close together with the sun in the sky.

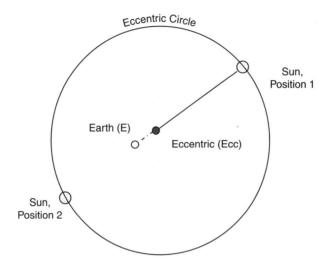

Figure 1.3 Eccentric motion. Explanation of apparent non-uniform motion on the assumption of uniform motion. The sun moves uniformly around point (*Ecc*). Seen from the Earth (*E*), however, the uniform motion looks non-uniform. At point 1 the sun appears furthest away from the Earth (apogee), while at point 2 it appears at its closest approach to the Earth (perigee)

The ordinary eastward path of planets seems interrupted – for a time, observers see the planets go westward. [Figure 1.4]

The appearance of nonuniform retrograde motion was solved by using the geometric device of epicyclic motion. The planets are carried on smaller circles (epicycles) moving on larger ones (deferents). Although the Greeks observed retrograde motion, it was only apparent, not real motion. The real motion of celestial objects required uniform circular motion. The task consisted in constructing models that fitted the observations without violating the presuppositions. Epicyclic motion is modeled by introducing a *deferent*, with the Earth at the center, and a smaller circle, an epicycle, mounted on the deferent. [Figure 1.5a] The radii of both epicycle and deferent move in the same direction. For an observer on Earth the planet performs a retrograde motion as it passes through the lower part of the epicyclic motion. This model, however, is too simple. It cannot account for the observational variations in retrograde motion of the planets. To explain the variations, Ptolemy invented a new device: the *equant*. [Figure 1.5b] This is an imaginary point on the other side of the center of the deferent as seen from the Earth. At the equant, an observer would see the planet move around its orbit through the sky at a uniform speed relative to the stars. But from a viewpoint on Earth away from the circle's center, the motion appears nonuniform.

For our later philosophical exploration we should note several points. Ptolemy was very well aware of the role of representational models in his theory. His usual method is to use geometric models but in order to represent the fixed stars he chooses a solid globe as a scale model. [Ptolemy 1984, Bk. VIII.3] At the same

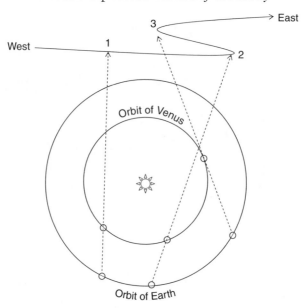

Figure 1.4   A simplified scheme of the appearance of retrograde motion of Venus as seen from by an Earth-bound observer. The observer "marks" the position of Venus against the background stars as the planet prepares to overtake Earth in its orbit – position 1. When Venus has overtaken Earth, the observer makes a second observation: as expected, Venus has moved from west to east – position 2. But at a later stage, a third observation reveals an apparent and abnormal retracing of the orbit of Venus toward the west. In a heliocentric view this is due to the relative position of the Earth with respect to Venus around the sun. [See Zeilik 1988, 40; Copernicus 1543, Bk. V, §35]

time, Ptolemy worried about the fit of his geocentric hypotheses. He was skeptical enough to warn his readers not to expect his geometric models to properly represent the celestial phenomena. [Ptolemy 1984, 600–1] In the spirit of Aristotle's two-sphere cosmology he cautioned that geometric models invented by an inhabitant of Earth could not do justice to the perfection of the heavenly phenomena. As we shall see, the question of how models manage to represent physical reality is of great interest to philosophers. Finally, Ptolemy agreed with the Greek tradition that the epicyclic and eccentric models were equivalent devices. [Ptolemy 1984, §§III.3, IV.5, XI.5; Rosen 1959, 37; 1984, 27; Dijksterhuis 1956, I, §71] Either of these two hypotheses will account for the appearance of irregular motion of the planet to the Earth-bound observer. Nevertheless, Ptolemy adopts the principle that only the simplest hypothesis be used. [Ptolemy 1984, §§III.1, XIII.2] The acceptance of equivalence raises interesting philosophical issues regarding explanation and representation. If the eccentric circle is as good a representation as the deferent–epicycle device, is there no way of deciding which one fits the actual physical system better than the alternative? Such concerns belong to the philosophical consequences of scientific theories.

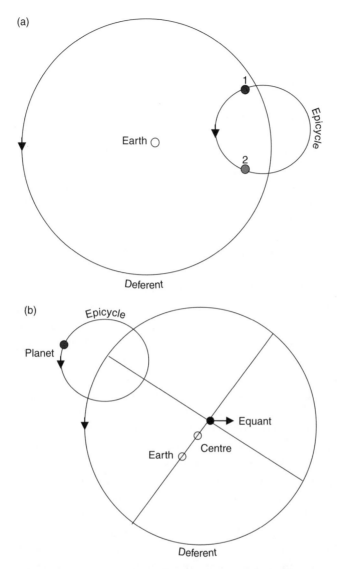

Figure 1.5 (a) Epicyclic motion. Retrograde motion occurs when the planet moves from P1 to P2 on its epicycle; (b) The equant. Explanation of retrograde motion with a new geometric device, the equant. [See Copernicus 1543, Bk. III, §15–16; Ptolemy 1984, §IX.6; Andersen/Barker/Chen 2006, Ch. 6.3] This representation is supposed to be a closer fit of the model to the data than the elementary model. From the point of view of the equant, the motion of the planet on the epicycle would appear uniform. Further flexibility is introduced by letting the Earth either sit at the center of the deferent or off-center, as indicated in the diagram

Before we mention some of the developments of the thirteenth and fourteenth centuries, which created the conditions for the emergence of the Copernican revolution, we should add another historical dimension. This is the synthesis between *Aristotelianism* and *Theology*. Only this further historical dimension could bring to a head the clash between Galileo and the Church in the seventeenth century. The synthesis was worked out by Thomas Aquinas (1225–74), Albertus Magnus (1206–80), and others.

According to Acquinas, real knowledge is based on sense experience. Albertus Magnus also stresses that the study of nature is based on sense experience, which provides the highest form of proof. Where we lack knowledge we have to appeal to revelation. The perfection of the heavens, postulated in the Greek tradition, is now identified with Divinity. Consequently our knowledge of the world is restricted to the sublunary sphere. The perfection of the heavens transcends our reasoning powers. Aquinas welcomes the systematic study of nature because he sees in it a means to acquire knowledge of the wisdom of God. To put it drastically, Aquinas hopes that a systematic study of nature will help to eradicate superstition. Couched in these terms, no conflict between Reason and Revelation is permitted to arise, for our reason is weak and faulty and in questions of doubt has to submit to the eternal Truth as expressed in the Revelation. This is a common attitude in the Middle Ages. Roger Bacon (c.1210–92) defends a similar idea. The value of science lies in its contribution to the interpretation of the Bible. It helps to glorify God. Once the Church had embraced Aristotelianism, all criticism directed at the geocentric worldview would also be a criticism of theology and the Church.

Nevertheless, progress was made and some developments took place toward the end of the Middle Ages. Progress, however, depended on the ability to overcome unquestioned presuppositions, which impose constraints on permissible models. This need to clear away presuppositions before progress could occur could be expressed in Kantian terms. Kant asks very generally in his Copernican turn in philosophy, "What are the conditions of the possibility of knowledge?" By analogy we can ask, "What are the conditions of the possibility of the Copernican, the Darwinian, and the Freudian revolutions?" Which new presuppositions were needed for the Copernican view to be able to arise? The questioning of ancient presuppositions happened in two stages: the Aristotelian theory of motion attracted critical scrutiny before the ideal of circular motion was questioned.

2.4   A philosophical aside: Outlook

Let us regard the Aristotelian theory of motion and his physical cosmology or the Ptolemaic devices as instructions in a toolkit, with which we try to build a model of the universe. Our building blocks are fixed stars and planets, circular spheres, a stationary Earth. Our instruction sheet contains a further restriction: the model we build must be as close as possible to naked-eye observations. Most strikingly, we observe the movement of the planets against a background of stars, the succession of the seasons, and the regular sequence of day and night. With these elements at hand we can build only a *geocentric* model. The sun, the planets, and even the fixed background of distant stars must parade before our eyes. The Earth must therefore be located at the center of these movements, for otherwise we could not account for them. [Figure 1.1]

Our ability to build *one* type of model, if we follow the instructions, is at the same time an inability to construct a *different* model. We can think of the toolkit as creating a space, more precisely a *constraint space*, to accommodate cosmological models. Such a constraint space is a logical space, because it creates certain possibilities, whether they materialize or not, for model-building. Geometric-type models will be allowed to inhabit the space; other types of models are excluded. This play of possibilities and impossibilities is regulated by the constraints we accept. Aristotle operated under the constraints of his theory of motion and his two-sphere cosmology. If we change the constraints, a different logical space will appear, which will accommodate other types of models. **Constraints** can be understood as restrictive conditions, which symbolic statements must satisfy in order to qualify as admissible scientific statements about the natural world. This teaches us some lessons, which will interest the philosopher of science.

1. Scientific theories come with certain *constraints*: empirical and theoretical constraints, which can be further subdivided into mathematical, methodological, and metaphysical constraints. Scientific theories operate under such constraints. With the exception of empirical constraints, these constraints form *presuppositions*. Presuppositions are fundamental assumptions, which, at least for the time being, are protected from critical inquiry. They are accepted as "true." They serve as historical a priori. They are not unquestionable but they remain unquestioned for certain periods of time. Whether true or false, they channel research into particular directions. The Aristotelian toolkit contains such presuppositions. The two-sphere cosmology and his geometric devices, including the theory of motion, form the Aristotelian presuppositions. Presuppositions can be exposed to doubt. This happened when Aristotle's concentric shells were replaced by other devices. Under such scrutiny, constraints will be amended or rejected. Already a modification of the model instructions, keeping the elements, will change the possibilities for model construction. The adoption of epicycles, for instance, created the space for the Ptolemaic model. Sometimes a modification of a constraint is more far-reaching. A questioning of the Aristotelian theory of motion and its replacement by the so-called *impetus* theory liberated the constraint space for the development of new theories. It is difficult to imagine how heliocentrism could have emerged, if some fundamental presuppositions had not changed. [Blumenberg 1965, Ch. I] Copernicus, for instance, was able to reject some of the classic objections against the motion of the Earth, because he no longer shared the Aristotelian theory of motion. The development of the impetus theory allowed him to regard the motion of the Earth as natural.

2. We can also see that constructing a cosmological model is not a matter of simply reading it off the available observational data. It cannot be, if presuppositions are as much a reality of scientific thinking as its methods and established results. [Weinert 2004] So a simple inductive view of the scientific method will not do, at least not in the case of scientific revolutions. Let us call the view that sees science as a straightforward generalization from observations

and experiments *induction by enumeration*. Francis Bacon already criticized this view. There is a more sophisticated way, called *induction by elimination*. Francis Bacon recommended this view as a fruitful scientific method. It is called *eliminative induction* because alternative or rival models are confronted with empirical evidence and other forms of constraints. The model, which fares better in view of these constraints, will gain in credibility while the rival model will lose credibility. So this view requires that there are at least two models available, which face the evidence. As we shall see, the Copernican and Darwinian revolutions come about by a progressive elimination of unsuccessful models in the face of an increasing number of constraints. The difficulty with Freudianism is precisely that the available evidence is unable to credit some model at the expense of its competitors. Is the overwhelming method of science eliminative induction or the more familiar method of falsificationism, as proposed by Karl Popper?

Even at this early stage of the argument, it is good to raise these philosophical questions because – and this is one of the **central theses** of this book – philosophical issues are inseparable from more scientific and historical concerns. In other words, *scientific revolutions have philosophical consequences*. We shall witness this logic at many points along the road.

3. An immediate question springs to mind, not just for the philosophically inclined. Do these geometric devices actually represent physical spheres, while the nonuniform variations are just appearances? Do these geometric devices – the epicycle and the deferent, the equant and the eccentric – describe some physical mechanism, which exists in nature? This is the question of the *representational* force of scientific models, which already exercised Ptolemy. Does the distinction between appearance and reality, between how the planets appear to move according to naked-eye observations and how they are said to move according to the Greek presuppositions, correspond to a physical feature of the universe? If we are interested in what science is and does, such questions, although philosophical in nature, are inescapable. Whatever position we adopt in response to these questions, they actually do some real work. The proponents of the geocentric worldview were divided on this question. Aristotle thought that the spheres were real physical spheres. They possess a natural motion: circular rotation. The natural motion of these spheres drives all the heavenly bodies. They depend on an unmoved mover for their energy requirements. Ptolemy was much less certain of the physical reality of the crystal spheres, the epicycles and deferents, which he was employing as geometric devices. True, they served to *save the appearances*. But Ptolemy did not think that the geometric devices fitted the heavenly phenomena very well. [Ptolemy 1984, 600–1; Dreyer 1953, Ch. IX] The Greek models try to match naked-eye observations with a priori presuppositions about the physical world. The postulation of uniform circular motion, of the two-sphere universe, of geometric devices is not based on observations. On the contrary: the observations seemed

to contradict the presuppositions. So much worse for the observations! The separation between physical cosmology and mathematical astronomy did not encourage Greek astronomers to think of observations as tests for the mathematical models. The question is whether models can achieve more than saving the appearances. This question leads to considerations of *instrumentalism* and *realism*, *explanation* and *representation*.

The uncertainty as to the physical reality of the geometric models plagued the geocentric worldview until the moment of its definite demise. Especially the Arab world, which preserved the tradition of Greek astronomy throughout the Middle Ages, voiced much opposition against the "reality" of the geometric devices. [Rosen 1984] But they remained in use for some 1,400 years. They predicted planetary positions to the accuracy needed by astronomers who relied on naked-eye observation. And they conformed to Aristotle's theory of motion and cosmology.

As we can see, the description of the Aristotelian–Ptolemaic geocentric world-view points to some general philosophical issues, which are difficult to separate from the scientific material.

2.5   Shaking the presuppositions: Some medieval developments

(…) by the purpose of movement it is proved "that movement belongs to the Earth as the home of the speculative creature." [Kepler, Epitome of Copernican Astronomy *(1618–21), Bk. IV, Part I, §5, 75]*

In Brecht's play Galileo clashes with the representatives of scholastic learning. Galileo is a believer in heliocentrism, observations, and the independence of the scientific method. The Mathematician and the Philosopher represent a world in decline: they put their faith in bookish learning, in the authority of the ancients, and cling to their belief in geocentrism. Galileo attempts to shake his visitors' beliefs. But they are not shallow opinions. They are based on philosophical presuppositions, which define their constraint space. In this constraint space certain models can be accommodated but others cannot. In the fourteenth century, some outstanding scholars became critical of the established doctrines: Nicolas d'Autrecourt (died after 1350), Johann Buridan (c.1300–58), Nicolaus of Oresme (c.320–82) at the University of Paris, and William of Ockham (c.1295–1349) at the University of Oxford. Two developments are particularly noteworthy: 1) Nicolas d'Autrecourt argued that philosophy and theology should be kept apart – a suggestion later taken up by Galileo and Pascal. The general idea is that natural philosophy should investigate the natural world and theology the spiritual world. 2) The Aristotelian theory of motion comes under critical scrutiny. Oresme and Buridan suggested, as Copernicus was later to do, that the diurnal motion of the Earth cannot be disproved by arguments derived from the Aristotelian theory of motion. According to Aristotle, the Earth rests motionless at the center of the world, because it inhabits its natural place. If it were to move, Earth-bound observers should feel the effect of this unnatural motion. Jean Bodin, a famous sixteenth-century political philosopher, echoes this age-old reasoning in 1597, fifty years after the publication of Copernicus's book (1547):

> No one in his senses, or imbued with the slightest knowledge of physics, will ever think that the Earth, heavy and unwieldy from its own weight and mass, staggers up and down around its own center and that of the sun; for at the slightest jar of the Earth, we would see cities and fortresses, towns and mountains thrown down. [Quoted in Kuhn 1957, 190]

As we have seen, Ptolemy advanced similar arguments involving the fall of objects and the destruction of buildings. The rejoinder in all these cases was to make the daily and yearly rotation of the Earth, to which Bodin refers, a *natural* motion. If we participate in this motion, Oresme and Buridan argue, then we do not perceive it. It is not a violent motion, says Copernicus, as Ptolemy thought. True, violent motion has the effect of breaking things up. But the rotation of the Earth "accords naturally with its form," so that every part of the Earth, "the clouds and the other things floating in the air or falling or rising up" take part in this natural motion of the Earth. [Copernicus 1543, Bk. I, §8] Copernicus employs impetus ideas to rebut the ancient, commonsense argument. If the air envelope travels along with the Earth and shares in its natural motion, the lack of violent winds is to be expected. Today we no longer accept the impetus theory. We are, however, all familiar with such phenomena. In constantly moving vehicles, our actions – drinking coffee, playing cards, reading books – happen as if the vehicles were stationary. Galileo's relativity principle serves as our explanation. The impetus theory was an important step toward the modern explanation of motion.

The *impetus* theory of motion was developed in the fourteenth century as an alternative to the Aristotelian theory of motion. According to this theory, as Oresme and Buridan explained it, a motive force is impressed upon an object, which carries it along. Then the argument against the motion of the Earth falls flat. Buridan first argues against the Aristotelian view of motion. If both a blunt and a sharp object are propelled along the same parabola, the air could not push in the same way on the sharp object as on the blunt one. It is better to say that a projector (internal propellant) impresses a certain *impetus* or motive force onto the moving body and that the projectile moves in the direction of the impetus. But air resistance and the "gravity" of the projectile decrease the impetus "till it is so diminished that the gravity of the stone wins out over it and moves the stone down to its natural place."[2]

The impetus theory played an essential role in the Copernican revolution: it was one of the conditions that made it possible. Buridan's pupil, Oresme, also based his refutation of Aristotle's central argument for the immobility of the Earth on the

---

[2] Quoted in Kuhn [1957], 120; according to Jeans [1943], 106, Hipparchus (c.140 BC) already held an impetus theory; on the impetus theory see the studies of Wolff [1978], Mittelstraß [1962] and Drake [1975].

impetus theory. He turned his attention to the first argument against the mobility of the Earth. It was claimed by the ancients that if the Earth moved eastward on its own axis, then an observer who threw a stone straight up into the air would see the stone return to the ground to the west of his feet. In the absence of the impetus theory, this argument seemed to make sense. On the ancient view the stone would be forced from its natural position and strive to return to it. But while the stone was in the air, the Earth would turn to the east. As the stone could not accompany the moving Earth, it must fall to the west from its point of departure. But Oresme argued that the moving Earth endows the stone with an eastward impetus. It will cause the stone to follow the moving Earth. [Kuhn 1957, 121; Mason 1956, §II.11; Wolff 1978, Pt. II, Ch. 7]

Buridan and Oresme extended this argument to the motion of the Earth. There was no need for "angelic intelligences" to move the celestial bodies. There was no need to postulate friction between the crystalline spheres to keep them moving in their 24-hour rhythm. There was no need for Aristotle's Unmoved Mover. There was no need for an energy-deficient universe. In creating the Earth God imparted a motive force to it, which sustains it in its motion. Unlike projectiles on Earth, which, according to the impetus theory, slow down because they encounter wind resistance and the Earth-bound force of their own gravity, no such forces interfere with the eternal motion of the Earth. More generally, the impetus theory suggested self-sustaining circular motion for the planets too. [Kuhn 1957, 121–2; Dijksterhuis 1956, II, §§111–5]

On the conceptual level, the impetus theory had important consequences. It lifted the ban on the possibility of the mobility of the Earth. The logic of the Aristotelian view immobilized the Earth. The arguments against its motion – falling objects, howling winds, tumbling houses – seemed to make sense. The impetus theory showed that it was conceptually possible for the Earth to move. The impetus theory also hinted at a unification of terrestrial and celestial physics. For it explained the trajectory of objects on Earth and in the heavens according to the same principle. It therefore led to the potential destruction of the two-sphere universe. Aristotle had made a distinction between rotatory locomotion, reserved for heavenly bodies, and rectilinear motion, for earthly objects. He regarded rotation as primary. [Aristotle 1952a, Bk. VIII, §9] The impetus theory held out the prospect of a dissolution of the dichotomy between supralunary and sublunary spheres.

However liberating the impetus theory was, arguments in its favor were never pushed to their logical conclusion. Fourteenth-century thinkers were content to investigate logical alternatives to Aristotelianism. They were not in the business of overthrowing it.

If the impetus theory was one of the conditions of the possibility of Copernicanism, the rise of *humanism* in the Renaissance was another. Renaissance humanism was directed against medieval scholasticism. As the Mathematician and Philosopher in Brecht's play show, the scholastic attitude viewed the Aristotelian tenets as sacrosanct. Scholastic scholarship consisted in the interpretation of Aristotle's texts. The Philosopher in Brecht's play reminds Galileo that "the universe of the divine

Aristotle is an edifice of (...) exquisite proportions." In the spirit of modern science Galileo counters that "the authority of Aristotle is one thing, tangible facts are another." But this objection only provokes the Philosopher into an indignant outburst: "if Aristotle is going to be dragged in the mud – that's to say an authority recognized not only by every classical scientist but also by the chief fathers of the church – then any prolonging of this discussion is in my view a waste of time. I have no use for discussions, which are not objective. Basta."

With humanism emerged a renewed interest in the mathematical and geometrical regularities of the phenomena of nature. This was important because Copernicus was among the first to revive the full Hellenistic tradition of mathematical astronomy, which had flourished in Ptolemy's time. [Blumenberg 1957; 1965; 1981] Humanism also put a new emphasis on human attitudes toward the cosmos. It reversed the age-old tradition, ever present since Ptolemy, that human knowledge could not extend as far as the heavens. Humanism elevates the astronomer to the status of a "contemplator caeli." [Blumenberg 1957, 77] It emphasizes that humans can understand the workings of the cosmos. This emphasis shifts the focus from understanding by way of observation to understanding by way of rational thinking. The emphasis on rational understanding on the basis of perspectival, Earth-bound observation had important implications for the heliocentric worldview.

3   The Heliocentric Worldview

And why not admit that the appearance of daily revolution belongs to the heavens but the reality belongs to the Earth? [Copernicus, De Revolutionibus *(1543), Bk. I, Ch. 8 (17)]*

Nicolaus Copernicus died on May 24, 1543. Only a few weeks later his great book *De Revolutionibus* was published. The original title of the book had been *De Revolutionibus orbium mundi.* This intended title was changed, by Andreas Osiander, to *De Revolutionibus orbium caelestium: On the Revolutions of Heavenly Spheres.* [Blumenberg 1957, 79; 1981 Vol. II, 344] Osiander, a theologian and preacher based in Nuremberg, oversaw the publication of *De Revolutionibus.* He also added an anonymous, philosophically significant preface to Copernicus's work. Kepler later identified Osiander as the author of the anonymous preface. It is philosophically significant because Osiander tries to interpret *De Revolutionibus* as a treatise which, contrary to first impressions, does not challenge the accepted worldview. Copernicus had been working on his masterpiece for years but had hesitated to publish it. Like Darwin after him, Copernicus feared that his ideas would meet with a hostile reaction. Nevertheless, prior to the publication of *De Revolutionibus* handwritten copies of his "Sketch of his Hypotheses for the Heavenly Motions," known as *The Commenariolus,* had been circulating. It was written between 1502 and 1514. [See Rosen 1959, Introduction] In these works Copernicus worked out a heliocentric model of the solar system.

3.1 Nicolaus Copernicus

Copernicus's achievement was not something forced by fresh observations, but rather was a triumph of the mind in envisioning what was essentially a more beautiful arrangement of the planets. [Gingerich, The Book Nobody Read *(2004), 116]*

On closer inspection *De Revolutionibus* falls into two parts. In the first chapter Copernicus introduces the general idea of a heliocentric system. He argues that

Nicolaus Copernicus (1473–1543)

Greek objections to the concept of a moving Earth do not hold water. He points to a number of Greek predecessors of heliocentrism. He claims that heliocentrism provides a simpler or more coherent explanation of the solar system. For Copernicus, as for the Greeks, the solar system, with the fixed stars, constitutes the universe. The second part of the book contains the mathematical determinations of planetary motions. It is much more technical. But Copernicus uses the same geometric devices as the Greeks (eccentrics and epicyles).

Since Kant, it has become customary to describe the result of Copernicus's labor as a **Copernican turn**. [Dijsterhuis 1956, Part IV, I, §§9–10, 18; Blumenberg 1981, Part V, V] This term is very useful: it marks the Copernican achievement without elevating Copernicus's work to a scientific revolution.

The Copernican turn is the conception of a heliocentric universe, in which the planets are carried on their spheres, not around a central Earth, but around a central (mean) sun. This in itself was not an original idea, since it had existed since antiquity. The Greek astronomer Aristarchus of Samos had constructed a heliocentric world system, according to which the Earth rotates daily around its own axis and annually around the sun. Other thinkers, both in antiquity (Herakleides) and in the fourteenth century (Buridan, Oresme, and Nicolaus of Cues), had conceived of a diurnal motion of the Earth. So what is original in Copernicus? Since Aristarchus's work has not survived, he became the first astronomer to have constructed a mathematical system of planetary motion from a heliocentric perspective. Copernicus attempts to derive all the celestial phenomena from a few basic assumptions. [*Commentariolus* 1959, 58–9] All the observations can be explained from the assumption of a nonstationary Earth. Copernicus assumes that the sun is stationary but Kepler later corrects this view: the sun turns on its own axis.[3] [Kepler 1618–21, Pt. II, §1] Copernicus was the first to develop a detailed account of the astronomical consequences of the Earth's motion. [Kuhn 1957, 142, 144] He claims that it accounts for the phenomena and creates a *coherent* system of the orders and magnitudes of all spheres and stars.

Copernicus was well aware that the *interrelatedness* of natural phenomena would lead to a coherent model of the universe. In his Dedication to Pope Paul III he spells out how his reasoning took him from the correlation of natural phenomena to a more adequate heliocentric model:

> And so, having laid down the movements which I attribute to the Earth farther on in the work, I finally discovered by the help of long and numerous observations that if the movements of the other wandering stars are correlated with the circular move- ment of the Earth, and if the movements are computed in accordance with the revolu- tion of each planet, not only do all their phenomena follow from that but also this correlation binds together so closely the order and magnitudes of all the planets and of their spheres or orbital circles and the heavens themselves that nothing can be shifted around in any part of them without disrupting the remaining parts and the universe as a whole. [Copernicus 1543, 6]

Copernicus and his disciple Georg Joachim Rheticus (1514–1576) claim that the heliocentric hypothesis has many advantages over the Ptolemaic hypothesis. The advantages derive from treating the planets and their motions as a *system*:

- According to Copernicus, the concept of a moving Earth – its daily and annual rotation – naturally explains all the celestial observations. For instance, the two great problems, inherited from antiquity, seem to dissolve in a heliocentric model. The retrograde motions of the (inner and outer) planets become a natu- ral consequence of the motion of the Earth around the (mean) sun. An inner planet, like Mercury, has a shorter orbital period than the Earth. It overtakes the Earth in its annual orbit. For an Earth-bound observer its motion appears as retrograde motion. [Figure 1.3] The second problem was the nonuniform motion of the planets. Planets seemed to require different times to complete their successive journeys around the ecliptic. Part of the solution derives from placing the planets at their correct distances from the sun. The outer planets need longer for their annual journeys than the inner planets. But Copernicus's solution is only partially successful because he still assumes uniform circular motion. Still, the two "appearances" can be explained without the use of major epicycles. The major irregularities of the planetary motions are only apparent. [Kuhn 1957, 149, 166–71; Zeilik 1988, 49] These appearances are produced by the orbital motion of the Earth. As the sun is stationary in the heliocentric

---

[3] Strictly speaking, the sun is not the physical center of the Copernican system; it is placed near the center of the orbit of Earth. It was only Kepler who attributed to the sun a "vital physical role in keeping the planets in motion." [Gingerich 1993, 42]

Box 1.1   The order of planets in heliocentrism
Sun) Mercury) Venus) Earth) $_{Moon}$**) Mars) Jupiter) Saturn) Fixed Stars**
compared with the geocentric order:
Earth) Moon) Mercury) Venus) Sun) Mars) Jupiter) Saturn

---

system, it does not have a retrograde motion. [Figure 1.1; Box 1.1] [See Kuhn 1957, 66, 69; Copernicus 1543, Bk. III, §§15–6]

- Copernicus also determines the *relative* distances of the planets from the sun, using techniques known from antiquity. [Copernicus 1543, Book I, §10; Neugebauer 1968, §2; Kuhn 1957, 142, 175; Zeilik 1988, 40–1] If the Earth–sun distance is taken as 1 unit, then Mercury is at 1/3 of the Earth–sun distance, Mars at 1½, Jupiter at 5 and Saturn at 9 Earth–sun distances. Copernicus argues that "the magnitude of the orbit circles should be measured by the magnitude of time." [Copernicus 1543, Bk. I, §10] By this he means that the distance of a planet from the sun is to be determined from its orbital period. Thus he rejects the medieval practice of deriving cosmic distances from Ptolemy's method of nesting celestial shells within each other according to certain proportions. Copernicus argues that only the heliocentric model satisfies the distance–period relationship. In the heliocentric system, the order of the planets is determined by observation of the orbital period of the planets. Copernicus treats the planets as a coherent system. [Figure 1.6]

- Although the assumption of a moving Earth allowed Copernicus to abandon major epicycles, he still needed minor epicycles. Major epicycles were employed to explain the qualitative appearance of retrograde motions. Minor epicycles are small circles, which are needed to eliminate minor quantitative discrepancies between the observations and the geometric models. [See Kuhn 1957, 68] Copernicus needed these minor epicycles because he endorsed the Greek principle of circular motion for the planets. The motion of celestial bodies is "regular, circular and everlasting." [Copernicus 1543, Book I, §4] In fact Copernicus desires to rescue the Greek tradition from Ptolemy. He wants a "system in which everything would move uniformly about its proper center as the rule of absolute motion requires." [Copernicus, *Commentariolus* 1959, 57–8] He swaps the geometric position of the Earth but still clings to the Platonian ideal of the uniform and circular motion, which he attributes to the planet-carrying spheres. [Figure 1.6] He criticizes Ptolemy for his introduction of the equant, although his model used a mathematically equivalent device, an epicyclet. [Gingerich 1993, 36, 175; Neugebauer 1968]

An important aspect of modern science is that observations are regarded as tests of scientific theories. But the Greeks sought to fit the appearances they observed to their prior beliefs about celestial phenomena. Copernicus claims that his work is based on long and numerous observations, his own and those of the Greek tradition. [Copernicus 1543; *Letter Against Werner* 1524; Rheticus 1540; see also

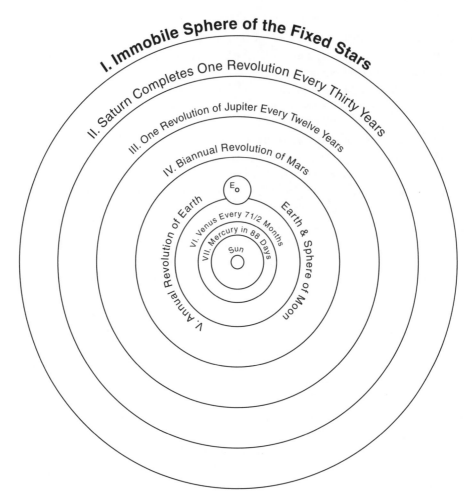

Figure 1.6   The Copernican model of the solar system

Koestler 1964, 203, 581 n. 20] We do not need to doubt Copernicus's sincerity. His book contains long tables of astronomical data, which are largely derived from ancient observations. But Copernicus was also aware that some of these ancient observations were out of date, when compared with modern values.[4] *De Revolutionibus* contains a long discussion of what he calls "artificial" instruments. Such observational instruments serve to determine "the distance between the tropics," the "altitude of the sun," and "the positions of the moon and the stars." [Copernicus 1543, Bk. II, §§2, 14] Nevertheless Copernicus's observations do not establish any *new* facts. The Copernican observations do not go beyond the discoveries of the Greeks. They do not cast in doubt Greek presuppositions about circular

---

[4] See for instance his discussion of the precessions of the solstices and equinoxes, *Revolutions* [1543], Bk. II, §14, Bk. III, §1.

motion. It is therefore fair to say that from an observational point of view, the Copernican and Ptolemaic systems were equivalent. [De Solla Price 1962; Gingerich 1982; Heidelberger 1980] No observation available at the time could trench in favor of one of these competing models. When Tycho Brahe and Galileo Galilei provided new observational discoveries, they had significant implications for Copernicanism. Copernicus's main achievement lay in his awareness of the need to treat the solar system as a coherent system. And he worked out the mathematical consequences of a heliocentric universe.

Although the Copernican treatise *De Revolutionibus orbium caelestium* (1543) has many defects, it arguably set in motion the rise of modern science, whose first phase culminated in the publication of Newton's *Principia Mathematica* (1687). Despite its defects, the Copernican model has greater explanatory power than its rival. Representing the solar system as a *coherent* system, it shows the correlations between many celestial phenomena and relates them to one underlying cause. We can see its explanatory power in the explanation of the seasons.

3.2   The explanation of the seasons

For the sun is not inappropriately called by some people the lantern of the universe, its mind by others, and its ruler by still others. [Copernicus, De Revolutionibus (1543), Bk. I, Ch. 10, quoted in Rosen, Copernicus and the Scientific Revolution (1984), 132]

Any human being is aware of the seasons. Any astronomical model must explain this most obvious of phenomena. But if the Earth sits stationary at the hub of the universe, with the sun orbiting it in a concentric circle, the gliding variations of the seasons cannot be explained. A uniformly moving sun would always remain at the same distance from the Earth, resulting in unchanging seasons. The Greeks were aware of this problem. Ptolemy knew from Hipparchus's observations that "the interval from spring equinox to summer solstice is 94½ days, and that the interval from summer solstice to autumn equinox is 92½." [Ptolemy 1984, Bk. III, §4; Kuhn 1957, 67] Ptolemy employed the eccentric or displaced circle to solve the problem. [Figure 1.3] The seasons have unequal lengths, but they are also asymmetrically distributed across the globe. When it is summer in the northern hemisphere, it is winter in the southern hemisphere and vice versa. Let us fix our attention on two cities, Madrid (Spain) and Wellington (New Zealand). The choice of these two cities can easily be explained. If we could drill a hole through the center of the Earth from Wellington we would reemerge in Madrid. How do you achieve the simultaneous asymmetry between the seasons in the northern and southern hemispheres on a geocentric model? As the Greeks observed that the sun rises high in the sky in the summer and remains low in the winter, and as they took the Earth to be stationary at the center of the universe, they assumed that the annual orbit of the sun around the Earth is tilted. They knew that the tilt was approximately 23.5°. The solution of the puzzle of the seasons results from the tilt of the eccentric circle of the sun. It explains nicely why the sun rises high in the sky in the summer and remains low in winter. The tilt of the ecliptic circle explains the sun's variation in latitude in the same location – like Alexandria, where Ptolemy lived – or in two different locations, like Wellington and Madrid. Consider two Greek expatriates, one living in Madrid, the other in Wellington, in AD 150. For them the Earth is stationary, spherical, and the sun, riding on its ecliptic circle, performs an eccentric, tilted motion around the central Earth. When it is summer for the Greek in Madrid, the tilt of the eccentric will raise the sun high in the sky. For his compatriot in Wellington, it will be winter. The tilt of the ecliptic means that the sun rises high above the equator, leading to short days and long nights in the southern hemisphere. For both, the sun occupies the same location on the ecliptic. Six months later the seasons are reversed. When this device is accurately employed, it seems that "the sun's motion on the eccentric can exactly match the unequal length of the seasons." It can also show why the sun's passage from vernal equinox to autumn equinox takes six days longer, according to modern values. [Rosen 1984, 26; Neugebauer 1968, 91, Kuhn 1957, 67]

How does Copernicus explain the seasons? There are a number of phenomena to be explained, which, as Copernicus insisted, are related to each other. Can the Copernican system solve the problem? Apart from the familiar motions of the Earth: the daily rotation and the annual motion, Copernicus stipulates what he calls the "deflexion of the axis of the moving Earth." [Copernicus 1543, Bk. I, §2, §11] It attributes a third motion to the Earth. The third motion has the function of explaining the change of seasons. Rheticus calls it "the motion of its poles":

> The third motion of the Earth produces the regular, cyclic changes of the season on the whole Earth; for it causes the sun and the other planets to appear to move on a circle oblique to the equator (…). [Rheticus 1540, 150–1]

Why does the Copernican system need to assume a rotation of the Earth's axis to explain the seasons? According to Copernicus the Earth is a planet but it is attached to a sphere, which carries it round the sun. This means, however, that the Earth's axis does not remain parallel to itself. An easy experiment will convince the reader that the axis will not keep its fixed orientation in space. All we need is a pen, a rubber band, and a cup. Attach the pen at an angle to the cup and rotate the cup slowly anticlockwise. Let us say that at the start the pen points from northeast to southwest. We now rotate the cup by 90°. The pen will point from northwest to southeast. The rotation of the cup, which corresponds to the second motion of the Earth in the Copernican system, does not keep the orientation of the Earth's axis constant. Copernicus therefore assumed a third, conical motion, which returns the axis to its original orientation in space. [Kuhn 1959, 165] [Figure 1.7b]

A curious situation confronts us. Both the geocentric and the heliocentric models are able to explain the seasons. Yet the Ptolemaic account seems simpler, since Copernicus needs to postulate a third motion of the Earth. Formally, it makes no difference whether we assume a tilted eccentric circle around a stationary Earth or a tilted axis of a moving Earth around a stationary sun. A simpler explanation is, however, not necessarily the most adequate explanation.

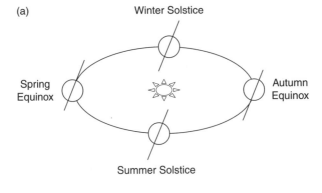

(a)

Winter Solstice

Spring
Equinox

Autumn
Equinox

Summer Solstice

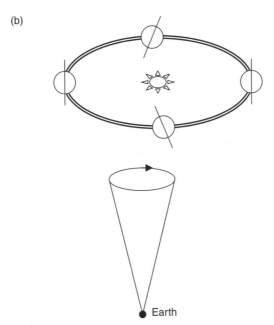

(b)

Earth

Figure 1.7 (a) The seasons as seen from the viewpoint of an observer in the northern hemisphere, according to the modern, Copernican view. The central axis of the Earth is inclined at 23.5° and remains constant with respect to the plane of the orbit around the sun. The model illustrates how the summer and winter solstices are linked and result in the different lengths of the day. [Copernicus, *Commentariolus* 1959, 63]; (b) The seasons as seen from the viewpoint of an observer in the northern hemisphere, according to the Copernican view. The central axis of the Earth is inclined at 23.5° but it does not remain constant with respect to the plane of the orbit around the sun. The Earth is carried on a sphere (double lines) around the "central" sun. As the experiment with the cup shows, this leads to a change of the orientation of the axis, which Copernicus calls "the deflexion of the axis" (of the Earth). The motion performs a small circle, in the opposite direction to the motion of the Earth, to compensate for the changing "tilt" of the axis. In his modification of the Copernican system, Kepler dispenses with the third movement.

The Ptolemaic explanation derives from a model that captures little of the physical reality of the solar system. Unlike the Ptolemaic model, the Copernican model represents the solar system as a proper system. We would expect that the Copernican system succeeds better at explaining the asymmetry of seasons across the hemispheres. With the stipulation of the three motions of the Earth Copernicus hopes to explain the phenomena conjointly, not separately. (Ptolemy can actually ignore the daily rotation of the Earth, since it plays no part in the explanation of the seasons.) As the Copernican model deals with the planets as a system, it has no difficulty in explaining the asymmetry of the seasons and the varying lengths of the days. In this sense the Copernican system has greater explanatory power: by adopting the mobility of the Earth, it naturally explains retrograde motions, the seasons, and the relative distance of the planets from the sun. But the Copernican system is more unwieldy than the Ptolemaic model, at least in this respect. Nevertheless, it retains a closer fit to the solar system than the Ptolemaic model. It was amended when Kepler pointed out that the third motion of the Earth is not needed. Kepler can dispense with it because there is no need for spheres. Astronomy can easily do "without the useless furniture of fictitious circles and spheres." [Kepler 1618–21, Bk. V, Part I, 124] The Earth moves freely around the sun, always keeping its axis constant with respect to an axis drawn through the center of the sun. [Figure 1.7a]

3.3   Copernicus and the Copernican turn

This transformation of the planetary loops from a physical reality to an optical appearance was an invincible argument for the validity of the astronomy of Copernicus. [Rosen, Copernicus and the Scientific Revolution *(1984), 115–16]*

It has become customary to speak of the *Copernican turn* since Kant referred to the Copernican hypothesis in his *Critique of Pure Reason*. [Kant ²1787, Preface]. Kant proposed that philosophy needed a change of *perspective*. Empiricism had regarded the mind as a blank sheet, a *tabula rasa*. Through observation and inductive reasoning humans acquired sense impressions of the material world. From these impressions the mind forms ideas, which slowly fill the *tabula rasa*. Rationalism had equipped the human mind with innate capacities. Through pure thinking the human mind could understand the basic structure of the natural world. Observation was needed only to confirm the postulations of the mind. Kant argued that each approach to knowledge was mistaken on its own terms. What was needed was a synthesis. Empiricism was right to insist on the importance of empirical knowledge. Rationalism was right to insist on the importance of rational principles. The synthesis could be brought about by a Copernican turn in philosophy. Do not look at knowledge either from the perspective of the world–mind relationship, like Empiricism, or the mind–world relationship, like Rationalism. Change your perspective. Knowledge is not the result of an active world etching its stamp on a passive mind (Empiricism). Nor is knowledge the result of an active mind putting its seal on a passive world (Rationalism). Human knowledge comes about through a partnership between an active mind and an active world. The mind already comes equipped with basic principles about causality, substance, space, and time. But they are too abstract to constitute empirical knowledge; they are presuppositions to objective knowledge. They need the encounter with the empirical world to give rise to empirical knowledge. The rational mind seeks a union with the empirical world. This is the Copernican turn in philosophy.

Copernicus himself provides us with a sense, a very primary sense, of what we mean by the Copernican turn. *It is a shift in perspective.* Copernicus invites the reader to change the focus of the explanation. Consider an object, which appears to move. And consider an observer, who appears to be stationary. Sitting on a train at a platform, the passenger is often momentarily unclear as to whether her/ his train has started to move or whether it is the train on the neighboring rail which has begun to pull out of the station. If we exchange the perspective between object and observer the motion remains invariant. We can describe it as motion of our train in the forward direction or as motion of the other train in the opposite direction. What is true for the train is true for the planets. If the sun appeared to move past the stationary observer on Earth from east to west, it is now the observer who must move past the "stationary" sun from west to east. In this change of perspective, some features must remain *invariant*. As we saw in the explanation of the seasons, Copernicus exchanges the tilt in the sun's ecliptic against a tilt in the Earth's axis. The tilt ($23.5°$) remains invariant but the tilted sphere passes from the sun to the Earth. How will this change of perspective work in the case of an observer on Earth? From the point of view of an apparently stationary Earth-bound observer, the fixed stars seem to move from east to west, while the planets generally move from west to east, with the exception of retrograde periods. If we change the perspective and make the Earth-bound observer move from west to east in the daily rotation of the Earth, the movement remains but the direction changes. The sun appears to us to rise in the east and to set in the west. If we hold the sun fixed and make the Earth turn on its own axis from west to east, the orbit of the sun through the sky remains the same but then its direction changes. In fact, all the properties of the apparent movement of the sun through the stellar constellations – the *ecliptic* – remain constant, only the perspective of motion changes. This, as Copernicus argued, is altogether more economical. It is more rational for the motion of the Earth to produce the apparent rapid motion of the fixed stars than it is for the fixed stars to rotate rapidly once on their sphere in a 24-hour rhythm. Copernicus announces his change of perspective very early on in the book:

> Although there are so many authorities for saying that the Earth rests in the center of the world that people think the contrary supposition inopinable and even ridiculous; if however we consider the thing attentively, we will see that the question has not yet been decided and accordingly is by no means to be scorned. *For every apparent change in place occurs on account of the movement of either of the thing seen or of the spectator,* or on account of the necessarily unequal movement of both. For no movement is perceptible relatively to things moved equally in the same directions – I mean relatively to the thing seen and the spectator.

This passage describes the change of perspective, which will have to leave the observations invariant:

> Now it is from the Earth that the celestial circuit is beheld and presented to our sight. Therefore, if some movement should belong to the Earth it will appear, in the parts of the universe which are outside, as the same movement but in the opposite direction, as though the things outside were passing over. And the daily rotation in especial is such a movement. For the daily revolution appears to carry the whole universe along, with the exception of the Earth and the things around it. And if you admit that the heavens possess none of this movement but that the Earth turns from west to east, you will find – if you make a serious examination – that as regards the apparent rising and setting of the sun, moon, and stars the case is so.

Copernicus concludes the argument with a slightly veiled appeal to Ockham's razor. This is the principle, stated very liberally and without respect for its original context, that simplicity of explanation is a great virtue in science. Of two different explanations concerning the same phenomenon, the simpler explanation is generally to be preferred. A simpler explanation is not a simplistic explanation. It is an explanation that leaves fewer things unconnected and explains more things with fewer principles.

> And since it is the heavens, which contain and embrace all things as the place common to the universe, it will not be clear at once why movement should not be assigned to the contained rather than to the container, to the thing placed rather than to the thing providing the place. [Copernicus 1543, Bk. I, §5; Bk. III, §1]

Copernicus hints at a *second* feature of the Copernican turn. The shift in perspective, which occurs on the background of some invariant feature, must be accompanied by some explanatory gain. If it were not, we would have a mere exercise in *perspectivism*. We would have different perspectives, all equally valid, without recourse to an adjudication between them. But such perspectivism would not be true of the history of science. Copernicus takes great pain to argue that the Copernican hypothesis gives us explanatory advantages. He uses the movement of the Earth as the more plausible principle. [Copernicus 1543, Bk. I, §1; see also the seven principles in *Commentariolus* 1959, 58] From this perspective the relative distance of the planets can be determined. Retrograde motion is not a problem of geometry. It is a physical reflection of our position in the solar system. It is true that the explanation of the seasons is more cumbersome from the Copernican perspective. But this could easily be amended by abandoning the spheres. The Copernican explanation is then a better approximation to the appearance of the seasons than the Ptolemaic account.

A shift in perspective is an important feature of the Copernican turn. Many great scientists began with a shift in perspective. Darwin, as we shall see, argued for a shift in perspective with respect to the great problem of his time: the "origin" of species. The scientists argued for a change in perspective to increase the explanatory gain, while keeping other things invariant. What about influential thinkers in the social sciences, like Marx and Freud? These thinkers, too, brought about a change in perspective. But it is a matter of dispute whether the explanatory gain, for which they claimed credit, really does accrue to their shift in perspective. This shadow of doubt also hangs over Copernicus. It explains why Copernicus is not seen as a true scientific revolutionary. A scientific revolution requires a change in perspective. But a mere change in perspective does not constitute a scientific revolution. An important ingredient in a scientific revolution is some "explanatory gain." If doubts remain about the explanatory gain achieved by a turn in thinking, there are doubts about a true revolutionary impact. This was Copernicus's problem, as we shall see. Thinkers like Copernicus and Freud, however, remind us of a further consequence of a scientific revolution: it has a significant impact on the way people begin to perceive the world. In the case of Copernicus this leads to a loss of centrality. In the case of Darwin it brings about a loss of design. And in the case of Freud it results in a loss of transparency.

A scientific revolution may need time to unfold. It is possible for one thinker to introduce a change of perspective and for others to complete the picture. The history of astronomy from Copernicus to Newton illustrates this point. In its etymological sense, the term "revolution" indicates an uprooting, a reversal and an overthrow of old established views or conditions. Working with the notion of "turn of ideas" or "shift of perspective" allows us to focus on scientists who completed the shift of perspective. The Copernican turn consisted in a realignment of the geometric arrangement of the planets. Astronomers built models out the existing material: the six planets known from antiquity to the eighteenth century. Once the components are at hand an immediate question imposes itself. How are these elements to be arranged with respect to each other? The Greeks started a long tradition of model-building in the history of astronomy. It consisted of two tasks: first, to determine a *topologic* structure of the model, which would arrange the planets in a geometric or spatial order. Most Greek astronomers opted for a geocentric arrangement. Copernicus reversed this tradition by choosing a heliocentric arrangement. Once a topologic structure is chosen, an *algebraic* structure for the model must be found. The algebraic structure determines the quantitative relationships between the components in the models. The Greeks worked with various geometric devices: eccentric circles or deferents and epicycles. Copernicus changed the topologic structure of planetary models. But he retained the geometric assumptions of his Greek predecessors. For this reason Copernicus never achieved the explanatory gain associated with a scientific revolution. Any explanatory advance to which Copernicus can lay claim accrues to the topologic structure of the heliocentric model. Copernicus made no contribution to the algebraic structure of planetary models. The explanatory gain in algebraic structure was achieved slowly through the work of Kepler, Galileo, and Newton.

We can applaud Copernicus for his introduction of a shift in perspective, and yet credit his brilliant successors with the completion of the Copernican turn. In a scientific revolution, a change of perspective against the background of invariant elements must be augmented by an explanatory gain in the algebraic structure. We shall see that Darwin's theory was able to offer the explanatory gain, while Freud failed as much as Copernicus. We do no harm in considering Copernicus's work as the dawn of the Copernican revolution and modern science. Copernicus is more a major figure in the history of scientific ideas than in the history of scientific revolutions. [Blumenberg 1957; 1981, Part I, I; Part III, II; 1965, II] Copernicus had a major impact on the way humans placed themselves in the wider cosmos. We shall see that major philosophical questions arise from the Copernican turn and the Copernican revolution. But let us first complete the story of the Copernican revolution.

3.3.1   A philosophical aside: From empirical adequacy to theoretical validity   The preceding sections harbor some philosophical lessons. The explanation of the seasons in geocentrism and heliocentrism, respectively, shows that in scientific explanation we require more than an agreement of the model with the empirical data. Let us say that two incompatible models, which agree with the empirical evidence, enjoy *empirical adequacy.* Both the Ptolemaic and the Copernican models can explain the available data equally well, but they do so on the assumption of different structures. Both are possible explanations with respect to the available evidence. The Copernican model reveals, however, a better topologic structure than the geocentric model. In order to mark the difference in fit we shall say that it gains *empirical validity.* We also require of a model that its mathematical structure must be in agreement with the structure of the target system. In order to achieve this fit the model must become a structural model or theory. [See Section 6.5.1] As the mathematical structure explains the observable phenomena, we shall say that an explanatory theory must acquire *theoretical validity.* We see the need for theoretical validity in the history of planetary models. The Greeks strived to "save the phenomena." They tried to match their sense observations with their presuppositions about planetary motion. The geocentric model was fairly accurate in its predictions of planetary motion but it was based on a mistaken structure: devices like eccentric and epicylic motion. As these devices do not reflect any physical mechanism, they have no theoretical validity. Although Copernicus also employs these devices, his model arranged the planets in a spatial order, which is close to the spatial (topologic) structure of the solar system. In this respect it was empirically valid. The heliocentric model, in its Keplerian form, enhances the approximation of the model to the reality of the solar system, because it replaces the traditional geocentric devices by a new algebraic structure. With Newton, it finally becomes a theory. As we shall see in the following section, Kepler discovered mathematical laws to describe planetary motion which no longer require the planets to be carried on spheres in circular orbits. The upshot is that we want the model assumptions to be more than instrumentalist hypotheses. The model assumptions must be in agreement with the structure of the natural system. [See Section 6.5] This requirement points us toward a discussion of instrumentalism and realism. [Section 6.2]

3.4   Copernicus consolidated: Kepler and Galileo

Kepler's marvelous achievement is a particularly fine example of the truth that knowledge cannot spring from experience alone but only from the comparison of the inventions of the intellect with observed fact. [Einstein, "Johannes Kepler" (1930), 266]

Figure 1.8 Tycho Brahe's observatory on the Island of Hven. *Source*: *Nature* **15** (1876–7), p. 407

Tycho Brahe (1546–1601)

Copernicus launched a new research program, whose completion relied on some groundbreaking contributions. The next central figure to enter the stage is Tycho Brahe, a Danish astronomer (1546–1601). Brahe was a lifelong opponent of Copernicanism. Nevertheless he occupies a pivotal position in the history of heliocentrism. For Brahe developed ingenious observational methods and collected a wealth of new data: [Figure 1.8]

- In 1572 he discovered a new star, which at first shone very brightly in the sky but later disappeared. Brahe had in fact discovered a supernova. This is the appearance in the sky of a very bright light, owing to the momentous explosion of a massive

star. The light increases its brightness hundreds of millions of times in the span
of only a few days.

- Between 1577 and 1596, Brahe discovered comets in the sky, the orbits of
which had to be located beyond the moon's sphere. Perhaps the most famous
comet is Halley's comet, named after the Astronomer Royal, who used Newton's
theory to predict its orbit. More recently, Earthlings had the visit of comet
Hale–Bopp, whose closest approach to Earth occurred on March 22, 1997 at
123 million miles. [Figure 1.9]

These discoveries were highly significant, because they raised serious questions
about the immutability of the heavens, a feature of the supralunary sphere in
Aristotelian cosmology. The appearance of a supernova far beyond the sublunary
sphere was not compatible with the dogma of its never-changing nature. The orbits
of comets are highly elliptical. For instance, the orbit of Halley's comet crosses the
orbits of the outer planets, reaching almost as far as Pluto before it returns to
Earth. Comet Hale–Bopp traversed the solar system from outer space, returning
from its last visit in 2214 BC. On the Aristotelian–Ptolemaic view, such orbits should
simply not be permitted. Recall that for these reasons Galileo's visitors refused to
contemplate the existence of Jupiter moons. An alternative attitude is a kind of

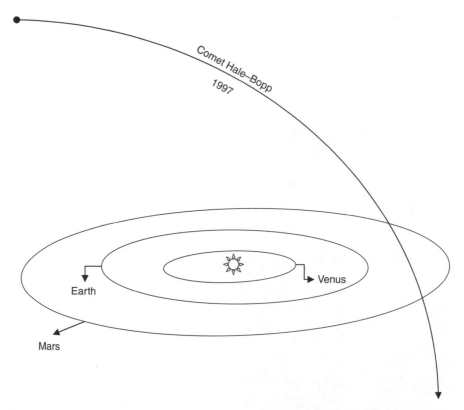

Figure 1.9 A schematic view of the orbit of comet Hale–Bopp between Venus and Mars
on its way through the solar system in June 1997. The orbits of Mars and Venus are inclined
with respect to the ecliptic

defeatism, which we find in the concluding part of Ptolemy's *Almagest*. Our knowledge of celestial bodies is so limited that what is impossible according to our model – comets smashing through the spheres – may turn out to be possible in the heavens. [Ptolemy 1984, Bk. XIII, §2]

Tycho Brahe proposed a compromise between the Copernican and Ptolemaic systems. In his system the Earth is at the center of the universe with the moon and the sun circling around it; but the other planets circle around the sun. Brahe's system was important to those astronomers who wanted to use the Copernican system because of its calculational advantages but could not adopt the motion of the Earth for philosophical reasons. [Kuhn 1957, 202] It leads to better astronomical tables.

Johannes Kepler (1571–1630)

Without Brahe's astronomical data, Johannes Kepler could not have discovered his three astronomical laws. They are the first mathematically precise laws in astronomy:

1. The **first law** states that the orbit of planets is not circular but elliptical. With this law, the ancient ideal of circular motion is consigned to the dustbin of the history of ideas.

2. The **second law** gives up the notion of a uniform motion, which had still been assumed by Copernicus. It states that the orbital period of each planet varies in such a way that "A line drawn from a planet to the sun sweeps out equal areas in equal times." A planet near the sun moves faster than a planet further away, but a line joining each planet to the sun sweeps through equal areas of the ellipsis in equal intervals of time.

3. The **third law** establishes a relation between the speeds of planets in different orbits, P, and their average distance from the sun, A: $A^3 \propto P^2$.

Some believe that the statement of these laws makes Kepler the true revolutionary in the history of astronomy. [Koestler 1964, Part IV] Recall the distinction between *topologic* and *algebraic* structure. Kepler rejects much of the Greek tradition to which Copernicus still adhered. The fictitious circles and spheres (Kepler 1618–21, Bk. IV, Part I, 124), and even more importantly the doctrine of circular motion, are rejected as elements of the topologic structure. With his three laws, Kepler makes a major contribution to the improvement of the algebraic structure of the heliocentric model. Further, Kepler wants to build an astronomy based on the physical causes of planetary motion. He makes a proposal that solar heat and light may keep the planets in their elliptical orbits. [Kepler 1618–19, Bk. IV, Pt. II] His intention is to appeal to natural powers, rather than "intelligences," to "move the planets." He attributes a "motor soul" to the sun. Unsurprisingly, Kepler's proposal failed. Several more steps were needed before the Copernican revolution was completed. The completion required that the proponents of heliocentrism shared some but not all of the basic convictions.

This point is well illustrated in the work of Galileo Galilei. Galileo is a Copernican who hardly takes any notice of Kepler's achievement. He ignores Kepler's discovery of the elliptical orbit of all planets and embraces the notion of circular movement as the natural movement of all bodies. Yet his importance for science cannot be doubted. It is threefold:

Galileo Galilei (1564–1642)

1. He defends the Copernican worldview and provides new evidence, through his work with the telescope, which discredits the Aristotelian–Ptolemaic view, and bestows credit on the heliocentric hypothesis. As we have already noted, Galileo starts with new presuppositions. What is important, he argues against scholastic scholars, like the Mathematician and Philosopher in Brecht's play, is the use of observations and the mathematical description of nature. All of Galileo's observations provide evidence that the heavens are not immutable.

    (a) The Jupiter moons, which he would like his guests to observe through the telescope in lieu of a scholarly dispute, provide a visible scale model of the Copernican solar system. The moons orbit Jupiter as their center. If Jupiter were carried on a crystal sphere, the moons would break through it. Contrary to the Philosopher's opinion, there are celestial bodies which "orbit around centers other than the Earth."

    (b) The study of the topography of the moon shows the similarity between the Earth and the moon. It casts into doubt the rationale of the two-sphere universe.

    (c) The observations of the sunspots, like the moon's surface, conflict with the assumed perfection of the celestial region of the universe.

    (d) The phases of Venus provide direct information about the shape of Venus's orbit. As Venus lies inside the orbit of the Earth, Earth-bound observers see it illuminated in different orientations. It provides direct proof that at least Venus orbits the sun. [Koestler 1964, 431–5; Kuhn 1957, 222–4; Copernicus 1543, Preface; DeWitt 2004, 156–64]

    (e) The study of the Milky Way hints at the potential infinitude of the universe.

2. Galileo develops the science of mechanics. It paves the way for a modern theory of motion, which dispenses with "pushes" and "impulses." Galileo develops his fall law, according to which all objects fall at the same rate, given by the gravitational acceleration near the surface of the Earth. Galileo also formulated a *principle of relativity*. [Galileo 1953, 199–201] A system at rest and a system in constant motion are equivalent from the physical point of view. The systems are invariant to this change of viewpoint. Galileo offers a famous thought experiment to demonstrate the equivalence of inertial systems. In a cabin below the deck of a large ship, observe the behavior of "flies," other "small winged

creatures," and "fish in a bowl." At first the ship is at rest. When the first set of observations is completed, let the ship proceed with uniform speed. The observations will reveal no difference in the behavior of the creatures. [Galileo 1953, 199–201] Anything that happens to objects in these systems happens according to the same laws. Thus, although we change the perspective, the regularities are invariant. Galileo's discovery of the principle of relativity was vital for the understanding of the most pressing question of astronomy: "How do planets move in their orbits?" It was left to Newton to provide the final answer. [Section 5]

3. Finally, he becomes an ardent defender of the freedom of scientific inquiry against the interference of the Church. Like Roger Bacon in the Middle Ages, Galileo pleads for a separation of theology and natural philosophy. Passages in the Bible may not literally mean what they appear to say. For this reason we should not use biblical passages to call in question what observations or mathematical reasoning teach us.

Kepler and Galileo developed the heliocentric model; standing on their shoulders Newton completed it. We begin to see more clearly why Copernicus was not a scientific revolutionary.

4 Copernicus was not a Scientific Revolutionary

Therefore, since the sun is the source of light and eye of the world, the center is due to it in order that the sun (...) may contemplate itself in the whole concave surface (...) and take pleasure in the image of itself, and illuminate itself by shining and inflame itself by warming. [Kepler, Epitome of Copernican Astronomy *(1618–21), Bk. IV, Part I, §2 (20)]*

Ever since Copernicus effected his Copernican turn, the question has been asked whether he was a scientific revolutionary. Copernicus himself and his disciple Joachim Rheticus were aware of the explosive nature of even a Copernican turn. In his Dedication to Pope Paul III, Copernicus admits that his heliocentric hypothesis will strike many of his contemporaries as absurd. Rheticus seems to find it necessary to emphasize that Copernicus was not "driven by lust for novelty." [Rheticus 1540, 187] But the geocentric view is unable to explain the "remarkable symmetry and interconnection" of planetary motions. The ancients failed because they did not regard the planets and their motions as a system. [Rheticus 1540, 138] As we have seen in the foregoing discussion, it is important for a scientific theory to explain all the phenomena that fall into its domain. Rheticus appeals to this criterion when he holds that only those hypotheses that can explain both apparent anomalies of planetary motion are acceptable. [Rheticus 1540, 168]

There was a clear perception that the Copernican turn bore the seeds of a new worldview. But was the Copernican turn revolutionary? Many scholars have considered this question. Some will give Copernicus very little credit. Copernicus's book is "little more than a re-shuffled version of the *Almagest*." [De Solla Price 1962, 215] The heliocentric system is not an improvement in computation over the geocentric system but "it is more pleasing to philosophical minds." [Neugebauer 1968, §10; Koestler 1964, Part III] The Copernican system has aesthetic advantages. [Kuhn 1959, 171–81] It also explains two gross planetary irregularities without resorting to major epicycles: retrograde motion and the varying times planets need to complete their orbits around the sun. [Kuhn 1959, 165–71] As we have seen, it also explains the seasons, although this explanation is aesthetically less pleasing than the Ptolemaic attempt. Most historians of science agree that Copernicus did not accomplish a scientific revolution.[5]

There are many reasons for this judgment. *Firstly*, Copernicus is still committed to the Greek ideals of circular motion. His main objection against Ptolemy is the use of the equant, which violates the ideal uniform circular orbs.

Secondly, there is an inconsistency in the Copernican treatment of planetary motion, which reveals itself in a discrepancy between the first part of the *De Revolutionibus* and the rest of the book. In Part I, Copernicus starts confidently with an assertion of the annual motion of the Earth around the sun. He believes that the motion is real and that it has explanatory value. But in the technical sections of his book, we encounter what Ptolemy called the "equivalence of hypotheses." Different geometric techniques are regarded as equivalent for the description of planetary motions. It is true that they may be "sufficient for the appearances" but they do not provide real explanation. Copernicus's indifference toward different methods reveals that he is not concerned with a physical explanation of the appearances. Such a physical explanation is, however, required to advance astronomy beyond a mere description of planetary orbits. Copernicus agrees with the Greeks that "planets are not carried on homocentric circles." [Copernicus 1543, Bk. V, §3] This geometric device fails to account for apparent irregularities in planetary motions. But he relies on the techniques to which the Greeks had already resorted: the use of deferents and epicycles. He regards these alternative techniques as equivalent and as "sufficient for the appearances." [Copernicus 1543, Bk. V, §4] In this respect Copernicus made no progress over Ptolemy. Kepler rightly complained that his predecessors had sought the "equipollence of their hypotheses with the Ptolemaic system." [Kepler 1618–19, Bk. IV, Pt II, §5] For this reason we need to distinguish between empirical adequacy and theoretical validity.

Thirdly, there are more dynamic reasons why Copernicus is not regarded as a scientific revolutionary. Copernicus employs the impetus theory to confer natural circular motion on the Earth. This explains why buildings do not crumble to the ground when the Earth turns but it does not answer the central question of sixteenth-century astronomy: why planets orbit the sun at varying speeds and distances. Copernicus offers geometric devices, which Kepler had to replace by physical laws. But Copernicus has no concept of inertia or gravitation. The concept

[5] See Dreyer [1953], 342–4; Dijksterhuis [1956], Part IV, I; Koestler [1959], 148–9, 213; Mittelstraß [1962], IV.6; Neugebauer [1968], 92, 103; Rybka [1977], 171; Wolff [1978], Part III, 8; Gingerich [1982]; Blumenberg [1981], Part I, VI, 99; Rosen [1984], 133.

of planetary motion around the sun allows Copernicus to abandon the major epicycles. But his concept of steady circular motion forces him to adopt minor epicycles. From a mathematical point of view his system is not much simpler; from the physical point of view it leaves unanswered "why"-questions.

Nevertheless, Copernicus initiated a Copernican *turn*. [Dreyer 1953, 342–3] His change of perspective brought some noteworthy advantages to astronomy. The most important is, as Rheticus repeatedly stresses, that Copernicus binds the planets into a coherent system. Move one sphere out of place and you disrupt the entire system. [Rheticus 1540, 147; Copernicus 1543, Preface] As we have seen, Copernicus was very aware of the importance of coherence:

> (The) Mobility of the Earth binds together the order and magnitude of the orbital circles of wandering stars. [Copernicus 1543, Bk. V, Introduction]

Kepler also perceived this advantage very clearly:

> Ptolemy treats planets separately; Copernicus and Brahe compare the planets with one another. [Kepler 1618–9, Bk. I, Part I, §5]

The conception of the coherence of planetary phenomena obliges the Copernicans to build a model of the planetary system which must accommodate all the known empirical data. In this respect the Copernican model is partially successful. By correlating the movement of the "wandering stars" with the "circular movement of the Earth," "all" phenomena follow, so Copernicus claims. [Copernicus 1543, Preface] Although they do not all follow, the Copernican system naturally explains the appearance of retrograde motion of the planets and the seasons; it correctly determines the order and relative distances of the planets from the sun. [Copernicus 1543, Bk. I, §10] It also makes the daily and annual motion of the Earth around the sun a reality, rather than a computational device. [Copernicus 1543, Bk. I, §11] The successes and failures of the Copernican system provide useful indications as to the criteria of scientific revolutions.

4.1   The Copernican method

In the center of all rests the Sun (…) as if on a kingly throne, governing the family of stars that wheel around. [Copernicus, De Revolutionibus (1543), Bk. I, Ch. 10, quoted in Gingerich, The Eye of Heaven (1993), 34]

Although Copernicus relied to a large extent on ancient observations, he was no stranger to making his own observations. At the same time Copernicus was aware of the theoretical work of his predecessors. He shows much respect for Ptolemy. Unsurprisingly, a particular mention is reserved for Aristarchos of Samos who anticipated a heliocentric system. In his appreciation and awareness of the work of his forebears, Copernicus in turn anticipates Charles Darwin. There are two noteworthy elements in these stories of discovery. Copernicus – and this is true of

Darwin, as we shall see – enters a conceptual space in which some theoretical accounts already vie for attention. These theoretical accounts claim to be able to account for the "appearances." The Copernican model arrives in an inhabited niche. This conceptual space already accommodates an elaborate system of geocentrism, a sketchy report of heliocentrism, ancient observations, and the impetus theory of motion. As we know, Copernicus made his own observations, which did not, however, lead to new discoveries. The existence of a conceptual space allows us to infer two points. Copernicus did not arrive at his heliocentric system by way of an inductive generalization over the available observations. And secondly, we find in Book I of *De Revolutionibus* and in *Narratio Prima* an explicit consideration of the virtues and vices of contrasting models of the solar system.

Rheticus has left us a brief statement of the Copernican method. First, he reports, Copernicus compared the ancient and medieval observations with his own findings, "seeking the mutual relationship which harmonizes them all." [Rheticus 1540, 163] He then compared these observations with the "hypotheses of Ptolemy and the ancients." The examination shows that the ancient hypotheses do not stand up to the test. Copernicus was forced to adopt new hypotheses, elements of which, as he himself acknowledged, he found in the existing store of astronomical knowledge. Rheticus embellishes the situation. Copernicus found the geocentric hypothesis wanting for reasons of economy and simplicity, not because it was in direct contradiction with the observations. In his Sketch of the Heliocentric System (*The Commentariolus*), of which only handwritten copies existed during his lifetime, Copernicus admits that the Ptolemaic system is "consistent with the numerical data." However, it also postulates the geometric device of the "equant," which Copernicus finds aesthetically objectionable. It violates his belief in heavenly uniformity and regularity. [Copernicus, *Commentariolus*, 1959, 57; see Rosen 1959, 38; 1984, 67] Copernicus considers a heliocentric hypothesis on the background of the geocentric tradition. By applying mathematics, Rheticus continues, Copernicus

> geometrically establishes the conclusions which can be drawn from them [i.e. the new hypotheses] by correct *inference*; he then harmonizes the ancient observations and his own with the hypotheses which he has adopted; and after performing all these operations he finally writes down the laws of astronomy. [Rheticus 1540, 163; italics added]

The "laws of astronomy" are the circular uniform orbs, which Kepler replaced. What is important in the present context is the observation that Copernicus made *inferences*. He uses the available data to infer that the Ptolemaic model was inadequate. Simultaneously, he infers from the data that the Copernican hypothesis is more adequate. The available data do not just consist of observations. Copernicus employs impetus considerations to parry the traditional plausibility arguments against the motion of the Earth. The Copernican inference is double-pronged. The same observations, which discredit the geocentric models, lend some credit to the heliocentric model. Furthermore, deductive consequences follow from the heliocentric model, which are in better harmony with the observations. For instance, the correct order and the relative distances of the planets are deductive consequences of heliocentrism. It is also a deductive consequence of heliocentrism that retrograde motions are an artifact of geocentrism. As we shall see, inferential practices are of great importance in the history of science. These are not to be confused with simple induction by enumeration. Scientists like Copernicus faced available evidence and competing models. They use observational evidence and other criteria, like the probability of explanation, to infer that some models are more appropriate than others. Observational evidence, probability considerations, and the impetus theory of motion now act as constraints on the acceptability of competing models. In a later chapter we will treat this basic procedure – inferring the cognitive adequacy of one model from the available constraints and simultaneously discrediting a competing model – as the method of **eliminative induction**. We face a competition between rival models, which claim to explain the available evidence. Each model is based on different presuppositions – geocentric vs. heliocentric assumptions – but no model enjoys absolute validity. Rather, it is a question of explanatory weight. Given the observations and other constraints, which system provides the more *likely* explanation? We will find this attitude in Darwin. The hypothesis of natural selection is a more likely explanation of species diversity than the design argument. The Copernicans employed *probability* arguments in favor of heliocentrism. It is physically more probable, they said, that the Earth turns once on its own axis in 24 hours than that the sphere of the fixed stars moved "at incalculable speed," in the same period, around a stationary Earth. [Kepler 1618–21, Pt. I, §3] And so, Kepler continues,

> (…) it is more probable that the sphere of the fixed stars should be 2,000 or 1,000 times wider than the ancients said than that it should be 24,000 times faster than Copernicus said. [Kepler 1618–21, Bk. IV, Part I, §4 (43)]

The annual movement of the Earth around the sun gives us "a more probable cause for the precession of equinoxes." [Kepler 1618–21, Pt. II, Book IV, §5; Copernicus 1543, Book I, §6]

As Copernicus and Kepler clearly saw, some models are better at dealing with the evidence than others. An inference to a model, which is more adequate with respect to the available constraints, is not an inference to *the* true model. The constraints themselves are subject to critical scrutiny. Copernicus still considered uniform circular motion as an all-important constraint and demanded that the Copernican hypotheses save the appearances. [Rosen 1959, 29] The history of heliocentrism from Copernicus to Newton confirms that a better model is better relative to both the available evidence and more theoretical considerations. With his planetary laws, Kepler introduced important changes into the Copernican model. Tycho Brahe and Galileo Galilei recorded observations which are more consistent with heliocentrism than geocentrism. Toward the end of the seventeenth century, Newton combined the idea of inertia and gravitation to arrive at a plausible mechanical explanation of why planets stay in orbits. In 1687 the constraints on an adequate model of astronomy had changed considerably. But now a difficulty confronts us. Recall that the geocentric model explains the seasons as well as the Copernican model. Why should we prefer the latter, given its additional redundant assumption of a third motion of the Earth?

4.2   The relativity of motion

Galileo introduced into physics the principle of the relativity of motion. (Einstein later adopted this principle and generalized it.) Insects fly through the cabin in the same manner, irrespective of the inertial motion of the boat. According to the principle of relativity, the kinetic motion of an object can be described from either a stationary or a moving reference frame. As long as the motion is inertial (either at rest or moving at constant velocity), both views are equivalent. They must lead to the same numerical results. It is a matter of choice, which system we regard as the frame at rest and the frame in motion, respectively. This makes no difference to the physics of the situation.

From the point of view of relativity it should therefore make no difference whether we adopt a geocentric or a heliocentric view. [Born 1962, 344; de Solla Price 1962, 198; Rosen 1984, 183–4] We can follow Ptolemy: regard the Earth as a stationary frame and the sun as a moving frame. Or we can follow Copernicus: regard the Earth as a moving frame and the sun as a stationary frame. According to the principle of relativity our choice makes no difference to the physics of the situation. And so it appears to be. The Earth turns on its own axis once in a 24-hour rhythm to give us day and night. If the sun turned around the stationary Earth once in a 24-hour rhythm it would give us day and night. The seasons result from either a tilted orb of the sun around the Earth or a tilted Earth around the sun. However, there is more to a description of the solar system than mere kinematics. From a strictly kinematic point of view, the models are equivalent. The kinematic point of view is concerned only with pure motion, without regard to its causes. [Dijksterhuis 1956, I, §83; IV, §18, IV, C] This is the Ptolemaic and Copernican perspective. But there is also the question of dynamics: What causes the planetary bodies to move? Imagine you sit on a train that has stopped at a station. Through the window you observe a train moving slowly along the rails. Your intuition tells you that you are stationary and the other train is moving. But physics informs us that your train can be regarded as moving and the other train as stationary. The kinematics will be the same. But now imagine that the locomotive has been removed from your train. The dynamic situation is no longer equivalent. The moving train clearly has a locomotive which causes its motion. Your train has lost its cause of motion. Kepler was preoccupied with the question of physical causes. He suspected that energetic rays from the sun drove the Earth around its elliptical orbit. When a planet shows its "friendly face" to the sun, its magnetic lines attract it. When a planet shows its "unfriendly face" to the sun, its magnetic lines repulse it. The game of attraction and repulsion constrains the planet to its orbital motion around the sun. [Kepler 1618–21, Pt. II, §93] As Newton showed, this dynamic explanation was mistaken. Nevertheless, Kepler advanced dynamic arguments in favor of the orbital motion of the Earth. Once Newton showed why the planets stay in their elliptical orbits around the sun, the heliocentric model gave a better representation of physical reality than the geocentric model. Newton improved the algebraic structure of the model. He provided a dynamic explanation of planetary orbits in a heliocentric model. And even if we focus only on the spatial arrangement of the solar system, the heliocentric model captures the topologic structure of the solar system better than the geocentric model. We suspect that the model structures will correspond differentially to the structure of the physical system.

5   The Transition to Newton

The telescope was a curiosity on display at the annual fair before, in Galilei's hand, it became an instrument of theory. [Blumenberg, The Genesis of the Copernican World *(1987), 648]*

Isaac Newton (1642–1727)

Newton had many reasons to believe that he was standing on the shoulder of giants. But he went one step further and produced the Newtonian synthesis. Newton's physics is not just a body of laws of mechanics, which govern the world of macro-objects both on Earth and in the heavens. It encapsulates a whole new view of the universe – a whole new image of how humans are to conceptualize the material world around them.

We can characterize the Scientific Revolution by two closely connected features: (A) the destruction of the ancient cosmos and the disappearance of all considerations based on its presuppositions; (B) the mathematization of nature and science. [Koyré 1957, 2, 29, 43, 61–2; 1965, 6–8] Let us look at these features in more detail.

(A)   *The destruction of the ancient cosmos.* We have encountered some of the features of the traditional cosmic world-order:

- its hierarchical two-sphere structure between the perfection of the supralunary sphere and the imperfection and decay of the sublunary sphere;
- its distinction between terrestrial and celestial physics;
- its finite and closed nature;
- its energy-deficiency.

We have seen how both astronomical observations and theoretical constructions began to chip away at the traditional cosmic world-order. Associated with the destruction of the traditional world-order is the disappearance of all considerations based on its presuppositions. Material causes replace final causes. It is not the aim of the stone to strive back to its natural place in the universe. The stone is subject to a downward accelerating force. Planets do not stray from their orbits because they obey physical laws. The stars in the firmament do not twinkle for the enjoyment of humankind. [Burtt 1932, 17–24]

Galileo conveys to his pupil Andrea the liberating elation of breaking out of the "walls and spheres and immobility." Once "the breaking of the circle" or "the bursting of the sphere" (Koyré 1965, 7 n. 1) had taken place, the new universe could take on infinite dimensions.

(B)   The second feature of the scientific revolution – the *mathematization of nature* – had equally important consequences for the development of Western civilization. It inspired a model of the universe that runs in accordance with deterministic laws. This image of the universe as a clockwork reached as far as Darwinism. [See Burtt 1932, 202, 206; Weinert 2004, Ch. I; Wendorff 1985, 144]

The language of mathematics applies to natural processes. It offers the great advantage of *algorithmic compressibility*. This means that a great number of data can be compressed in a precise mathematical equation. For instance, Kepler's third law establishes a relation between the orbital period of a planet around the sun, P, and its average distance from the sun, A (expressed in units of the Earth–sun distance AU). The law states that the square of the orbital period is directly proportional to the cube of its average distance from the sun:

$$A^3 \propto P^2.$$

This law can be used to find, for any body orbiting the sun (even a spacecraft), either the average distance from the period or the period from the average distance. For example, if $A = 4$ AU, then $P = 8$ years. Thus, Kepler's third law compresses into one neat formula a great multitude of data. All objects orbiting the sun, from planets to satellites, are subject to this law. The law expresses the structure of the orbits around the sun.

For Galileo and Newton, the book of nature was written in the language of mathematics. Newton's great achievement was to have provided a synthesis between the mechanics of the heavens (Kepler) and of the Earth (Galileo). Newton destroyed the two-sphere universe. Whether or not the apple fell on his head, the lesson from this episode is correct. The same force that makes the apple fall on his head keeps the planets in their orbits. Newton was able to formulate three fundamental mechanical laws to which many terrestrial phenomena – from accelerating cars, to colliding balls, moving elevators, and orbiting planets – were subjected:

1.   The law of inertia states that objects retain the same state of motion or rest unless some external force interferes.
2.   The force law states that inertial motion can be subject to the application of a force, which changes its direction and momentum.
3.   The law of interaction (action–reaction) states that to every action there is a reaction, equal in force and opposite in direction.

The growing mathematization of nature, to which Kepler, Galileo, and Hooke contributed, culminates in Newton's axiomatization of classical mechanics. He formulated a few fundamental laws and principles, from which other laws (like Kepler's) could be deduced. Philosophically, Newton's world consisted of four elements:

- **Matter**: an infinite number of mutually separated hard and unchangeable particles, called corpuscles. They possess primary and secondary qualities but only the primary qualities matter to physics. [Burtt 1932, 235–6]
- **Motion**: the motion of the corpuscles can be described by the laws of mechanics.
- **Absolute space**: an imaginary cosmic vessel, within whose walls the corpuscles (and the bodies built out of them) perform their lawful motions; Newton takes absolute space to exist even when there is no matter to fill it.
- **Absolute time**: an imaginary river, whose constant flow sets a unique time metric by which all natural processes can be measured; all observers throughout the whole universe assign the same time to events, however far apart they are. [Weinert 2004, Ch. 4]

There is a distinctly philosophical side to Newton's reasoning.

5.1   On hypotheses

Like most great scientists, Newton demonstrated philosophical awareness. He reflected on the philosophical dimensions of physics. Newton is famous for his statement: "Hypotheses non fingo." This Latin phrase can be rendered alternatively as "I do not feign hypotheses"; "I do not make use of fictions"; "I do not use false propositions or premises or explanations."[6] Historians of science have identified several senses in which Newton uses the word "hypothesis." Sometimes he meant a plausible though not provable conception. In his later years he came to regard a hypothesis as a gratuitous fiction. [Koyré 1965, 36–7]

> That which cannot be derived from phenomena is called a *hypothesis* and these do not belong to experimental philosophy. [Quoted in Dijksterhuis 1956, 537]

Newton was not the first to worry about the term "hypothesis." Copernicus and Rheticus had corresponded about the usefulness of hypotheses in astronomy with a figure who will soon come to greater prominence in the discussion: Andreas Osiander. Copernicus and his pupil considered that certain astronomical hypotheses were more probable than others. More probability accrued to the heliocentric hypothesis than to the geocentric hypothesis. Acceptable hypotheses in astronomy had to explain all the observable phenomena. They had to explain the phenomena in a coherent way. The Ptolemaic hypothesis, says Rheticus, does not suffice to establish the harmony of celestial phenomena. [Rheticus 1540, 132; see also

---

[6] Koyré [1965], 35; Dijksterhuis [1956], 541; Crombie distinguishes three senses of "hypotheses": improvised propositions, heuristic aids, illegitimate fictions; Crombie [1994], Vol. II, 1071.

Correspondence reprinted in Rosen 1959, 31–2; 1984, 125–6, 193–4, 198–205]
Kepler later agreed that the Copernican hypothesis enjoyed more probability than
the Ptolemaic hypothesis. The notion of hypothesis had great repercussions
throughout the next 140 years. The ambiguity of the term, as reflected in Newton's
views on hypotheses in science, invited opposing interpretations of the Copernican
model. In his *Dialogue Concerning the Two Chief World Systems* (1632), Galileo
epitomizes the ambivalent status of hypotheses in the sixteenth and seventeenth
centuries. The Preface states that his spokesman, Salviati, will defend the Copernican
system but only as a purely mathematical hypothesis. But as the dialogue unfolds,
Salviati is drawn toward probability arguments. Eventually he adopts the Copernican
position that the acceptance of the dual motion of the Earth as a physical assumption leads to a more coherent explanation of the appearances. Note that these
probability arguments invoke belief in a model, because its physical assumptions
are more probable. It is not believable, says Kepler, that the "fixed stars move at
incalculable speed."[7] [Kepler 1618–21, Pt. II, §5] The Copernican hypotheses are
more like conjectures than useful fictions. They have a much closer association with
the phenomena than Newton would later accept. They form, as Rheticus tells us,
the basis of inferences.

By contrast, labeling hypotheses as "useful fictions" in astronomy reassured
Copernicus's adversaries that his heliocentric model did not force them to abandon
their cherished geocentric beliefs. Cardinal Bellarmine reminded Galileo that
Copernicus had always spoken *hypothetically*: it is possible to use the motion of the
Earth as a mathematical device to render the calculations more economic, since
fewer epicycles and eccentrics are needed. However, to affirm the centrality of the
sun as a physical hypothesis is in conflict with the Scriptures.[8]

In order to soften the clash between the Church and heliocentrism, Osiander
inserts his Preface in an attempt to present the Copernican hypotheses as mere
calculating devices. They have the license to be false or replaceable as long as "they
reproduce exactly the phenomena of the motions." [Osiander, Letter to Copernicus
April 20, 1541, quoted in Rosen 1984, 193–4] By the time Newton appeared on
the scene, hypotheses did not command a respectable tradition. Rejecting them,
Newton claims to be an inductivist. The laws of motion are deduced from

[7] Kepler's probability argument states that we should attach more plausibility to the heliocentric view
because the evidence – the apparent motion of the "fixed" stars in a 24-hour rhythm about the Earth – is
more probable on the view that the Earth rotates on its own axis. These probability arguments can be
supported by a consideration of the angular velocities involved under the two scenarios. Under some
simplifying assumptions, the angular velocity of the rotating Earth for an observer at the equator is
$464 \, \text{m/s} = 1670 \, \text{km/h}$ The geocentric view, by contrast, has to assume an angular velocity of the
"fixed" stars about the stationary Earth. A calculation produces a value of $5.45 \times 10^6 \, \text{m/s} = 1.96 \times 10^7$
km/h It is such an enormous rotational velocity of the stars – 19.6 million kilometers per hour, compared to 1670 km per hour for the Earth at the equator – which the Copernicans consider implausible
on mechanical grounds. By comparison, the orbital velocity of the Earth around the sun is 30 km/h and
the velocity of the sun around the galactic center is 225 km/h. The evidence – the observable rotation of
the sphere of fixed stars – is more likely on account of heliocentrism than on account of geocentrism.

[8] See Koestler [1959], 454; similar statements, reflecting Osiander's instrumentalist attitude, are found
in Kuhn [1957], 191, 194; Crombie [1994], Vol. I, 599–600.

Phenomena and made general by Induction, he declares, and this is the highest evidence that a proposition can have in Philosophy. [Koyré 1965, 36–7; Dijksterhuis 1956, 544, 546–7] Phenomena are (reliable) observational or experimental data, from which are derived laws or axioms. Newton rejects any explanation of natural phenomena that appeals to metaphysical hypotheses, for which no evidence can be cited.

This does not mean, however, that the unobservable must automatically be suspect. The interior of the sun, for instance, is unobservable yet it is perfectly possible to make quite definite inferences about the chemical composition of the interior by use of spectral analysis. We have to distinguish *direct* from *indirect* observation. Directly observable phenomena are accessible through our eyesight or through the use of instruments. The directly observable is not necessarily the most reliable. Retrograde motion, as Copernicus reminds us, is a mere deception of sight. Indirectly observable phenomena are inferences from observations, and the use of reliable techniques, to unobservable parts of nature. We cannot directly observe the cause of planetary orbits but we can infer it from our observations and the heliocentric hypothesis. We begin to see that the story of heliocentrism is laced with philosophical lessons.

6   Some Philosophical Lessons

Copernicus reflects the cosmological differentiation between the parochial perspective of his terrestrial "corner" and the central point of construction from which the universe cannot, indeed, be viewed but can be thought. [Blumenberg, The Genesis of the Copernican World *(1987), 38]*

Copernicanism creates a problem situation from which a number of philosophical consequences follow. As we shall see in later chapters, philosophical consequences also follow from Darwinism and Freudianism. A **problem situation** in science occurs when a number of competing explanatory accounts propose solutions to a perceived scientific problem. The solutions are proposed on the background of a number of accepted presuppositions, techniques, and models. The presuppositions and techniques define acceptable problems and a set of possible solutions to the problem. Consider two famous problems in the history of science: the motion of planets and the existence of different species. In 1543 Copernicus proposed a solution to the first problem. In 1859 Darwin offered a solution to the second problem. Both solutions entered a conceptual space in which certain presuppositions, techniques, and models had already taken root. Copernicus and Darwin proposed rival models. They involved a set of solutions, which differed from previous solutions. They also gave rise to a set of presuppositions and techniques, which diverged from previously accepted presuppositions. The divergence was striking in the case of Darwin, but only partial in the case of Copernicus. At least Copernicus worked with a non-Aristotelian theory of motion. The set of presuppositions, techniques, and models renders certain solutions acceptable, others unacceptable. Certain solutions are possible solutions because they are compatible with the set of accepted presuppositions and techniques. The set also renders other solutions impossible. For instance, a mature form of Copernicanism renders all Greek presuppositions – the circle, the two-sphere universe – and techniques – eccentric circles and epicycles – obsolete. Hence geocentrism can no longer be regarded as an acceptable solution. It is important to distinguish *possible* from *actual* solutions. A certain solution, *S*, may be possible with respect to a cluster of presuppositions. We cannot accept it as the actual solution because other solutions will also be possible with respect to this cluster. For a possible solution to transform itself into an actual solution, it needs to prove its mettle. The actual solution needs to solve some old and some new problems.

Scientific problem situations have an impact on philosophical issues, to which we now turn. Note that there is a difference between the *deductive, inductive*, and *philosophical* consequences of a theory. The deductive consequences follow mathematically or logically from the principles of the theory. Deductive consequences can occur in the form of novel predictions or the accommodation of already known facts. In both cases they are often compatible with one theory but, ideally, not its rivals. If this situation obtains, we will later speak of *supportive* evidence. Inductive consequences follow from the theory with degrees of probability. For instance, if a theory is statistical in nature its consequences follow with higher or lower degrees of probability. Consider the difference between "All ravens are black" and "Most ravens are black." If "All ravens are black" and the observation is made that "this is a raven," it follows deductively that "this raven is black." But if the statement merely is that "Most ravens are black," then it follows only inductively that "this raven is black." Philosophical consequences are conceptual issues, as they are dear to the philosopher. Although they are often taken to follow from scientific theories, they are rarely subject to direct empirical testing. Hence they do not command the expert consensus which deductive consequences typically induce. Given one theory, *T*, incompatible philosophical consequences are often drawn from it. Copernicanism, Darwinism, and Freudianism, for instance, raise questions regarding an instrumentalist or realist interpretation of some of their fundamental assumptions. The fact that two incompatible philosophical views (instrumentalism or realism) are compatible with one theory does not exclude the possibility that one view is more in agreement with the principles of the theory than the other. In connection with Copernicanism we first note its impact on general worldviews. Then there arise lessons for epistemological attitudes: realism and instrumentalism and the question of underdetermination. Copernicanism also raises philosophical issues concerning models, theories, and laws. The Copernican turn also calls for an analysis of criteria of scientific revolutions. And finally we tackle the Anthropic Principle and ask whether it constitutes a reversal of the Copernican turn.

6.1   The loss of centrality

Copernicus, through his work and the greatness of his personality, taught man to be modest. [Einstein, "Message on the 410th Anniversary of the Death of Copernicus" (1953), 359]

In *De Revolutionibus* Copernicus sets out to convince his readers that the heliocentric hypothesis is not as absurd as it may sound. He managed to convince a number of his contemporaries, like Rheticus and Maestlin in Germany. He found some followers in England, like Thomas Digges and William Gilbert, and Italy, like Giordano Bruno. [Dryer 1953, Ch. XIII] Yet at the beginning of the seventeenth century Kepler still reports that many people were shocked by heliocentric ideas. [Kepler 1619, 175] His contemporary Francis Bacon (1561–1626) steadfastly refused to accept Copernicanism.

In many quarters the Copernican treatise was greeted with opposition and hostility. Copernicus's contemporaries did not find discomfort in the mathematical details of his work. Their objections were more philosophical. The Copernican view, if taken literally, displaced the Earth from the hub of the universe. The inhabitants of the Earth suffered a loss of centrality. According to Copernicus, they felt, it was no longer true that the universe had been created for the sake of humankind. Copernicus's heliocentric hypothesis presented more than a mathematical treatise in the esoteric science of astronomy. It was an attack on what people believed about the structure of the world. Especially amongst the Protestant clergy and theologians the heliocentric idea gained little favor. Martin Luther, citing Scripture, dismisses Copernicus as a fool. His chief lieutenant, Philip Melanchthon, calls him simply insolent. The Roman Catholic Church had embraced the geocentric view through the work of Thomas of Aquinas. The resistance of the Catholic Church against the physical motion of the Earth around the sun was partly due to ecclesiastical pressures. The Catholic Church saw its authority under threat from the rise of Protestantism. Copernicanism posed an additional challenge to Catholic dogma.

In their endeavor to cushion the shock of heliocentric ideas, Kepler and Rheticus were eager to employ teleological arguments. While physical centrality had been lost, humankind had not sunk to cosmic insignificance. The heavenly phenomena had been invested with a purpose. The purpose of movement, Kepler proclaims, is to prove that "movement belongs to the Earth as the home of the speculative creature." [Kepler 1618–21, Bk. IV, Part I, §5 (75, 77)] Rheticus even asserts that "the sphere was studded by God for our sake with a large number of twinkling stars." [Rheticus 1540, 143]

The early Copernicans did not accept that a mere physical removal of humans from the hub of the solar system to its third sphere represented a hurtful demotion. The celestial phenomena have a purpose, which remains unaffected by the physical position of the Earth among the planets. Their purpose resides in their service to humankind. Aristotle held that "Nature is a cause that operates for a purpose." [Aristotle 1952a, Bk. II, 8] And "as nature makes nothing purposeless or in vain, all animals must have been made by nature for the sake of men." [Aristotle 1948, Bk. 1, §11, 1256b] Copernicus conceives of his job as understanding the "machinery of the world, which has been built for us by the Best and Most Orderly Workman." [Copernicus 1543, Preface, 6] The dogma of teleology – that "Nature does nothing without a purpose" – reverberates through the history of human ideas about Nature. Rheticus even turns teleology against the Scholastics.

The Wise Maker would have created a heliocentric model, for He would have "shirked from inserting in the mechanism any superfluous wheel." [Rheticus 1540, 137]

It took until Newton to remove teleological thinking from the physical sciences. It took even longer in the biological sciences, as we shall see in Chapter II. It is easier to convince people that stars are not made to shine for their amusement than it is to convince them that eyes are not designed for them to see.

Nietzsche found that "since Copernicus man has been rolling from the center toward x." He saw himself as a second Copernicus who branded the self-deprecation of "European Man" as the greatest danger. [Kaufmann 1974, 122, 288; Nietzsche 1887, Bk. I, §12] Freud, too, interpreted the Copernican turn and the Darwinian revolution as serious blows to the self-image of humankind. The Oxford physicist David Deutsch observes that

> the prevailing view today is that life, far from being central, either geometrically, theoretically or practically, is of almost inconceivable insignificance.

Deutsch disagrees with this assessment. Yet it is a physical fact that

> the solar system is a negligible component of our Galaxy, the Milky Way, which is itself unremarkable among the many in the known universe. So it seems that, as Stephen Hawking put it, "The human race is just a chemical scum on a moderate-sized planet, orbiting round a very average star in the outer suburb of one among a hundred billion galaxies." [Deutsch 1997, 177–8; cf. Weinberg 1977, 148; Blumenberg 1981, Pt. I, VI; 1965]

There is, however, a difference between physical and rational centrality.[9] From a terrestrial perspective humans observe the universe from a particular physical angle, which is defined by the location of the Earth in the Milky Way. This angular perspective has led to misconceptions. The Greeks constructed from the appearances a geocentric worldview. Copernicus does not abandon the tight connection between observational appearances and geometric constructions. But he holds that a heliocentric model accounts better for the appearances, on account of its greater plausibility. Copernicus's change of perspective has two implications. One is that physically humans no longer occupy the geometric center of the universe. Another is that the heliocentric hypothesis affords humans a much better grip on the observational appearances. The change of perspective offers humans a more coherent model of the solar system. The Copernican turn replaces *physical* with *rational* centrality. Through rational thinking humans can construct an accurate model of the universe. Their perspective on the universe is predicated on a particular physical position in the universe. Their centrality is due to a rational comprehension of the universe, which far exceeds what their eyes will allow them to see. "Eyesight," muses Kepler, "must learn from reason." [Kepler 1618–21, Bk. I, Part I, §1]

---

[9] The distinction between physical and rational centrality runs through the work of Hans Blumenberg on Copernicanism – see Blumenberg [1955]; [1957]; [1965]; [1981], Part I, III; Part II, III, IV; Part VI, I.

Observations confirm that humans do not dwell in a privileged geometric location in the vast cosmos. There is no evidence that the other bodies in the solar system were especially designed for the purpose of human life. In this sense the Copernican hypothesis has led to a loss of centrality. The Copernicans, however, demonstrated that physical centrality is not of utmost importance. The human mind soars far above the physical limitation of its bodily habitat. What humans lost in geometric centrality they gained in rational centrality. Knowledge replaces location, reason enhances eyesight.

A certain symmetry exists between Copernicus and Darwin. Copernicus removed humans from the physical center of the universe. Darwin removed humans from the pinnacle of creation. These philosophical implications of the Copernican turn and the Darwinian revolution have recently been contested. Some modern cosmologists reject what they call the "Copernican dogma." According to them this dogma states that there is nothing special about humans and their habitat. The Earth is one of many planets orbiting a solar body of average size. The solar system itself can claim no central position in the Milky Way. And the Milky Way is just one of billions of galaxies. Darwinian evolution seems to support the Copernican "dogma." Evolution, as Darwin taught, has produced an offshoot of the evolutionary tree, which humans call their home. But evolution is contingent. It might never have brought forth intelligent life.

Some cosmologists argue that the "Anthropic Principle" needs to replace the Copernican dogma. [See Section 8] In evolutionary biology, "intelligent design" is set against Darwin's natural selection. Intelligent design scenarios seek to reinstate teleological thinking in evolutionary biology. [Chapter II, Section 5.4] The Anthropic Principle rejects the implication that human existence is not in any way special. The Anthropic Principle affirms that humans live in a very special epoch of cosmic history. It is special since it has permitted the "evolution of carbon life." [Barrow/Tipler 1986, 601] Before we consider this principle, a number of other philosophical concerns require attention.

6.2   Was Copernicus a realist?

Earlier we found that Copernicus was not the author of a scientific revolution. With his shift in perspective against the backdrop of some invariant features, Copernicus planted the seeds of a scientific revolution. The Copernican turn – a shift in perspective with some explanatory gain – was a significant opening move, which enabled the rise of modern science. Copernicus stood at the gate of modernity.

When we think about modern science, three features stand out:

- systematic observation;
- controlled experiment;
- mathematization.

Copernicus reports a number of his own observations, made at Frauenburg, in Prussia. [Copernicus 1543, Bk. III, §2; Rheticus 1540] Otherwise, he relied on

numerous Greek observations. Kepler formulated his planetary laws on the basis of
Tycho Brahe's discoveries. And Galileo made significant additions to the corpus of
astronomical data. They all used observations in a *systematic* fashion. That is, they
used them to establish the Copernican hypothesis. **Systematic observation** means
that observational data are employed to test the adequacy of a particular model. All
these scientists found that, compared to the Ptolemaic model, the heliocentric
model was more probable. In the case of astronomy **controlled experiment** does
not apply. Controlled experiment is the deliberate manipulation of selected parameters in scientific laboratories. That is, it involves a deliberate exclusion and inclusion of parameters in the experiment. In their famous scattering experiments
(1909–11), for instance, Rutherford and his collaborators used ionized helium
atoms, fired at gold atoms, to discover that the atom possessed a nucleus. In these
experiments Rutherford deliberately neglected electrons because they would not
interfere with the trajectory of the heavy helium atoms inside the gold atoms. The
experimenters concentrated exclusively on the interactions between the nuclei. We
have already seen that **mathematization** offers algebraic compressibility. [Section 5]
Ancient astronomy made extensive use of geometry. Angles and circles were the
main tools in the hands of the astronomer, even Copernicus. Geometry limits the
usefulness of mathematics in the description of nature. The use of geometry made
it impossible for Copernicus to provide an accurate quantitative theory of planetary
motion, let alone a dynamic analysis. His explanatory gain was limited to the topologic structure of his model.

But does this explanatory gain mean that a better astronomical explanation
is at hand? This question lies in the logic of the Copernican problem situation.
By changing the perspective between stationary and moving Earth, Copernicus
claims to achieve a better explanation of the observable phenomena. A philosophical issue immediately arises, of which his contemporaries were aware: Granted that
Copernicus achieved some explanatory gain, does this explanatory gain tell us
merely something about the structure of our theories or more informatively about
the structure of the physical world itself? Was Copernicus a realist? Osiander raised
this question in his anonymous Foreword to *De Revolutionibus*. This question also
lurks behind the ambivalent use of the term "hypotheses." We are thus dealing
with the philosophical issue of **realism** and **instrumentalism**.

6.2.1 Lessons for instrumentalism and realism   The most famous testimony to
the presence of this philosophical issue in the minds of Copernicus and his contemporaries is buried in Osiander's Preface. It presents the *Revolutions* to the European
world of 1543. Osiander saw it fit to add some introductory notes for the benefit
of The Reader Concerning The Hypotheses of This Work. [Copernicus 1543, 3–4;
Koestler 1964, 169–78; Rosen 1984, 195–6] Note, first, that Osiander follows
Copernicus in speaking of *hypotheses*. As we have seen, this innocuous-seeming
term developed its own divided pedigree in the span from 1543 to 1687. Osiander,
however, employs the term in only one of its senses. Reminding the reader of the
newness of the heliocentric hypothesis, he spells out the astronomer's dilemma. On
the one hand, the astronomer cannot know the "true causes" of the celestial

movements. On the other hand, the astronomer can establish fairly accurate descriptions of "the history of the celestial movements." How is this dilemma to be resolved? Osiander's recipe is the *locus classicus* of instrumentalist philosophy. The astronomer can establish *how* the planets move but not *why*. Yet the human mind is exercised by theoretical curiosity. Even though no true explanation can be given, any explanation is better than no explanation. It is then the job of the astronomer

> to think up or construct whatever causes or hypotheses he pleases such that, by the assumption of these causes, those same movements can be calculated from the principles of geometry for the past and for the future too.

It is therefore not necessary for the hypotheses to be true or even probable. Distancing himself directly from the probability arguments advanced in the main text, Osiander holds that:

> it is enough that they [the hypotheses] provide a calculus, which fits the observations.

Why should the reader then even read the Copernican tract? Osiander makes an appeal to simplicity. Some hypotheses render the calculations simpler, make the observations easier to understand. They may even give rise to more reliable predictions.

> Therefore let us permit these new hypotheses to make a public appearance among old ones which are themselves no more probable, especially since they are wonderful and easy and bring with them a vast storehouse of learned observations. As far as hypotheses go, let no one expect anything in the way of certainty from astronomy, since astronomy can offer us nothing certain, lest, if anyone take as true that which has been constructed for another use, he go away from this discipline a bigger fool than when he came to it. Farewell.

Osiander anticipated Newton's later skepticism regarding hypotheses in astronomy. He permitted heliocentrism as a mathematical hypothesis, but not as a claim about physical reality. As a reality claim it would be a thorn in the theologian's eye. By deflecting the Copernican hypothesis along instrumentalist lines, Osiander sought to remove its sting. It was another mathematical device, with no better claim to reality. It had as little probability as the established Greek hypotheses. The true causes of planetary motion cannot be known, because the human mind is too weak to apprehend the celestial sphere. In the absence of physical understanding, revelation takes its place.

What about Copernicus? Was he a realist? Just as there is no doubt that Aristotle believed in the physical centrality of the Earth, there is no doubt that Copernicus believed in the annual and daily motion of the Earth. Aristotle also believed in the reality of solid spheres, whose function was to carry the planets. But Copernicus was "unsure whether they were real or imaginary." [Rosen 1959, 11–21] Copernicus was uncertain about the physical significance of his geometric constructions. He observes that Ptolemy employed various geometric devices. Unsurprisingly, he finds himself unable to say which one of them corresponds to reality. His solution is to endorse the equivalence of hypotheses:

> It is not easy to determine which of them exists in the heavens (...) except that the perpetual harmony of numbers and appearances compels us to believe that it is some one of them.[10]

Copernicus was not a realist about his geometric devices. He had some difficulty believing that the theoretical structures employed in heliocentrism – the eccentric and epicyclic motions, which serve to account for the observable phenomena – have a counterpart in physical reality. Copernicus was not a realist about the algebraic structure of his model. Copernicus was a realist about celestial objects, their motions, and the system that holds them together. He believed that the place he had assigned to the Earth in the heliocentric model corresponded to a part of the structure of the solar system. Copernicus was a realist about the topologic structure of the heliocentric model. Copernicus's realist arguments are presented in his Preface and Dedication to Pope Paul III and Part I of his book.

First, Copernicus attributes a *causal* role to the movement of the Earth. It is the physical position of the Earth among the other planets that explains the appearance of retrograde motion, the seasons, and the natural length of the day. The observable appearances are causally explained by the physical motion of the Earth:

> We explained the appearances due to the movement of the Earth around the sun, and we proposed by that same means to determine the movements of all the planets. [Copernicus 1543, Bk. IV, Introduction]

As Copernicus believes in the planetary status of the Earth, he believes that the Earth's location is causally responsible for some of the observable phenomena. So Copernicus is not just a realist about the position of the Earth; he must be a realist about the physical consequences of this position.

The second move is Copernicus's *argument from coherence*. We have already noted that Copernicus was very well aware of the fact that natural phenomena are correlated in a number of ways. For Copernicus this meant that the movements of the Earth and the planets are correlated such that a change in one part leads to consequences in another part of the system. According to Copernicus we cannot arbitrarily change the order of the planets, without upsetting the whole cosmic picture. And if we correlate the orbits of the planets, their natural order is revealed to us. The correct choice of the initial position of the Earth – it is an orbiting planet

---

[10] Copernicus [1543], Bk. III, §§15, 20; Bk. V, §4. According to E. A. Burtt [1932, 49–51] Copernicus is not preoccupied with the question of the reality of the motion of the Earth but with the mathematical simplicity of the system, achieved through shifting the central reference point away from the Earth. This interpretation is true of the later chapters of *De Revolutionibus* but not of Book I; see Cushing [1998], 55.

rather than the stationary center of the universe – leads according to Copernicus to a more adequate model of the universe. Thus Copernicus argues from realism about the celestial objects, including the Earth, to realism about the scientific model he adopts. This realism is, however, restricted to the spatial distribution of the planets. The heliocentric model is a better model of the universe than the geocentric one, because it is more coherent. And its *coherence* provides a certain plausibility that it more correctly captures the structure of the universe than its rival, the geocentric model:

> Now we are turning to the movements of the five wandering stars: the mobility of the Earth binds together the order and magnitude of their orbital circles in a wonderful harmony and sure commensurability … [Copernicus 1543, Bk. V, Introduction]

On the hypothesis that the Earth moves, many observable consequences follow. Once the assumption leads to a coherent model, the coherence boosts the credibility of the original hypothesis. The heliocentric model, on account of its coherence, is a better representation of the interrelatedness of nature than the geocentric model. As Kepler and Newton later realized, the representation of the heliocentric model could be enhanced by abandoning many of the Copernican presuppositions.

6.3   Modern realism

If, therefore, there is a lesson which scientists should teach realists it is that all-or-nothing realism is not worth fighting for. [Psillos, Scientific Realism (1999), 113]

Some of Copernicus's friends and followers felt outrage at Osiander's instrumentalist tinkering with the Copernican model. The equivocation of the term "hypothesis" pointed them in the direction of realism. With hindsight we can have a more relaxed attitude. Copernicus does not improve on the algebraic structure of the ancient models. Copernicus could advance no striking observational evidence in favor of the motion of the Earth. Osiander's instrumentalist Preface sounded a note of caution. For the technical part of the Copernican treatise, with its acceptance of the equivalence of hypotheses, does not live up to the promise of the first part.[11]

The Copernican model was able to provide a coherent account of the observational data known during Copernicus's lifetime. It was also compatible with later observations. But the original Copernican model lacked a credible mechanism that could explain the observations. The Copernican model enjoyed empirical validity, owing to its topologic structure. But a more sophisticated model or theory must also satisfy the demand for an accurate algebraic structure. The mechanism that

---

[11] For a defence of Osiander see Dijksterhuis [1956], Part IV, §§14–15; see Rosen [1959] for the correspondence on hypotheses; Mittelstraß [1962], IV, 6, 199–204; Neugebauer [1968], §6, 100; Blumenberg [1957], VIII, 73; [1965]; [1975], Part III, II; Rosen [1984], 125.

explains the "appearances" must correspond to the structure of the real world. A theory that has an accurate algebraic structure enjoys theoretical validity. The philosophical dispute between Osiander and Copernicus, between instrumentalism and realism, has its modern equivalent. The modern-day instrumentalist is as hesitant about mechanisms and structures underlying the observable features as Osiander. The modern-day realist is as confident about the underlying mechanisms and structures as Copernicus was about the reality of the motion of the Earth.

Modern instrumentalists advance two reasons for their caution about unobservable structures. On the one hand, there is the underdetermination of theories by evidence. [Section 6.4] This is the view that the empirical evidence can never clearly decide between any two scientific theories which are empirically equivalent but structurally different. This problem already found its expression in Ptolemy's equivalence of hypotheses. On the other hand, there is the "pessimistic meta-induction." This is the view that many scientific theories, which were once regarded as "true," have since landed on the scrapheap of mistaken ideas. Geocentrism is a case in point. What reason do we have for trusting our current theories? Modern instrumentalism therefore concerns itself with "saving the appearances." Scientific theories can at best be empirically adequate: they fit their domain as far as the observable phenomena are concerned but we have to remain agnostic with regard to the underlying theoretical structures. Two scientific theories may stipulate incompatible, unobservable mechanisms, although they both account for the available evidence. Furthermore, it is always possible to explain the same evidence on the basis of different theoretical structures. The realist wants more. It is not enough for our models to be adequate as far as the observations reach. The underlying theoretical structure, which can explain the observations, must also represent the structure of reality. Copernicus was still hampered by accepting the equivalence of the geometric hypotheses, even though they render the observational data coherent. Kepler, however, was interested in physical causes. The planets move in certain regular ways, expressed in Kepler's three laws. The further question is why they move in this way. It was not until Newton combined the first law of mechanics with the law of gravitation that a viable physical explanation became available. [See Figure 2.12] For the realist such episodes show that our mature scientific theories constitute good approximations to a genuine explanation of physical processes.

The realist claims that realism is the "only philosophy that does not make the success of science a miracle." [Putnam 1975, 73] Yet the story of astronomy shows that the anti-realist seems to have a point. The geocentric and heliocentric models were at first observationally equivalent, both endorsing the "equipollence of hypotheses." Yet, they were structurally different. And much of the theoretical structure ended up in the wastebasket of wrong-headed ideas. Ptolemy made no exaggerated claims about the "realism" of his geometric devices, and Copernicus was only a realist with respect to the spatial arrangement of the planets. Kepler advanced probability arguments in favor of the Copernican model, while jettisoning some of its central presuppositions. Newton abandoned Kepler's "physical" causes, while completing the Copernican revolution.

How can realists accounts for this double aspect of continuity and discontinuity? Recently the thesis of **structural realism** has been advanced to stem anti-realist arguments.[12] [Worrall 1989; Ladyman 1998; Psillos 1999] To account for continuity amid conceptual change, structural realism focuses on the structural aspects of systems. We can approach the idea by reflecting on the fact that science is concerned, quite generally, with natural systems (say, solar or biological systems). To obtain a system it is not enough to juxtapose elements (relata). A collection of planets does not constitute a system. To turn it into a system, the components must be interrelated. Thanks to the achievements of the Copernicans we know that the planets are related in a systematic fashion. The planets are bound into a system by the relations (laws) of the system. What constitutes a system in the natural world is the interaction of the components of the system. Planetary systems have planets as components, and Kepler's laws as their "glue," which holds them together. Biological systems have species or individuals as their components. What hold them together are the laws of evolutionary biology. Human societies have individuals and social groups as their components. What hold them together are the values, norms, and legal rules of particular societies. Apart from the relata of the systems, there must be appropriate relations between the components. It is hard to imagine that a human society could be held together by Kepler's laws.

In natural systems several components combine in a regular fashion. Natural systems therefore display a structure, consisting of the relata of the system and the relations between them. Science attempts to construct theories with models representing such systems. The models must have a model structure, which represents, in symbolic form, the structure of the natural system. If the scientific enterprise is preoccupied with the description and explanation of natural and social systems, structural realism is the thesis that the model structure represents, in approximation, well-confirmed structural aspects of the target system. It is concerned with the theoretical validity of model structures. So a structural realist will want to claim that the models of science aim at representing the (topologic and algebraic) structures of natural or social systems. The model structure, for present purposes,

---

[12] Structural realism comes in two flavors. *Epistemological* structural realism (ESR) claims that "all we know is structure." ESR stresses the continuity of the mathematical equations as scientific theories undergo drastic changes in their ontology and vocabulary. ESR tends to remain agnostic regarding the question whether there is more to the world than what the mathematical structure reveals about it, since it is conceivable that some elements of reality remain hidden from our view. *Ontological* structural realism (OSR) stresses that "all there is, is structure" and there exist no further constituents of reality beyond this structure; and the job of scientific theories is to capture this structure in symbolic form. OSR is divided over the question whether relations enjoy ontological primacy over objects (relata), in which case objects are just "nodes" in the relational structure, or whether relata and relations are taken to constitute the structure in a union. [See Rickles et al. 2006] The author's own inclination is to adopt a strong version of OSR, according to which all that exists are structured natural systems. OSR, on this strong view, is committed to both the reality of the relata and the relations, which are best captured in structural models. The relations are typically expressed in the laws of science, of which we will in a later section encounter a structural interpretation. Through the employment of equations, models, and theories, science expresses structural aspects of the material world. The existence of natural systems, like planetary and organic systems, shows that there is much structure for science to describe and explain.

consists of the components and their interrelatedness, the relata and the relations. To support this view the structural realist will want to show how to accommodate both the continuities and discontinuities in scientific theories at the level of structures. These features can be considered differentially with respect to both the topologic and the algebraic structures. In the transition from Ptolemy to Copernicus the topologic structure of the models changes but the algebraic structure remains the same, with the exception of the equant and major epicycles. In the transition from Copernicus to Kepler the topologic structure remains unchanged, but the algebraic structure undergoes dramatic changes. From Kepler to Newton, the algebraic structure experiences refinements, but no longer major changes. Since Newton the structural model of the solar system has no longer experienced drastic changes. It is a fact that scientific representations (in terms of models) always face limits in approximation and idealization. For instance, the circle is an idealization of an ellipse. The structural realist will emphasize that there is enough incontestable evidence in the history of science for the postulation of continuities in the underlying structures. The continuities extend to both the algebraic and the topologic structures. Ultimately, it is only mature models that can be fully representative of the structure of the system modeled. Furthermore, these elements change differentially and usually for good scientific reasons. We may therefore suspect, as we will discuss in Section 7, that even in revolutionary periods in science, a certain chain-of-reasoning process links the transitions from old to new theories.

6.4   The underdetermination of theories by evidence

The inability to remove the equivalence of hypotheses was one of the reasons why Copernicus failed to become a scientific revolutionary. Realists must hold that the theoretical structure scientists assign to a set of observational or experimental data refers to some physical process, which can reasonably be taken to explain the observational data. Ideally this amounts to a causal explanation. [See Chapter II, Section 6.6] How can we make sure that our theoretical accounts are approximately true of the material world? This is in part the question of how scientific accounts manage to represent sections of the natural world. This is done, as we shall see, through the use of a variety of models. [See Sections 6.4 and 6.5] Before we turn to these concerns a stumbling block must be removed. There is a famous argument – the Duhem–Quine thesis – which attempts to show that the evidence is *never* strong enough to weed out all competing theoretical accounts. If the argument succeeds, there will always be perhaps infinitely many theoretical accounts, ontologically incompatible, which will be compatible with the evidence. The evidence will be unable to select one account as superior to another. All theoretical accounts will be underdetermined by the available evidence. The instrumentalists will have a powerful argument in their armory.

Consider some alien creatures that are able to utter the number "9" when shown a set of objects. How do they arrive at this answer? There are obviously several mechanisms we can ascribe to them: (1) 3×3; (2) $4.5 + 4.5$; (3) $\sqrt{81}$; (4) $1 + 2 + 6$; etc. If we learn nothing else about the creatures' abilities it will be difficult to decide. But imagine we learn that the aliens only possess mastery of natural numbers 1 to 10 and are not aware of fractions. Then we can exclude hypotheses (2) and (3). In this way we can build up evidence, which strongly suggests that a certain mechanism must be at work. This seems to have been the history of the Copernican hypothesis. In 1543 the observational evidence was simply not strong enough to favor the Copernican system over the Ptolemaic one. But then Tycho Brahe and Galileo made their significant empirical discoveries. They were difficult to reconcile with the Ptolemaic system. Through the discovery of his planetary laws, Kepler significantly improved the Copernican system. Finally, Newton crowned its success when he showed that inertia and gravitation could account for the elliptical orbits of the planets. As we shall explain in detail in the chapter on Darwinism, a process of elimination of rival accounts takes place. This elimination is possible, as the history of Copernicanism makes clear, because theoretical accounts run up against a number of constraints: the stubbornness of the phenomena is one such constraint, coherence and the probability of explanations are others. If it is a reality of the solar system that the Earth orbits the sun, and not the other way round, it will be hard for theoretical accounts to evade this fact. The Ptolemaic system survived for 1,500 years because of the paucity of the evidence. Once the evidence hardened, the Ptolemaic system floundered on the stubbornness of the facts and the implausibility of its assumptions.

A determined instrumentalist may not be swayed by such arguments. The instrumentalist will point out: (a) that it is always possible to dismiss the evidence as unreliable; the Mathematician and Philosopher in Brecht's play were right, it may be said, to be skeptical about the telescope, which in 1610 was not yet a reliable instrument; (b) that it is always possible to change certain background assumptions to save the theory; (c) that the history of science is full of cases of underdetermination. Until the beginning of the seventeenth century astronomical models were incompatible with each other, yet equally compatible with the evidence. A similar situation prevailed in evolutionary biology until the beginning of the twentieth century. And social-science models of societal phenomena still suffer from the scourge of underdetermination.

6.4.1 The Duhem–Quine thesis Consider Popper's falsificationist scheme: from a scientific theory, T, we derive a testable hypothesis, H_E; this hypothesis is then subjected to "severe" tests; if the hypothesis does not survive the tests, $\neg H_E$, the theory, from which it was deduced, will be falsified, $\neg T$. Against this falsificationist move Duhem and Quine hold that it is the theory, T, and background assumptions, A, together that face experience; if the tests fail to confirm the hypothesis there is some latitude of choice. Either we reject the hypothesis, H_E, or we change the background assumptions, A, to save the theory. [See Box 1.2 for a more logical statement]

Before we turn to an example, some preliminary remarks are in order. First, the disjunction, $\neg H_E \vee \neg A$, is not a definite result. Some reason should be advanced for retaining either H_E or A. Second, we should consider that the background hypotheses, A, and the hypothesis under test, H_E, may not have equal

---

Box 1.2   The Duhem–Quine thesis in logical terms
Popper's falsificationist scheme:

$$\left[(T \Rightarrow H_E) \,\&\, \neg H_E\right] \Rightarrow \neg T$$

Duhem–Quine Thesis:

$$\left\{\left[(T \,\&\, A) \Rightarrow H_E\right] \,\&\, \neg H_E\right\} \Rightarrow (\neg H_E \vee \neg A)$$

To save the theory, T, Duhem–Quine envisage that the background assumptions can be changed, say to A', so that from the conjunction of T and A' the negation of H_E can be deduced:

$$\left\{(T \,\&\, A') \Rightarrow \neg H_E\right\}$$

Key to symbols:
T = theory; H_E = testable hypothesis; $\neg$ = negation; $\vee$ = logical disjunction; $\Rightarrow$ = deductive consequence; A = background assumptions

---

epistemological weight. Often in science, there are fairly well-established theories, or results, which are presupposed and not currently under test. Third, the evidence, E, which is used to test the hypothesis, will often be fairly robust in the sense that it has been arrived at through independent methods. The Duhem–Quine thesis ignores these reservations. It holds that it is always possible to save T, despite $\neg H_E$, if we are prepared to make appropriate changes in the background knowledge, say from A to A'.

The history of astronomy provides us with nice examples to illustrate and evaluate the Duhem–Quine thesis.

Let us first enrich Popper's falsificationist scheme with some additional background assumption A. Let T be Ptolemaic astronomy, let A stand for the immutability of the supralunary sphere in Aristotelian cosmology; we take H_E from the Mathematician's assertion, in Brecht's play, that according to the ancients "there can be no stars, which turn round centers other than the Earth," that is, Jupiter can have no moons; finally let $\neg H_E$ stand for Galileo's discovery of the Jupiter moons. Galileo's discoveries, including the phases of Venus, and Tycho Brahe's observation of the appearance of the supernova of 1572 demonstrate the mutability of the heavens. Such empirical discoveries are difficult to accommodate in the geocentric model, which explicitly postulates the immutability of the heavens. It is difficult to see how the background assumption, A, could have been changed in order to accommodate the empirical results. Changing A to, say, A' – the mutability of the heavens – would have destroyed the structure of the geocentric model. The empirical discoveries were robust. "The universe of the divine Aristotle is an edifice of such exquisite proportions," declares the Philosopher, "that we should think twice before disrupting its harmony." The Philosopher's strategy is therefore to deny Galileo's evidence, $\neg H_E$. This is a legitimate move as long as the evidence is not robust. Once the evidence is fairly reliable, this strategy degenerates quickly into dogmatism. Galileo's visitors appear dogmatic because they insist on the "truth" of the Aristotelian dogma. Their skepticism would have been more justified had they questioned the reliability of the telescope.

There are, however, cases where it is perfectly reasonable to reject background assumptions. Let T be Copernican astronomy, let A stand for the assumption of uniform, circular orbs in the supralunary sphere, which Copernicus shared with the Greeks; we take H_E to stand for the circular motion of a planet around the central sun; finally let $\neg H_E$ stand for Kepler's discovery of the elliptical, nonuniform motion of this planet. We know that Kepler rejected the background assumption, A, of circular orbital spheres. He replaced it with nonuniform motion, A', as expressed in Kepler's laws. As Kepler considered the basic approach of Copernicus to be correct, he could not really reject the theory, T. To do so would have meant to return to earlier theories, like Ptolemy's geocentrism or Brahe's compromise system. These earlier theories held few attractions for Kepler, since they were not compatible with the observational evidence. It is true that Kepler saved the Copernican theory, T, by changing the background assumption, A. As required by the Duhem–Quine strategy, the conjunction of T and A' now had the deductive consequence $\neg H_E$: $\{(T \ \& \ A') \Rightarrow \neg H_E\}$. But note that it was the background assumption, A, that Kepler could not make compatible with the evidence. It was not a matter of saving T come what may. T was a relatively successful theory. But it could not account for the observations with the accuracy required by Kepler, since it was based on a background assumption which it shared with geocentrism.

These two examples suggest that we should distinguish a logical from a practical point of view. From a logical point of view it may indeed be possible, as Duhem and Quine suggest, to save a theory, T, by a number of stratagems: changing the background assumptions, denying the evidence. But from a practical point of view, scientists are usually faced with a limited number of theoretical accounts, which they assess, as Einstein showed, by submitting them to the power of constraints. [Einstein 1918; 1919; Weinert 2006]

→ *6.4.2   The power of constraints*   A practical solution to the Duhem–Quine thesis relies on the appeal to constraints. Constraints can generally be regarded as restrictive conditions on admissibility. They either control which parameters are to be admitted into a scientific theory or model or, more generally, which theories and models are admissible as scientific constructs. Consider a doorman outside a nightclub. This nightclub serves alcohol but only to punters who are over a certain age limit. The doorman must make sure that only punters who satisfy the age limit are admitted. If you are an under-age punter, the doorman imposes a restrictive condition on your admissibility to the nightclub. If he does his job properly, you will not be admitted. As we shall discuss now, a variety of constraints operate in scientific theorizing. Basically, there are empirical and theoretical constraints, which can be further subdivided.

Consider scientific models and theories as embedded in some *logical space*, which is structured by a range of empirical and theoretical constraints. The mathematician regards an equation as a **constraint** on the set of possible solutions. Consider the equation for a parabola: "$y = x^2 + 1$." The equation excludes as possible solutions all those numbers smaller than one (<1), since numbers <1 do not satisfy the constraint imposed by the equation. Scientific constraints are restrictive conditions on models and theories "such that out of a set of available parameters only those which satisfy the constraints constitute admissible inputs." With the development of science and the emergence of new discoveries, these restrictive conditions can change in various ways. [Weinert 1999, 308–13] Thus for the Greeks, as for Copernicus, circular motion was a powerful constraint on model building. As broad categories, we encounter in science empirical and theoretical constraints. Within these broad categories, further distinctions can be made. Under *empirical constraints* we understand the availability of stable, repeatable empirical data (experimental and observational results), but also the existence of fundamental physical constants (like h and c, see Weinert [1998]), which may appear across quite different models or theories. Scientific theories are to be testable against such empirical constraints. Under *theoretical constraints* we understand physico-mathematical principles (like the relativity principle); methodological norms: simplicity, unification, logical consistency, and the conceptual coherence of a theory (by which is meant here the maximization of the logical connections, mathematical derivations, and evidential relations); and finally metaphysical postulates (the uniformity of nature, causality, circular orbits, determinism, perfection and harmony in nature). A conjunction of these different constraints can delineate a number of different constraint spaces, in which models and theories can be embedded. Geocentrism and its constraints constitute one constraint space, while Kepler's heliocentrism constitutes another constraint structure. The idea of a set of constraints operating on scientific constructs is useful for the elimination of inadequate models: the latter founder on the rock of constraints. We have already alluded to the procedure of eliminative induction, which we shall discuss in Chapter II. Its strength lies in the fact that it can eliminate not just individual models but whole sets of models that satisfy a particular set of constraints. [Norton 1995; Earman 1996]

A *constraint space* is a structure, defined by various types of constraints, into which actual and possible models can be embedded. The constraints operate on admissible and inadmissible constructs. The constraints are always finite, but the constraint space permits an infinity of possible (unarticulated) models and theories. However, a finite number of constraints can govern a potentially infinite number of models or theories, just like an infinite number of specific cases, both actual and possible, falls under one scientific law. A small number of constraints can eliminate an infinite number of models and theories, just like one doorman can turn back a large number of punters. Thus a heliocentric structure, combined with Kepler's laws, eliminates all geocentric-type models as unsuitable constructs. The elimination succeeds because the geocentric models become incompatible with the new constraints. In other words, the attempt to insert old models into new constraint spaces produces inconsistencies.

For the purpose of discussing the Duhem–Quine thesis, in the light of constraints, we shall appeal only to empirical data, as examples of empirical constraints, and coherence, as an example of a theoretical constraint.

Let us first look at some examples of how the appeal to *empirical* constraints helps to alleviate the Duhem–Quine problem, at least from a practical point of view. Recall the situation of geocentricism around 1600. In 1572, Tycho Brahe had discovered a new star (supernova). In the period 1577–1596 he discovered comets. He proved that they were located beyond the moon's sphere. These observations posed a serious problem for one of the central dogmas of geocentrism: the immutability of the supralunary sphere. It cannot be excluded, from a logical viewpoint, that geocentrism may have found a way to accommodate these phenomena. In the later parts of his *Almagest* Ptolemy suggests that any precise knowledge of the supralunary sphere is beyond human understanding. A strange way of accommodation is to plead ignorance. It is important to emphasize that geocentrism would have *had* to accommodate the phenomena. They were stubborn phenomena, whose denial would lead to the dogmatism of the Philosopher. The question is which cost the accommodation would have incurred. [Kitcher 1993, 247–56; Quine 1990, 3–21] As we have seen, the accommodation was natural to heliocentrism. From a practical point of view, the cost of accommodation would have placed a severe strain on geocentrism. It is not coherent to postulate the immutability of the heavens and accept the evidence of mutability.

Let us also look at some examples of how the appeal to *theoretical* constraints, like the coherence of a scientific theory, helps to alleviate the Duhem–Quine problem. Coherence means that the elements of a scientific theory form a tight network. Coherence measures the number of interconnections of the components and deductive consequences of a theory. To make coherence act as a constraint means that only those elements which do not upset the coherence of the system are allowed to enter. The coherence of scientific theories can be compared to a crossword puzzle. As we fill the columns and rows of the puzzle with answers, we begin to see a tight fit. A crossword puzzle has only one solution, which determines which answers are permitted. If a column answer is correct, it exerts a constraint on all the row answers. With almost all columns and rows filled, the puzzle becomes a rigid system. The filled columns and rows impose severe constraints on entries in the remaining blank spaces. With this conception of coherence in mind, consider how an attempt to fit the Jupiter moons into the Ptolemaic model would fare. It would upset the coherence of the system, which was based on the metaphysical belief in supralunary perfection. The non-circular, elliptical orbit of comets would not only have posed considerable problems for the geometric construction of the system. [Figure 1.9] It would have destroyed its coherence. Why did Kepler's introduction of real, nonuniform velocities of planets not destroy the coherence of the Copernican system? Kepler accepted the spatial arrangement of the planets in the Copernican system. Circular orbits can be regarded, mathematically, as good approximations of the near-elliptical orbits of the planets. But Kepler had to abandon the metaphysical need for spheres and circular motion.

We are not arbitrarily imposing philosophical ideas on the Copernicans. We have seen that Copernicus expresses a strong belief in coherence in his Preface to *De Revolutionibus*. He stresses that if the movements of the "wandering stars" are correlated to the circular movement of the Earth, many observable phenomena follow. Taking up these thoughts on coherence, his pupil Rheticus is even more explicit on this score. The great astronomers of the past, he declares,

> fashioned their theories and devices for correcting the motion of the heavenly bodies with too little regard for the rule which reminds us that the order and motions of the heavenly spheres agree in an absolute system. [Rheticus, 1540, 145]

If it is true, however, that scientific theories tend to display a great amount of coherence (interconnectedness) between their components, then it is hard to believe, with Duhem and Quine, that a scientific system only faces the verdict of evidence as a whole. As we have just seen, the coherence requirement seems to limit the number of changes that can be made. It also limits the nature of the changes that are acceptable. We can target particular elements of a system, knowing that it will affect the whole system. The Copernican hypothesis targeted individual components of geocentrism: the topologic position of the earth; the algebraic tool of the equant. Kepler targeted the dogma of circular orbs.

→ 6.5   Theories, models, and laws

In the preceding pages we have spoken of astronomical theories, geocentric and heliocentric models, and planetary laws. The scientific enterprise rests on a number of pillars: theories and models, laws, and constraints. How are they related to each other? How does a theory differ from a model? What is a law of nature? The constraints, as we have emphasized, constitute a constraint space. We can enlarge the constraint space by introducing further constraints, as the history of astronomy illustrates. Tycho Brahe enlarged the observational basis on which astronomical models had to be built. When Kepler rejected the metaphysics of circular spheres he became the author of the most fundamental change in the constraint space of astronomy for two thousand years.

→ *6.5.1   Theories and models*   We can think of a scientific theory as a coherent conceptual system, linking a number of theoretical elements, which are important for the scientific exercise. A scientific theory applies to a domain. The domain comprises all the phenomena in all the possible systems to which the theory applies. The Copernican theory takes all inanimate planetary systems as its domain. The Darwinian theory takes all animate biological systems as its domain. The theories claim that they can account for all the relevant types of behavior in the systems which fall within their respective domains. The Copernican theory wants to account for the distribution of solar systems and planetary motions. The Darwinian theory wants to account for all evolutionary phenomena. To a certain extent scientific theories also provide worldviews. That is, they deliver a particular perspective on the natural or social systems that make up their respective domains. Worldviews tell people what the world is like and what place humans occupy in it. Heliocentrism, Darwinism, and Freudianism express such metaphysical views of the world. When a theory changes, these worldviews come under threat. Such moments in the history of science are the occasions for scientific revolutions. The resistance to a change of scientific theories can partly be explained by their association with worldviews. This process is well illustrated in the transition from geocentrism to heliocentrism. We shall see this process at work in the transition from pre-Darwinian to evolutionary accounts of biological systems.

Scientific theories usually embody a number of fundamental principles, which act as constraints. Copernicus, for instance, states in his *Commentariolus* seven principles of astronomy, including the movement of the Earth. [Copernicus, *Commentariolus* 1959, 58–9; Copernicus 1543, Bk. I, §11] There are *metaphysical principles*, like the belief in circular motion, the unity of nature, and the postulate that every natural event is determined. Galileo's belief that the book of nature is written in the language of mathematics is also a metaphysical conviction. There are *methodological principles*, like the belief in a coherent system, the simplicity of explanation, and the empirical confirmability of the theory. There are *mathematical principles*, like the geometric devices of the Greeks or, after Kepler, the use of algebra. Apart from these principles, most modern scientific theories contain a body of *mathematical laws*. This became evident, for the first time, in the work of Kepler. Later, as the heliocentric model became more sophisticated, the theory could show how the various laws are connected to each other. Newton showed how Kepler's laws could be derived from a more fundamental law, the law of gravitation. Finally, scientific theories must embody a body of *empirical hypotheses*. They face the empirical constraints, the empirical evidence. These must be derivable from the abstract principles. This can be seen at work in, say, heliocentrism. A heliocentric theory makes a very general statement that *all* planetary systems, not just the solar system, consist of a number of satellites, which orbit around a central gravitational body. In order to confirm such a universal theory it is necessary to derive testable statements about a particular system. The solar system was particularly convenient because it could be observed with the instruments available to Galileo and his contemporaries. We can treat these principles as constituting a constraint space. The constraint space consists of empirical and theoretical constraints. Scientific theories also comprise a number of models. The models allow the theory to represent particular aspects of the world. The job of a scientific theory is to throw a blanket of coherence over all these elements. The theory shows how all these elements fit together and how the models of the theory are connected. It shows how the observational and experimental data are deductive or inductive consequences of the principles of the theory.

Models are of particular interest, because they provide the theory with the means of representation. A theory needs models to represent particular aspects of the world. The idea can be quickly illustrated. The Copernican theory is easily confused with a model because it is a theory with only a very restricted domain. For Copernicus the solar system and the fixed stars constituted the universe.

Strictly speaking, the Copernican theory is only a Copernican model. But this is just an accident of discovery. The Copernican theory is not restricted to the solar system. Its intention is to include in its domain not only the known planets of the solar system, but all planetary systems, in any galaxy. If we extend it in this way, it becomes a theory. The Copernican theory includes in its domain all planetary systems in all galaxies. But its extension is much wider. It also includes all artificial systems, like satellites, which may be sent into orbit. One essential feature of models is that they cover only a limited domain of data. The solar system provides the data to construct various astronomical models.

The ability of models to bind selected parameters into a system is one of their most important functions. We may call this function *coherence* or *interrelatedness*. Copernicus was keenly aware that the heliocentric model must represent the planetary system. To enable the models to perform this role, they must serve three other functions: abstraction, idealization, and factualization.

Models concentrate on a few manageable parameters of the target system and *abstract* from a number of interfering factors. The interfering factors are neglected for the purpose of modeling. This operation is called *abstraction*. These interfering factors may be demonstrably negligible, in which case the model will justifiably ignore them. For instance, planetary moons are routinely neglected in the models. The model focuses on a central body and its satellites. However, closer scrutiny may reveal that the abstracted factors have a non-negligible influence on the relationship between the parameters, in which case they need to be incorporated in the model. The Earth moon has an important effect on the tides.

The real factors, which operate in the material world, may be too complicated to compute, in which case a model needs to introduce mathematical simplifications. The models *idealize* the parameters to make their relationships computable in the models. This operation is called *idealization*.[13] Once the dogma of circular motion has been cast aside, it becomes computationally easier to regard the circle as an idealization of the ellipsis.

Again, more complicated models may be able to reduce the idealization of the parameters. The inclusion of non-negligible factors and the elimination of idealized parameters are called *factualization*. In the history of astronomy the most important case of factualization is Kepler's introduction of his planetary laws.

There are also various types of model. Most models have *representational* functions. In this way most models in science serve a practical function. A distinction between various types of model will help to clarify what it means for models to represent. The job of models generally is to capture structural aspects of the natural systems modeled. Recall that models either emphasize the *spatial* ordering of the components in the system – as for instance the spatial distribution of the planets around the sun in the solar system – or place more emphasis on the *mathematical*

---

[13] There has been a considerable amount of literature on abstractions, idealizations, factualizations in science: Krajewski, *Correspondence Principle* [1977]; Nowak [1980]; McMullin [1985]; Brzeziński et al. [1990]; Brzeziński/Nowak [1992]; Herfel et al. [1995]; Cartwright [1999], §9.5; Sklar [2000], Ch. 3; see also Morgan/Morrison [1999].

relationships between the parameters – as for instance in the functional dependence of one parameter on another. When the models emphasize the spatial order, they represent the *topologic structure* of the system modeled. When the mathematical relationship between the parameters comes to the fore, the models represent the *algebraic structure* of the system modeled.[14] In sophisticated models, these two ways of representing will often appear combined, as in the mature Copernican model.

We will briefly distinguish different types of model:

- *Analogue models* represent the unfamiliar or unobservable in terms of the familiar or observable. This type of model suggests that there is an analogy between certain elements of already known systems and some elements of unknown systems. Analogue models are based on formal or material similarity relations. In order to consider a physical cause of planetary motion, Kepler uses the analogy of magnetic rays of the sun, ensnaring the planets. But the mere analogy does not assure that the real systems will resemble the analogue model. The sun does not "lead" the planets by magnetic rays; and planets do not display "friendly" or "unfriendly" faces. Analogies often exploit visual resemblances between the models and the system modeled. The sun seems to attract the planets like a magnet attracts a piece of metal. Analogue models are a useful, if limited, step in an attempt to achieve physical understanding. They suggest useful approaches to problem situations. However, we want more from models than just analogies. We want the models to represent structural features of the natural systems being modeled. To achieve real physical understanding we need more sophisticated models.

- *Hypothetical models* – or *as-if* models – incorporate idealizations and abstractions. They claim to represent the system modeled *as if* it consisted only of the parameters and relationships stipulated in the model. Graphic representations of the solar system are typical hypothetical models. [Figure 1.6] They represent the solar system *as if* it consisted only of, say, six planets, without moons, and *as if* they orbited the sun in circular orbits. However, we know that such idealized factors are mathematical simplifications and that abstracted factors are present in the real systems. (We shall later argue that hypothetical models play an important part in the social sciences.)

- *Scale models* represent real-life systems either in reduced size (the solar system) or in enlarged size (planetary models of atoms). Geocentric and heliocentric models are typical scale models, which represent, in different ways, the solar system. Scale models are usually three-dimensional and require a fairly precise knowledge of the operation of the system. The history of astronomy shows that an accurate representation of the solar system was difficult to obtain.

- *Functional models,* as the name suggests, represent the functional dependence between several parameters. They are widespread in science, ranging from the

---

[14] This distinction between *topologic* and *algebraic* structures is due to P. Roman [1969], 363–69; see Weinert [1999], 313–17.

Carnot cycle of an ideal gas to space-time diagrams and supply and demand curves in economics. There is no need to assign precise values to the symbols that stand for the parameters. What counts is the nature of the *functional* relationship between some parameters. We obtain a functional model, if the functional relationship between various parameters is represented in a diagram or graph. A functional relationship is captured in *Bode's law*. This relationship was discovered by Johann Titius. But it became better known through Johann Bode (1772). Bode's law states that the distance of the planets from the sun (measured in units of the Earth–sun distance, AU) follows the rule:

$$r_n = 0.4 + 0.3 \cdot 2^n \, (n = -1, 0, 1, 2....8)$$

Thus the distance, r, varies with the exponent n. When $n = 1$, for instance, we find $r_n = 1$, which is the distance between the sun and the Earth in the chosen units. When $n = 4$, $r_n = 5.2$ (AU), which is the distance of Jupiter from the sun. In these models, the basis of representation begins to shift from the topologic to the algebraic structure.

• *Structural models* typically combine algebraic and topologic structures in order to represent how some underlying structure or mechanism can account for some observable phenomenon. Structural models are very useful in the representation of macroscopic systems, like planetary systems, and microscopic systems, like atoms. Kepler's heliocentric model combines Copernicus's topologic structure of the solar system with an improved algebraic structure. As we have seen, Copernicus's geometric arrangement of the planets is structurally correct, but the failure of his model lies in the algebraic structure. Once the topologic structure is combined with Kepler's laws and later Newton's theory of mechanics, a fairly accurate structural model of heliocentrism emerges. As we shall see in Chapter II, structural models can also be used to provide structural explanations.

→ *6.5.2 Laws of nature, laws of science*   A scientific theory usually comprises a number of scientific laws. In the history of science, Kepler was one of the first to introduce mathematical laws of planetary motions. [See Ruby 1995] Kepler's work allows us to distinguish between the *laws of nature* and the *laws of science*. [Weinert 1995a, b] The laws of nature are the empirical regularities that exist in nature, irrespective of human awareness. The laws of science are symbolic expressions of the laws of nature. For instance, prior to Kepler's discoveries, the planets moved in near-elliptical orbits around the sun. They moved approximately according to Kepler's three laws of motion. But before Kepler, astronomers assumed that the planets moved in circles, which were modeled using eccentric or epicyclic motion. All these geometric devices were human artifacts. But when Kepler wrote down his three laws of planetary motion, he employed symbolic expressions, which encode the real motion of the planets. Recall Kepler's third law: $A^3 \propto P^2$. This symbolic expression tells us, in terms of averages, that the cube of the average distance of the planet from the sun varies as the square of its orbital period around the sun.

This law of science conceptualizes the phenomena. It provides what has been called "algorithmic compressibility." [Davies 1995] That means that all the observational data about planetary motions can be compressed into a succinct algebraic formula. The equation spares us the tedious task of recording or remembering all the data about these motions. In Kepler's law we have a polynomial formula, which expresses the relations in a structure: we expect the trajectory of a planet, any satellite, to follow a physical pattern according to this formula. Once the formula is at hand there is no need to observe and measure the position of *every* planet. Formulae like Kepler's laws or Bode's law inform us, in compressed algebraic form, of the trajectories of the planets. It is therefore not difficult to conceive of laws as structural constraints on objects. They lay down how a body like a planet must move. They prohibit any other type of behavior of such bodies. Of course, a scientific law may be wrong, as the impetus theory illustrates. But the point is that the regular behavior of natural systems can be expressed in the language of mathematics. If it is the case that planets and satellites behave according to Kepler's laws and that bodies fall according to Newton's laws, then the laws of science give us, in algebraic form, the structure of the behavior of physical systems. The mathematical relationship defines a graph, so to speak, on which the observational data of planetary motion can be arranged. If many of these data cannot be arranged along the prescribed path of the graph, then the mathematical formula is mistaken.

→ *6.5.3 Philosophical views of laws* The algorithmic compressibility offered by the laws of science is extremely convenient. Laws of science express systematic relationships between parameters, enabling us to make inferences from a known to an unknown case. Laws allow scientists to find answers to what looks like insoluble problems. For instance, Newton's laws allowed them to determine the mass of the Earth. The virtues of the laws of science are so great that philosophers have constructed a number of conceptual models about them.

→ *6.5.3.1 The inference view* According to the *inference-license* view, laws are licenses, which allow scientists to infer A from B. The scientist has a certain set of empirical data – the height of a projected ball, the orbital period of a planet – and with the help of appropriate law statements, the scientist is able to work out another set of data: the initial velocity of the ball, the average distance of the planet from the sun. Wittgenstein calls it an illusion "that the so-called laws of nature are the explanations of natural phenomena." [Wittgenstein 1921/1978, §6.371] He compares scientific theories, like Newtonian mechanics, with conceptual networks, which bring "the description of the universe to a unified form." (§6.341) Anticipating the Duhem–Quine problem, he adds that there can be different networks to which different systems of describing the world correspond. The fact that the world can be described by Newtonian mechanics "asserts nothing about the world," according to Wittgenstein. (§6.342)[15] The inference-license view expresses the gist of instrumentalism, since it holds: (1) that laws of nature are "laws

---

[15] Similar instrumentalist views on laws in Toulmin [1953], Ch. 3; Hanson [1958], Ch. 5; Watson [1938]; for a discussion of these different accounts, see Weinert [1995b]; Carroll [2003].

of our method of representing nature" (Hanson, quoted in Musgrave [1979–80], 69); (2) laws permit us to infer particulars from other particulars (for instance, from the position of a planet today, we can infer its position 300 years ago) (Toulmin, quoted in Musgrave [1979–80], 73); (3) they are just rules of inference, so that laws cease to be true or false "empirical" statements.

What are we to think of this view? It is true that scientists do not inductively read off the laws of science from the regularities of nature. As they scrutinize the natural world, they construct the laws of science. Nevertheless, the idea that laws of science are not about nature, but about the conceptual networks according to which we describe nature, is very unsatisfactory. The mathematical formulations, which the scientists construct, must fit the observational data. Even though the scientist may be regarded as free in their formulations of the laws of science, the laws must fit the constraints of the empirical world. In this sense the lawful regularities of nature act as a constraint on the formulation of the laws of science. This empirical check often leads to a modification of law statements. With the emergence of Kepler's laws, for instance, the ancient worry about the equivalence of the geometric devices lost its rationale. If laws of science were inference tickets, we would always face the question: "Why are some inference tickets better than others?" "Why are Kepler's laws better than epicycles?" The problem with the inference view is that rules of inference cannot be confirmed or disconfirmed. They are simply adequate within a certain scope of applicability. But a rule can be adequate without being valid. Ancient astronomy made many adequate predictions, although the planetary "law," on which they were based, was mistaken. The adequacy of a law of science is not exhausted by its ability to make successful predictions. The law must be valid. It must accurately describe, within acceptable limits of approximation, the underlying pattern of regularity. The law is the spine that holds the observational bones together. It states a structure, which governs the behavior of the observables. But inference rules cannot be refuted. They can only be shown to be inadequate for a task at hand. A hammer is an inadequate tool to fasten a screw but not to drive in a nail. Rules need not be eliminated. As science works by a process of elimination, instrumentalism cannot explain the progress of science. [Popper 1963, 112–14; Musgrave 1979–80, 97]

The conclusion has to be that the instrumentalist account of the nature of physical laws is inadequate. Laws of science express more than a license to draw an inference from one particular set of data to another.

→ *6.5.3.2 The regularity view*   The *regularity theory* of natural laws is a more ambitious program. According to this account, the statement,

"It is a law that Fs are Gs,"

must be analyzed as the statement,

"All Fs are Gs."

(Let F stand for "planets" and G for the predicate "circular motion." This statement then tells us that all planets move in circles.) If we look at Kepler's and Newton's laws, it may strike us at first that none of them look like universal propositions of logic:

"All Fs are Gs"

or, in logical symbols,

$$(\forall x)(Fx \supset Gx).$$

Newton's second law and Kepler's third law do not, admittedly, appear as universal propositions in the textbooks of science. But, say proponents of this view, they can be cast in logical symbols. All we have to do is let the symbols *F* and *G* be placeholders for the parameters involved in these two laws. The logical form expresses the universality. [See Hempel 1965, 25–30, 40, 271]

The distinctive feature of the regularity approach is its portrayal of the natural world as governed by uniformities in the Humean sense. That is, if we observe that all instances of *A* precede *B*, then we have reason to believe that whenever *A* occurs, *B* will follow. Thus we infer from the occurrence of sunrise in the past that the sun will rise in the future.

The regularity theory claims that the world is governed by *contingent* uniformities, which we express symbolically in our laws of science. The theory denies that any form of necessity is involved in natural laws. In a way, this approach seems to be quite plausible. Even though the sun has "risen" in the sky for thousands of years, this is not a sufficient reason for the assertion that the sun *must* rise tomorrow. All observed swans (*S*) may have been white (*W*) up to a certain moment in time. But this observation does not forbid the occurrence of black swans. Nothing forbids the non-occurrence of *W*, even when *S* occurs. But consider the two propositions:

(1) "All sodium salts burn yellow."
(2) "Nothing travels faster than the speed of light."

It is tempting to smuggle in a modal operator:

(1a) "All sodium salts *must* burn yellow."
(2b) "Nothing *can possibly* travel faster than the speed of light."

This temptation stems from the intuitive feeling that laws of nature comprise more than contingent uniformities. An intuition tells us not only that *A*, *B*, and *C* have a certain property *P*, but that if some objects have the properties *A*, *B*, and *C*, then they also must exhibit the property *P*. We feel that natural laws must not give us accidental generalities, but unrestricted, cosmic uniformities. It must not simply be the case that a certain number of objects under investigation (planets, sodium salts) share certain respective properties (elliptical orbits, yellow flames). They must possess these properties essentially. Can the regularity theory capture this intuition? For the regularity theory to be viable, it must endorse a crucial distinction between *accidental* and *cosmic* uniformities. Consider the difference between

(3) "All lumps of gold have a radius smaller than 1 mile"

and

(4) "All atoms have a radius smaller than 1 mile."

The first statement looks like an accident of nature. Nothing in nature seems to forbid the formation of lumps of gold with a radius larger than 1 meter. The second statement looks like a universal regularity of nature. It is not simply the case that no scientist has ever observed an atom of a larger size. Science tells us that no atom *can* exist if it exceeds a certain size. Only cosmic uniformities should count as laws of nature. The challenge is to formulate a set of criteria, which make that distinction. regularity theorists define a statement, *P*, as a law of nature by a number of conditions:

i.   *P* is either a universal or a statistical proposition.
ii.   *P* is true at all times and all places.
iii. *P* is contingent.
iv.   *P* contains only non-local empirical predicates, apart from logical connectives and quantifies. That is, *P* is purely descriptive.
v.   *P* takes the form of a conditional ("$\subset$").

[See Swartz 1985, 28–29; Molar, quoted in Armstrong 1983, 12] Physical laws, according to the regularity view, are descriptions of actualized, empirical, contingent connections between states and events in the physical world. [Swartz 1985, Chs. 2, 3]

Stated in this way, the regularity theory suffers from a number of weaknesses. A consideration of these weaknesses has led to a much stronger view, the necessitarian account of laws. What are the weaknesses?

One worry is that the regularity account cannot properly distinguish between accidental generalizations and cosmic uniformities. There are widespread accidental generalizations, which we would not be tempted to count as laws of nature. Statement (3) is an accidental generality, whilst statements (2) and (4) are true cosmic uniformities. But statement (3) satisfies the conditions imposed by the regularity theory on the laws of nature. On the other hand, certain statements, like (1), only express a restricted uniformity. Yet they qualify as lawful regularities, just like Ohm's law

(5) $V = IR$,

which states that voltage, V, is the product of current, I, and resistance, R. Ohm's law is only true at constant temperatures. Unrestricted universality is not a necessary condition for laws of nature.

Therefore we must make a distinction between *fundamental* and *phenomeno-logical* laws. Fundamental laws, like statements (2) and (4), are valid for all physical systems. Phenomenological laws, like statements (1) and (5), only apply to physical systems under certain restrictive conditions.

A further unwelcome consequence of the regularity theory is its exclusion of unrealized physical possibilities. Unrealized physical possibilities are an important feature of science, which often lead to technological innovations. For instance, 60 years ago, satellites were unrealized physical possibilities. One hundred years ago, lasers were unrealized physical possibilities. And two hundred million years ago most of today's species were unrealized biological possibilities. All these unrealized physical possibilities, had they been realized at the time, would have been subject to the same laws that govern them today. Satellites are subject to Kepler's laws. Lasers are governed by the laws of quantum mechanics. Species are subject to the laws of evolution. But according to the regularity theorist, unrealized physical possibilities are physically impossible. This account only accepts as laws of nature the actualized cosmic uniformities. The regularity view cannot deal with *counterfactuals*, as they occur in the contemplation of unrealized physical possibilities.

The regularity theory faces, perhaps surprisingly, a great number of problems. A philosophical account of laws should account for counterfactuals, the distinction between unrealized physical possibilities and genuine physical impossibilities, and the difference between accidental and cosmic uniformities. It should also justify our confidence in making inferences from known to unknown cases. A philosophical account should explain how scientific laws are employed in our explanations of the behavior of physical systems. This is a tall order. The instrumentalist and regularity views have not lived up to expectations. Can a stronger view, the necessitarian account, help?

→ 6.5.3.3 *The necessitarian view* The *necessitarian view* has a straightforward answer to all these problems. Natural laws are relations between universals. The necessitarian account transforms the formula, used by the regularity theorist,

"It is a law that Fs are Gs"

into the much stronger claim:

"It is physically necessary that Fs are Gs."

In terms of our previous distinction between the laws of science (scientific laws) and the laws of nature (natural laws), the necessitarian view states that

Laws are (expressed by) singular statements describing the relationships that exist between universal qualities and quantifiers.

To say that it is a law that Fs are G is to say that "All Fs are G" is to be understood, not as a statement about the extensions of the predicates "F" and "G" but as a singular statement describing a relationship between the universal properties F-ness and G-ness (where properties can be magnitudes, quantities, features). [Dretske 1957, 252–3; cf. Leckey/Bigelow 1995; see also Weinert 1995b]

The necessitarian approach to laws of nature is based on two essential notions: (a) laws involve universals and (b) laws are relations between universals.

Ad (a) According to David Armstrong, a proponent of the necessitarian view, "universals are either properties or relations of some, or holding between some real particulars." [Armstrong 1983, Ch. 6] Universals are the repeatable features of the spatio-temporal world.

Ad (b) If we accept this Aristotelian characterization of universals, we are left to specify the relation between universals. Given that some entity is *F* and we know that it is a law of nature that $(\forall x)(Fx \supset Gx)$, we can conclude that *F must be G*, unless there are some intervening factors. It is physically necessary for *F* to be *G*, given the law of nature. The relation, which is postulated as holding between universals, is one of nomic necessitation. Yet nomic necessitation is only contingently true as it may not hold in all possible worlds.

The logician's symbolic formulation highlights the difference between the necessitarian and the regularity views. From the postulation of nomic necessitation, $N(F, G)$, it follows that for all entities, *x*, if they have property, *F*, they must have property, *G*. The relation of nomic necessitation between F-ness and G-ness entails the corresponding cosmic uniformity:

$$N(F,G) \rightarrow (\forall x)(Fx \supset Gx).$$

But the converse does not follow. Each *F* may be *G* but this does not mean that F-ness necessarily entails G-ness, as the case of accidental regularities demonstrates:

$$(\forall x)(Fx \supset Gx) \overset{\neg}{\longrightarrow} N(F,G)$$

Accidental generalities allow exceptions, nomic necessity does not (barring domain restrictions). One criticism leveled against this theory is that nomic necessitation is not observable. Nothing in our observational evidence points to the existence in nature of physical necessity, which the necessitarian theory requires. Armstrong postulates nomic necessity as an unexplicable primitive notion of his theory. [Armstrong 1983, Ch. 6.4] Science observes repeatable features between spatio-temporal events. We are free to label such features – ravenhood and blackness – as Aristotelian universals. Even if we grant that science observes universals, we cannot leap to the conclusion that physical necessity holds between them. Cosmic universality does not imply nomic necessity. (Other necessitarians go beyond Armstrong and stipulate the existence of uninstantiated universals, thus embracing Platonism.)

We should distinguish two senses of physical necessity. On the one hand we have the necessitarian sense of nomic necessitation between universals. This physical necessity between universals constitutes the laws of nature. On the other hand we have the scientific sense of necessity. This sense states that certain physical systems are subject to the laws of nature. Planets must orbit the sun, because they are constrained by Kepler's laws to do so, given initial conditions. On the necessitarian account, all uniformities in nature reduce to just one connection: nomic necessitation between universals. But the laws of science teach us that there are many different mathematical relations between many different systems. They can be expressed in precisely defined parameters. Furthermore, these systems are interrelated. The laws of science express a number of different algebraic relations between relata of physical systems. For the laws of science to express the laws of nature, they must approximate the lawful regularities in nature.

The necessitarian view tells us very little about how laws are used in science. If we are interested in how the laws of science express nature's regularities, we should turn to an account of laws, which takes us closer to scientific practice. Let us call this view a structural approach. The basic idea behind this approach is that the laws of science encode the relations in the structure of physical systems.

→ *6.5.3.4 The structural view*   Karl Popper rightly points out that "natural laws" are logically stronger than true, strictly universal statements. This observation takes care of the need to distinguish natural laws from mere accidental generalities. But they are logically weaker than logical necessities. While logical necessities are true in all possible worlds, the laws of nature are contingent. We expect them to be true in the actual universe, not just on Earth, but they may not be true in all possible worlds. What logical character do the laws possess? Popper shifts necessity from the laws of nature to the laws of science.

> If we conjecture that "a" is a natural law – he writes – we conjecture that "a" expresses a structural property of our world; a property, which prevents the occurrence of certain logically possible singular events. [Popper 1959, 432]

Thus physical necessity means that the laws impose structural constraints upon the natural world. [Popper 1959, 430] According to Popper, scientific laws express certain structural properties about the physical world. Popper's *structural* view of laws stands in direct opposition to Wittgenstein's instrumentalist view. Popper stresses that the laws of nature forbid certain structural properties of the actual universe. The laws forbid circular planetary motions, perpetual motion machines, and superluminal velocities. The reverse of this characterization is that the laws will also enable certain physical events. The laws lay down structural constraints, according to which individual objects must behave. But the singular facts continue to enjoy a certain freedom within the structural grid that binds all the facts together. Autumn leaves, carried by the wind, "defy" the law of gravity. A planet on a collision course with a meteorite may be knocked out of its orbit. Evolutionary mutations are the result of genetic chance.

The structural view accounts quite well for some of the main features of scientific laws. It accounts for the unavoidable abstraction and idealizations involved in the scientific laws. It shows why we cannot deduce nomic necessity from the available evidence. It stays close to scientific practice. [Weinert 1993; 1995a, b] The laws of nature encode the structural properties of physical systems. The laws of science are the symbolic expressions of these structural aspects in mathematical language. The structural account makes the distinction between the laws of science and the laws of nature, which we find in scientific practice. It marks the distinction between accidental and cosmic uniformities. Accidental uniformities – "All solids of gold are smaller than 1 mile in diameter" – are not structural features of the natural world. There is nothing in the atomic structure of gold to forbid large solids of gold. Cosmic uniformities – "All atoms are smaller than 1 mile in diameter" – express the structure of atoms. There could be *no* atoms if they had these dimensions. It explains unrealized physical possibilities as possibilities allowed by the structures. It explains counterfactuals by relying on what the structures allow and what they forbid. "If there were a tenth planet, it would obey Kepler's laws" means: the structure of the solar system is such that a tenth planet may orbit the sun in accordance with Kepler's laws. It covers the distinction between fundamental and phenomenological laws. Fundamental laws express the structure of all or numerous systems; phenomenological laws are restricted to a few systems, requiring many boundary conditions.

According to the structural view of laws, the laws of nature constitute constraints on the possible trajectories of material objects in our universe. For mathematicians equations are constraints on the possibility of solutions. Imagine a mathematician who has taught algebra for many years. The mathematician has patiently explained to students the use of functions. A function $f(x)$ is an equation of a certain degree: (a) $y = 2x^2 + 3x + 4$ is a quadratic equation while (b) $y = 3x^3 + 2x^2 + x + 1$ is a polynomial expression. As Descartes discovered, functions can be represented in Cartesian graphs. Irrespective of the particular equation, a mathematical function displays typical algebraic structures, depending on the exponentials involved. Consider two material objects in our universe, one of which behaves according to (a) and the other according to (b). Even if we know nothing more about these objects, the *structure* of their trajectory is laid down in their respective graphs. (The details of the graphs change if more boundary conditions are included.) [Figures 1.10a, b]

These graphs represent the permitted trajectories of the object, under the constraint of the respective function. More precisely they represent a mathematical pattern, which says that any object, subject to the given algebraic structure, $y = x^2$, will display a behavior pattern of this structure. Similarly for objects, which are subject to the algebraic structure, $y = x^3$. The laws of science represent, in mathematical language, structural aspects of the world around us. The laws specify the relations in the structure, according to which material objects must behave. In this sense the laws of science impose constraints on the possible trajectories of objects through the material world. As we noted earlier (Section 6.3), a structure consists of relata and relations; the relations bind the relata together. In the case of planetary systems, the relata are planets, moons, comets, meteorites, and satellites; the relations are the mathematical relationships that hold between the relata; or the relata may be biological systems between which evolutionary relations hold. On the structural account, the laws of science determine mathematically specifiable correlations between the relata. As the graphs illustrate, the laws specify the

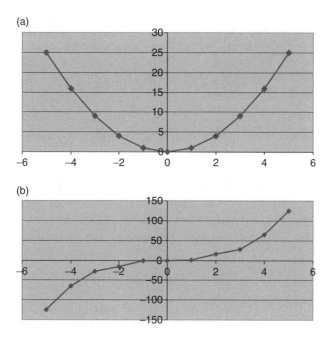

Figure 1.10 (a) The quadratic function leads to a parabola; (b) The polynomial function leads to a polynomial graph

trajectory of material objects as they weave their paths through a material world in the neighborhood of other objects.

We have now discussed a number of philosophical issues, which arose out of the problem situation of Copernicanism. Copernicus stood at the threshold of modern science. The Copernican turn pushed open the gates to the Copernican revolution. Can Copernicanism tell us something about the process of scientific revolutions?

7 Copernicus and Scientific Revolutions

One measure of the depth of a physical theory is the extent to which it poses serious challenges to aspects of our worldview that had previously seemed immutable.
[Greene, The Elegant Universe *(2000), 386]*

We have argued that Copernicus is the author of a change of perspective. Copernicus initiated the Copernican turn. Although his system had several advantages, it did not amount to a scientific revolution. On the one hand, Copernicus adheres too much to his Greek predecessors and their geometric methods. He adopts a purely kinematic view. He adds no significant new observations to the available catalogue of data. On the other hand, he leaves in place the equivalence of hypotheses.

He finds it acceptable to use a number of geometric devices interchangeably, without asking which one corresponds better to the physical mechanism of planetary motion. But his model is in better agreement with the topologic structure of the solar system. In this respect it enjoys greater empirical validity. However, its algebraic structure is deficient. For this reason the Copernican model fails to reach full theoretical validity.

If the Copernican turn was a first step in the direction of a scientific revolution, what criteria are available to judge an episode of science as a scientific revolution? Several models of scientific revolutions have been proposed.

A   *Kuhn's paradigm model of revolutions.* According to Thomas S. Kuhn, the history of science consists of a series of "normal" and "extraordinary" periods. [Kuhn 1970; Hoyningen-Huene 1993] A normal period of science is marked by the presence of a dominant paradigm. This paradigm is accepted as a valid framework for ongoing research. During periods of normal science, scientists are involved in problem-solving. The accepted problems and solutions are set by the ruling paradigm. Typical examples of paradigms are heliocentric astronomy, Newtonian mechanics, and Darwinian evolutionary biology. During normal periods of science, the practitioners of a scientific discipline accept the basic presuppositions of the paradigm. Their work consists in refining the representational force and explanatory power of the paradigm. Kepler, Galileo, and Newton all accepted the heliocentric hypothesis. Newton accepted Kepler's laws, although Galileo chose to ignore them. But Galileo's observations made significant contributions to heliocentrism. In their own way they all refined the paradigm and improved it. Eventually, however, any period of normal science faces a crisis. It then enters a period of extraordinary science. A crisis in science can happen for a number of reasons. It is, according to Kuhn, mostly associated with the failure of a paradigm to deal with all the phenomena in its domain. A crisis occurs because a paradigm faces a significant *anomaly*. An anomaly is not just the failure of a prediction or a discrepancy between theory and observations. Such discrepancies are unavoidable: the observational devices are not perfect and the theory always makes a number of abstractions and idealizations. An anomaly occurs when there is a persistent disagreement between a theory and its predictions. An anomaly arises when the theory claims that the world is one way and our observations tell us that it is another way. The Greek theory of concentric circles quickly faced an anomaly. The theory implied that the planets are always at a constant distance from the Earth. Observations told the Greeks otherwise: the planets change their distances from the Earth. The geocentric view forbids the appearance of comets beyond the lunar sphere. The Mathematician and Philosopher in Brecht's play insist that Jupiter can have no moons. Yet Brahe's and Galileo's observations told them otherwise. These theories faced anomalies. They could not accommodate the observations, which clearly fell within their domains. When such events happen, the once dominant paradigm enters a crisis. The practitioners try to solve the problem in a number of ways. If they succeed the paradigm may continue its rule. But if they fail, the scientific discipline enters a *revolutionary period*. During a revolutionary period, according to Kuhn, the old paradigm is dismantled and replaced by a new one. A scientific revolution is a replacement of a paradigm.

A paradigm is a conceptual scheme which mediates the interaction between the scientist and the world. It facilitates the mapping of symbolic structures onto the empirical world. Later, Kuhn preferred to speak of a "disciplinary matrix": an ordered set of elements. A matrix comprises a number of conceptual elements: symbolic generalizations, like fundamental laws; exemplary problems, on whose solutions students can practice the techniques of the discipline; scientific values, like consistency, coherence, testability, unification; and metaphysical convictions, like a belief in an ordered universe, an independent, external reality, and deterministic laws. Kuhn renamed a paradigm of science "a disciplinary matrix" to indicate that the conceptual constructs form a coherent network. A distinctive feature of Kuhn's paradigm model of scientific revolutions is that scientists are taken to be fundamentally committed to their paradigms. A paradigm gives them a firm foothold that enables them to approach the natural world in a principled manner. According to Kuhn, the scientists begin to see the world in terms of the ruling paradigm. It is as if a scientist was wearing glasses – Newtonian spectacles, Darwinian spectacles – through which alone s/he could see the world. According to Kuhn, scientists can inhabit only one paradigm at a time. If this is the case then paradigms which spell out completely different problem situations make the scientists see the world completely differently. They find it difficult to talk to each other, because they inhabit "different worlds." The Aristotelians in Brecht's play embrace geocentrism and inhabit a geocentric world. It makes sense to them to refuse to look through the telescope, because Jupiter's moons "cannot exist." By contrast Galileo defends heliocentrism. He inhabits a heliocentric world. His appeal to the observational evidence cuts little ice with the Mathematician and Philosopher. He speaks from the platform of a paradigm, whose language they cannot understand. There is an abandonment of critical discourse.

If this scenario describes the history of science correctly, Kuhn faces the question of how scientific change is possible at all. For both parties in the dispute seem to be absolutely convinced of the truth of their respective paradigms. Kuhn uses the term "incommensurability" to describe the stalemate. The term indicates that it is not possible to translate each element of a paradigm into the elements of its rival paradigm. It is possible to compare paradigms globally but it is not possible to map each conceptual element of one paradigm onto another. [See Andersen/Barker/Chen 2006, Ch. 5] We may ask why this should be important or even a problem. If there is no equivalent of an "epicycle" in Kepler's model, this seems rather a gain than a loss. Kuhn, however, considers the incommensurability of paradigms to be an important feature of the history of science. So he has to explain how paradigms can change at all, if scientists are chained to them like worldviews. Kuhn's answer is that the seeds of revolutionary change are built into each paradigm. Each paradigm eventually encounters an anomaly. It is the persistent non-agreement between theory and empirical data that precipitates a paradigm into a crisis.

Eventually a new paradigm emerges. With the new paradigm, a new period of normal science sets in. Most practitioners are now in fundamental agreement with the presuppositions of the new paradigm. Again it imposes a vision on the scientist, which differs markedly from the previous vision. "What were ducks in the scientist's world before the revolution are rabbits afterwards." [Kuhn 1970, 111] According to Kuhn, the quantum jump from one paradigm to another has drastic consequences. First, the whole conceptual network changes. There are vast differences in ontology – of the things which are supposed to exist in the world – between the old and the new paradigms. Kepler, for instance, replaced the solid spheres with free-floating planets. The Greek geometric devices are replaced by Kepler's planetary laws. The abrupt change also affects the range of acceptable problems and the accepted techniques. From Kepler to Newton the dominant problem became the question: "Why do planets move in orbits?" Geometrical methods became obsolete and were replaced by more sophisticated mathematical techniques. Second, a communication breakdown occurs, as Brecht's play seems to illustrate. Kuhn is not always clear about the extent of the communication breakdown. After criticism he seemed to accept that the communication breakdown was only partial. [Kuhn 1983; Nola 2003, Ch. 1.4.1] The scientists will at least partly agree on some continuity between the old and new paradigms. But rational reasons alone are not compelling enough to convince a doubting scientist of the virtues of the new paradigm. Kuhn explains the adoption of the new paradigm as a case of conversion and persuasion. The reality is more complicated, as Kuhn eventually admitted. There are continuous and discontinuous links between the paradigms. An intensive discussion took place between the Copernicans, the Ptolemaists, and the theologians. The correspondence between Osiander, Copernicus, and Rheticus has survived. The dispute lasted for 150 years. This was enough time for many different models to be marshaled. The Ptolemaists provided arguments against the motion of the Earth. Copernicus adopted the impetus theory to reject their arguments. Tycho Brahe adopted a compromise position. An inveterate Copernican like Galileo simply ignored Kepler's laws. Some converted to Copernicanism, like Rheticus in Germany and Digges in England. Until the beginning of the seventeenth century there was insufficient evidence of the heliocentric hypothesis so that some accepted only parts of the Copernican system. [Rybka 1977; Dreyer 1953, Chs. XIII, XIV] The Copernicans finally prevailed when Newton combined the law of inertia with the law of gravitation.

The Copernican turn does not really fit the Kuhnian paradigm model. [Heidelberger 1980] It did not constitute a paradigm shift. The Copernican version of heliocentrism is hardly incommensurate with geocentrism because of the large overlap between the two systems. Copernicus uses many of the Greek observations. He invents no new method. On the contrary, he wants to be purer than Ptolemy, since he objects to the use of the equant. As the discussion around the term "hypothesis" shows, it is hard to detect as much as a partial breakdown of communication. Of course, the opponents did not agree. The ambiguity of the term "hypothesis" seems to have provided a bond between proponents and detractors of Copernicanism. Later on, the Copernicans did not even agree on some of the central elements of the disciplinary matrix. Galileo continued to believe in circular motion. The Copernican position is compatible with both instrumentalism and realism. By concentrating on the equivalence of hypotheses, Osiander argued in favor of instrumentalism. By focusing on the topologic structure of his model (in Book I of *De Revolutionibus*), Copernicus arrived at a realist position.

The conceptual network did change but by a slow transition rather than quantum leap. Tycho's observations, Kepler's calculations, Galileo's findings provided powerful arguments in favor of the Copernican hypothesis. These achievements constituted powerful constraints, which the geocentric system could not accommodate. Let us therefore consider some alternative models of scientific revolutions.

B   *Cohen's four-stage-model.* Bernard I. Cohen, a historian of science, proposed a model, according to which a scientific revolution unfolds in four stages. [Cohen 1985] The *first stage* consists of a conceptual innovation, as we find in Copernicanism and Darwinism. We discussed this stage earlier as a change in perspective. A change in perspective is, however, not sufficient to constitute a revolution.[16] The *second stage* consists of new methods and techniques. We have found that Copernicus did not introduce new techniques, so he would fail this second criterion. In this regard both Kepler and Newton made the greatest advances. Darwin, too, proposed a new method: the Darwinian inferences. His principle of natural selection, as we shall see, incorporates a new algebraic structure. The *third stage* Cohen calls "dissemination." In this stage the revolution occurs on paper. It is given a voice in the public arena. In modern terms, we would say the work finds a publisher. There are some famous publication dates in science, which illustrate this stage: Copernicus's *De Revolutionibus* (1543); Darwin's *The Origin of Species* (1859); Einstein's special theory of relativity (1905). These publications submit the revolutionary ideas to public scrutiny. The *fourth stage* consists in the adoption of the new ideas by the scientific community. This is not a question of an instant conversion. The new ideas usually find some proponents but they also meet with skepticism and opposition. In such a situation intensive debates ensue. They do not have to take the turn of the unfortunate encounter of Galileo with the Mathematician and the Philosopher. Attempts to persuade are often based on arguments. In his *Narratio Prima* (1540) Rheticus tries to convince his contemporaries of the superiority of the Copernican system. At the beginning of the seventeenth century Kepler wrote the first textbook of astronomy. The *Epitome of Copernican Astronomy* (1618–21) is a long argument in favor of a modified Copernican system. Darwin conceived his *Origin of Species* as a "long argument" in favor of evolutionary explanations. If there is no instant conversion, most successful scientific theories experience a *convergence* of

---

[16] It may be asked whether a change in perspective is also a necessary condition for a revolution. This will be too strong a requirement for it amounts to the thesis that a scientific revolution cannot occur without a change in perspective. But even if this were true of all past revolutions in science, we cannot be committed to the view that it will also be a condition of all future revolutions in science.

expert opinion. After some period most scientific practitioners become convinced of the validity of the new theory. According to the evidence gathered in this and later chapters, this is mostly a matter of persuasion by argument. For the evidence, as we have seen, also converges on to one model or theory. What is more, the same evidence that increases the creditability of one theory begins to discredit rival theories.

Cohen's four-stage model sees scientific revolutions as temporally extended processes. They are transitions rather than abrupt conversions. The first two stages – the conceptual innovation and the new techniques – intimate an important aspect of scientific revolutions, namely the problem-solving abilities of new paradigms. This aspect was already emphasized in Kuhn's paradigm model. It becomes the focus of the chain-of-reasoning model.

C   *The chain-of-reasoning model.* Seen against the background of Copernicanism (and Darwinism), Kuhn's model suffers from two defects. With its thesis of incommensurability it overemphasizes the disruptive discontinuity between successive paradigms. And it often portrays a breakdown of communication between scientists, as parodied in Brecht's play, when the reality of scientific revolutions is often more sophisticated. Cohen's model may satisfy a historian of science but it does not emphasize sufficiently the conceptual continuity between successive revolutions. Successive revolutions display a transitional nature, which can be made more precise. Let us speak of *traditions*, rather than paradigms, to designate the conceptual networks in the history of science. Their respective elements can change *differentially*. That means, as the history of Copernicanism has shown, that the geocentric tradition did not collapse all at once. Rather, the Copernican spatial rearrangement of the planets was preceded by an abandonment of the Aristotelian theory of motion. Only later was the notion of circular notion discarded. This differential surgery means that we can look for *reasoned* transitions between the conceptual components of the network. [Shapere 1964; 1966; 1989; Chen/Barker 2000] We can follow the career of, say, the concept of circular orbs from Greek to post-Copernican astronomy. We will observe that it became obsolete and ask why this happened. Or we can study the role of theories of motion in the history of astronomical models. The theorists of the Parisian School replaced the Aristotelian theory of motion with the impetus theory, because they found that the latter gave a better explanation of motion. The impetus theory helped the Copernican model along the way. But it was not until the middle of the seventeenth century that the concept of inertia was developed. [Drake 1975; Wolff 1978; Dijksterhuis 1956] These are examples of *reasoned* transitions. They are reasoned transitions because they arise from problem situations, in which attempted solutions are evaluated through chains of reasons and arguments. In the history of astronomical models we can follow the career of constituent presuppositions. They are judged as to their ability to provide solutions to the problem of planetary motion. The transitions lead to the reorganization of at least part of the conceptual scheme. The transitions are part of problem-solving attempts. These attempts leave traceable lines of descent between astronomical models. There are *deletions*: the Aristotelian theory of motion; there are *modifications*: the circle gives way to the ellipsis; there are *replacements*: mathematical analysis replaces geometric methods. We can trace the lines of arguments in the development of conceptual traditions and their components: of theories of *motion* and the arrangement of *planets* from geocentrism to heliocentrism. We can evaluate these operations in the chains of reasoning. Chain-of-reasoning transitions emphasize the slow transformations of conceptual networks, traditions, through the weight of arguments and evidence. Integrating these insights we arrive at an analytic four-stage model of a scientific revolution as a series of successive events:

1. a turn or switch of perspectives, which often involves a questioning of existing presuppositions;
2. the introduction of new methods and techniques with problem-solving ability;
3. the emergence of a new tradition through differential chain-of-reasoning transitions, as a result of the problem-solving success of the emergent tradition;
4. convergence of expert opinion on to a new tradition; we shall note in Chapter II that this convergence does not exclude the coexistence of alternative models within the new tradition.

There is no guarantee that once a component is secured within a tradition, its position will remain unchallenged by further arguments. Ironically, this fate befell the Copernican principle itself. If we follow this particular chain of reasons into contemporary debate, it will lead us to a questioning of the Copernican principle: that we do not occupy a privileged position in the universe.

8 The Anthropic Principle: A Reversal of the Copernican Turn?

To paraphrase Descartes, "cogito ergo mundus talis est." [Carter, "Anthropic Principle" (1974), 294]

Freud considered that human self-pride had suffered serious blows from Copernicanism and Darwinism. It is true that the Copernican hypothesis dethrones humans from their imaginary physical center in the universe. And Darwin seems to have robbed humans of their "crown of creation." But the Copernican change in perspective also demonstrates the power of the human mind. Copernicus and his successors showed that the supralunary sphere was not forever veiled from human comprehension. The human mind, as Kepler insisted, is more penetrating than the human eye. What humans can see is not the limit of what they can comprehend. As we seem to be the only intelligent species in our immediate cosmic neighborhood, spanning several light years in all directions, humans can at least claim rational centrality. The Copernican turn shows that rational centrality does not depend on physical centrality. Once this had been understood, modern science could get off the ground. Humans will not be able to leave the solar system in the foreseeable future. Humans cannot even visit all parts of the solar system. Yet cosmologists know much more from scientific analysis than they would learn from cosmic sightseeing. Cosmologists have mapped the structure of the solar system, the Milky Way, and the cosmos to an extent that could not be matched by mere observations.

For some cosmologists this rational centrality is not enough. They have proposed an *Anthropic Principle* (AP), which attempts to restore some of the pre-Copernican pride to the human species. The Anthropic Principle seeks a reversal of the Copernican turn. It reasons from the existence of intelligent life on Earth to the special physical conditions that render intelligent life possible. Our position in the universe may not be central, but it is privileged. [Carter 1974, 291] There are in fact two versions of the Anthropic Principle.

- **The Strong Anthropic Principle (SAP)**: *The Universe must have those properties which allow life to develop within it at some stage in its history.* [Barrow/Tipler 1986, 21; Carter 1974, 294]

    This is a very stringent requirement, since it postulates that the physical layout of the universe is such that it inevitably becomes a self-cognizant universe. The universe will eventually lead to the creation of intelligent human observers. [Barrow/Tipler 1986, 248, 523; Breuer 1991, Ch. I] The strength of this requirement is also its weakness. On the one hand, the Strong Anthropic Principle embodies a determined anthropocentrism, which is no longer part of a physical explanation. Since the Scientific Revolution of the seventeenth century there has been a radical tendency to eradicate anthropocentric reasoning first from the physical sciences, and subsequently from the biological sciences. On the other hand, the SAP runs directly counter to evolutionary explanations of life. Biological principles, like natural selection, do not claim that new species evolve along a linear path, whose terminus is the emergence of human beings. It was one of the consequences of Darwin's work that any teleological thinking in evolutionary biology, which treated humans as the telos of evolution, was a misconception. Most cosmologists therefore prefer a weak version of the Anthropic Principle.
- **The Weak Anthropic Principle (WAP)**: *We must be prepared to take account of the fact that our location in the universe is necessarily privileged to the extent of being compatible with our existence as observers.* [Carter 1974, 293; Dicke 1961; Breuer 1991, 8]

    This postulate states that the existence of human life can be used to explain the delicate values of the fundamental physical constants:

    The observed values of all physical and cosmological quantities are not equally probable but they take on values restricted by the requirement that there exist sites where carbon-based life can evolve and by the requirement that the Universe be old enough for it to have already done so. [Barrow/Tipler 1986, 16; Dicke 1961]

    The idea is that the intricate balance between the fundamental physical constants cannot be a cosmic accident. Even the slightest changes in these values would destroy the possibility of human life. The precise numerical values of the fundamental constants are held to be necessary for human life to become possible. The stability of matter and three-dimensional space, for instance, is essential for life. A slight change in these numerical values would have launched the universe onto a completely different trajectory. Even more dramatically, there are key properties of the universe, without which biological evolution would have been impossible. Our discovery of these key properties "may in some sense be necessary consequences of the fact that we are observers" of them. [Barrow/ Tipler 1986, 383] The basic idea is that human existence imposes severe constraints on the numerical values of the fundamental constants. The constants must have the values they possess *because* we are here. Couched in counterfactual terms, if the fundamental physical constants had acquired slightly different values, human beings would not have evolved. As humans have evolved, the constants must possess these specific values.

Consider, for instance, the age and size of the universe. Proponents of the Weak Anthropic Principle maintain that these properties can be explained by considerations of the condition of human existence. Clearly the Earth must have been physically hospitable to the evolution of organic life. The transformation of hydrogen and helium into life-conducive molecules happens primarily in stars. But the production of the heavier molecules, on which life depends, takes billions of years. The universe must be sufficiently old for life forms to be present. Therefore, proponents of the WAP argue, the existence of human life can explain the age of the universe. And the age explains its size. "Many observations of the natural world (...) can be seen in this light as inevitable consequences of our own existence." [Barrow/Tipler 1986, 219]

For proponents of the WAP this reasoning restores a form of special centrality and reverses the Copernican turn. If only very special conditions produce a self-cognizant universe, a close link between the human race and the cosmic environment is reestablished. The Anthropic Principle puts constraints on the structure of a self-cognizant universe. It asserts that intelligent life in some ways selects the actual universe. [Barrow/Tipler 1986, 510; Breuer 1991, Ch. I] Solely such a universe gives rise to intelligent observers who can recognize their privileged position.

One problem with the Anthropic Principle is that it is at best only approximate. It invites the observer back into physical theory. But even then it can only state the order of magnitude of the fundamental constants, not their precise values. The WAP masquerades as a physical explanation when it is no more than an unobjectionable inference. From the fact that human observers inhabit a small corner of the universe it is inferred that this place must be hospitable to life. But our existence does not retrospectively explain why the universe possesses the physical conditions that have made intelligent life possible. [Salmon 1998, 396] Even proponents of the AP admit that the principle may ultimately be replaced by a physical explanation. [Carter 1974, 292, 295; Carr/Rees 1979, 612]

There is a certain similarity between arguments advanced by proponents of the AP and those of modern advocates of Intelligent Design. Advocates of Intelligent

Design argue that many biochemical processes are too fine-tuned for natural selection to explain them. [See Chapter II, Section 5.4]. They consider that such organs as the eye are so complex that they could not be the result of slow evolutionary adaptation. Advocates of the AP point to a similar delicate balance in the fundamental constants. They argue that the existence of intelligent observers can explain the fine-tuning of the fundamental constants. In both cases teleological explanations play an important part. Proponents of Intelligent Design infer design from the complexity and improbability of biological systems. Proponents of AP infer a special place and time for humans from the specific values of cosmic parameters. The problem is that the fundamental constants do not possess their particular values, *because* we are here. A slight change in the value of the fundamental constants in the distant past might indeed have rendered the evolution of life on Earth impossible. But our presence *now* does not make it necessary that in the past the constants acquired their particular values. It is true that we can infer special values of the fundamental constants from our presence. But this does not explain the values. The SAP seems to imply that observers are the goals of evolution. [Barrow/Tipler 1986, 28] Such teleological thinking, as Chapter II will show, has been prevalent for most of the history of ideas. It is strongly contradicted by the history of Darwinism. Even the WAP fails to show that the sizes of stars, planets, and people are the necessary consequences of the constants of nature. [Barrow/Tipler 1986, 387] One reason is that cosmic evolution is contingent. The Earth is not shielded from the rest of the solar system. For millions of years the Earth was subject to the bombardment from outer space. The cosmos is a vast system. According to the Alvarez hypothesis, the extinction of the dinosaurs, 65 million years ago, was the aftermath of a collision of the Earth with an asteroid. When they disappeared their lifespan had already stretched over 200 million years. Had they continued to thrive, humans might never have seen the day of light. During the dinosaurs' reign conditions were favorable to organic life. But this does not mean that human observers had to appear. There is another reason to be suspicious of anthropic reasoning, which follows from the first criticism. Anthropic reasoning ignores chains of causation. We cannot causally explain the occurrence of an earlier event through the occurrence of a later event. Yet anthropic reasoning leaps from a distant event in the past to present-day events. It answers questions like, "Why is the universe isotropic?" with statements like "because we are here." [Barrow/Tipler 1986, 426] However, the contingency of physical events forbids us from skipping several links in the chain of causation. It would be faulty reasoning to claim that I am here *because* my great-grandparents met in the 1870s. Physical thinking tells us that interferences can divert an event from its "predetermined" path. Evolutionary thinking tells us that the tree of life sprouts in a contingent manner. If we accept these insights, then humans are not a necessary consequence of evolution. Such an explanation sounds suspiciously like Lamarck's progressive evolution. As we shall see, Darwin's revolution led to the loss of rational design.

The Copernican turn, we may conclude, led to a loss of physical centrality. But human existence is still precious in a dual sense. We are the only intelligent species within our cosmic neighborhood. As such we do not depend on any physical centrality. Through the force of abstract reasoning human minds crisscross the universe. We know more from thinking than from seeing. This is a worthier kind of centrality. It is rational centrality.

Reading List

Andersen, H./P. Barker/X. Chen [2006]: *The Cognitive Structure of Scientific Revolutions*. Cambridge: Cambridge University Press

Aristotle [1948]: *Politics*, transl. E. Barker. Oxford: Oxford University Press

Aristotle [1952a]: *Physics*. Great Books of the Western World, The Works of Aristotle, Vol. I. Chicago: Encyclopaedia Britannica

Aristotle [1952b]: *On the Heavens*. Great Books of the Western World. Chicago: Encyclopaedia Britannica

Armstrong, D. [1983]: *What is a Law of Nature?* Cambridge: Cambridge University Press

Barker, P. [2002]: "Constructing Copernicus," *Perspective on Science* **10**, 208–27

Barrow, J. D./F. J. Tipler [1986]: *The Anthropic Cosmological Principle*. Oxford: Oxford University Press

Blumenberg, H. [1955]: "Der kopernikanische Umsturz und die Weltstellung des Menschen," *Studium Generale* **8**, 637–48

Blumenberg, H. [1957]: "Kosmos und System," *Studium Generale* **10**, 61–80

Blumenberg, H. [1965]: *Die kopernikanische Wende*. Frankfurt a./M.: Suhrkamp

Blumenberg, H. [1981]: *Die Genesis der kopernikanische Welt*. 3 vols. Frankfurt a./M.: Suhrkamp [English translation: *The Genesis of the Copernican World*, transl. R. M. Wallace. Cambridge, MA: MIT Press, 1987]

Born, M. [1962]: *Einstein's Theory of Relativity*. New York: Dover

Brecht, B. [1963]: *Leben des Galilei*. Berlin: Suhrkamp [English translation: *The Life of Galileo*. London: Eyre Methuen, 1963]

Breuer, R. [1991]: *The Anthropic Principle: Man as the Focal Point of Nature*. Boston: Birkhäuser

Brzeziński, J./F. Coniglione/Th. A. F. Kuipers/L. Nowak *eds.* [1990]: *Idealization II: Forms and Applications*. Atlanta, GA: Rodopi

Brzeziński, J./L. Nowak *eds.* [1992]: *Idealization III: Approximation and Truth*. Atlanta, GA: Rodopi

Burtt, E. A. [²1932]: *The Metaphysical Foundations of Modern Physical Science*. London: Routledge & Kegan Paul

Carter, B. [1974]: "Large Number Coincidences and the Anthropic Principle in Cosmology," in M. S. Longair *ed.*, *Confrontation of Cosmological Theories with Observational Data*. Dordrecht: D. Reidel 1974, 291–8

Carr, B. J./M. J. Rees [1979]: "The Anthropic Principle and the Structure of the Physical World," *Nature* **278**, 605–12

Carroll, John W., "Laws of Nature," in *The Stanford Encyclopedia of Philosophy* (Fall 2003 Edition), Edward N. Zalta *ed.*, URL = <http://plato.stanford.edu/archives/fall2003/entries/laws-of-nature/>.

Cartwright, N. [1999]: *The Dappled World: A Study of the Boundaries of Science*. Cambridge: Cambridge University Press

Chen, X./P. Barker [2000]: "Continuity through Revolutions," *Philosophy of Science* **67**, S208–S233

Cohen, I. B. [1985]: *Revolution in Science*. Cambridge, MA: Belknap Press at Harvard University Press

Copernicus, N. [1543/1995]: *On the Revolutions of Heavenly Spheres*. Amherst, NY: Prometheus Books

Copernicus, N. [1959]: The *Letter Against Werner*, in E. Rosen *ed.* [1959], 93–106

Copernicus, N. [1959]: *The Commentariolus*, in E. Rosen *ed.* [1959], 55–90

Crombie, A. C. [1961]: *Augustus to Galileo*. London: Mercury

Crombie, A. C. [1994]: *Styles of Scientific Reasoning in the European Tradition*. London: Duckworth

Cushing, J. T. [1998]: *Philosophical Concepts in Physics*. Cambridge: Cambridge University Press

Davies, P. [1995]: "Algorithmic Compressibility, Fundamental and Phenomenological Laws," in F. Weinert *ed.* [1995a], 248–67

Deutsch, D. [1997]: *The Fabric of Reality*. London: Penguin

DeWitt, R. [2004]: *Worldviews*. London: Blackwell

Dicke, R. H. [1961]: "Dirac's Cosmology and Mach's Principle," *Nature Letters* **192**, 440–1

Dijksterhuis, E. J. [1956]: *The Mechanization of the World Picture*. Oxford: Clarendon 1961

Dirac, P. M. [1961]: "Dicke's Cosmology and Mach's Principle," *Nature* **192**, 440–1

Drake, S. [1975]: "Impetus Theory Reappraised," *Journal of the History of Ideas* **36**, 27–46

Dreyer, J. L. E. [1953]: *A History of Astronomy from Thales to Kepler*. New York: Dover

Dretske, F. I. [1977]: "Laws of Nature," *Philosophy of Science* **44**, 248–68

Duhem, P. [1962]: *The Aim and Structure of Physical Theory*. New York: Athenaeum

Einstein, A. [1918]: "Prinzipien der Forschung," in A. Einstein [1977], 107–10; English translation in Einstein [1954], 224–7

Einstein, A. [1919]: "Was ist Relativitätstheorie?," in A. Einstein [1977], 127–131; English translation in Einstein [1954], 227–32

Einstein, A. [1930]: "Johannes Kepler," in A. Einstein [1977], 147; English translation in Einstein [1954], 262–6

Einstein, A. [1953]: "Message on the 410th Anniversary of the Death of Copernicus," in Einstein [1954], 359–60

Einstein, A. [1954]: *Ideas and Opinions*. London: Alvin Redman

Einstein A. [1977]: *Mein Weltbild*. Hrsg. von Carl von Seelig. Frankfurt a./M: Ullstein

Galilei, G. [1953/1632]: *Dialogue on the Great World Systems*. G. de Santillana *ed.* Chicago: Chicago University Press

Gingerich, O. [1982]: "The Galileo Affair," *Scientific American* **247** (August), 118–27

Gingerich, O. [1993]: *The Eye of Heaven: Ptolemy, Copernicus, Kepler*. New York: American Institute of Physics

Gingerich, O. [2004]: *The Book Nobody Read*. London: Heinemann

Greene, G. [2000]: *The Elegant Universe*. New York: Vintage Books

Gutting, G. *ed.* [1980]: *Paradigms & Revolutions*. Notre Dame, IN: University of Notre Dame Press

Hacking, I. *ed.* [1981]: *Scientific Revolutions*. Oxford: Oxford University Press

Hanson, N. [1958]: *Patterns of Discovery*. Cambridge: Cambridge University Press

Heidelberger, M. [1980]: "Some Intertheoretic Relations between Ptolemean and Copernican Astronomy," in G. Gutting *ed.* [1980], 271–83

Hempel, C. [1965]: *Aspects of Scientific Explanation*. New York: The Free Press

Herfel, W. E. et al. *eds.* [1995]: *Theories and Models in Scientific Processes*. Amsterdam: Rodopi

Hoyningen-Huene, P. [1993]: *Reconstructing Scientific Revolutions*. Chicago: Chicago University Press [German original: *Die Wissenschaftsphilosophie Thomas S. Kuhns*. Braunschweig: Vieweg & Sohn 1989]

Jeans, J. [1943]: *Physics and Philosophy*. Cambridge: Cambridge University Press

Kaufmann, W. [⁴1974]: *Nietzsche*. Princeton, NJ: Princeton University Press

Kepler, J. [1619]: *The Harmonies of the World*, in J. Kepler [1995], 1–164

Kepler, J. [1618–21]: *Epitome of Copernican Astronomy*, in J. Kepler [1995], 165–245

Kepler, J. [1995]: *Epitome of Copernican Astronomy & Harmonies of the World*. Amherst, NY: Prometheus Books

Kitcher, P. [1993]: *The Advancement of Science*. Oxford: Oxford University Press

Koestler, A. [1964]: *The Sleepwalkers*. London: Penguin

Koyré, A. [1957]: *From the Closed World to the Open Universe*. Baltimore, MD: John Hopkins University Press

Koyré, A. [1965]: *Newtonian Studies*. Chicago: University of Chicago Press

Koyré, A. [1978]: *Galileo Studies*. Hassocks (Sussex): The Harvester Press

Krajewski, W. [1977]: *Correspondence Principle and Growth of Science*. Dordrecht: D. Reidel

Kuhn, T. S. [1957]: *The Copernican Revolution*. Cambridge, MA: Harvard University Press

Kuhn, T. S. [²1970]: *The Structure of Scientific Revolutions*. Chicago: The University of Chicago Press

Kuhn, T. S. [1983]: "Commensurability, Comparability, Communicability" (PSA 1982). P. D. Asquith/T. Nickles *eds*. East Lansing, MI: Philosophy of Science Association, 669–88

Ladyman, J. [1998]: "What is structural realism?" *Studies in History and Philosophy of Science* **29**, 409–24

Leckey, M./J. Bigelow [1995]: "Necessitarian Perspective: Laws as Natural Entailments," in Weinert [1995b], 92–118

Mason, S. F. [1956]: *Main Currents of Scientific Thought*. London: Routledge

McMullin, E. [1985]: "Galilean Idealization," *Studies in History and Philosophy of Science* **16**, 247–73

Mittelstraß, J. [1962]: *Die Rettung der Phänomene*. Berlin: Walter de Gruyter

Morgan, M./M. Morrison *eds*. [1999]: *Models as Mediators*. Cambridge: Cambridge University Press

Musgrave, A. [1979–80]: "Wittgensteinian Instrumentalism," *Theoria* **45/6**, 65–105

Neugebauer, O. [1968]: "On the Planetary Theory of Copernicus," *Vistas in Astronomy* **10**, 89–103

Nietzsche, F. [1887]: *On the Geneology of Morals*. New York: Vintage 1967 [Translation of *Über die Geneologie der Moral*, 1887]

Nola, R. [2003]: *Rescuing Reason*. Dordrecht: Kluwer

Nowak, L. [1980]: *The Structure of Idealization*. Dordrecht: D. Reidel

Popper, K. [1959]: *The Logic of Scientific Discovery*. London: Hutchinson

Popper, K. [1963]: *Conjectures and Refutations*. London: Routledge & Kegan Paul

Psillos, St. [1999]: *Scientific Realism*. London: Routledge

Ptolemy, C. [1984]: *Ptolemy's Almagest*. G. J. Toomer *ed*. London: Duckworth [German translation: *Des Ptolemäus Handbuch der Astronomie*. Bd. 1, 2, übersetzt von Karl Manitius. Leipzig: B. G. Teubner, 1912]

Putnam, H. [1975]: *Mathematics, Matter and Method*. Philosophical Papers, Vol. 1. Cambridge: Cambridge University Press

Quine, W. v. [1990]: *The Pursuit of Truth*. Cambridge, MA: Harvard University Press

Rheticus, J. [1540]: *Narratio Prima*, in E. Rosen [1959], 107–96

Rickles, D./S. French/J. Saatsi *eds.* [2006]: *The Structural Foundations of Quantum Gravity*. Oxford: Clarendon Press

Roman, P. [1969]: "Symmetry in Physics," in R. S. Cohen/M. W. Wartofsky *eds.*, *Boston Studies in the Philosophy of Science*, Vol. V. Dordrecht: Reidel, 363–9

Rosen, E. *ed.* [1959]: *Three Copernican Treatises*. Mineola, NY: Dover

Rosen, E. [1984]: *Copernicus and the Scientific Revolution*. Malabar, FL: Robert E. Krieger

Ruby, J. [1995]: "Origins of Scientific 'Law'," in F. Weinert *ed.* [1995a], 289–315

Rybka, E. [1977]: "The Scientific Reception of the Copernican Theory," *Studia Copernicana* **17**, 158–71

Salmon, W. [1998]: *Causality and Explanation*. Oxford: Oxford University Press

Shapere, D. [1964]: "The Structure of Scientific Revolutions," reprinted in G. Gutting *ed.* [1980], 27–38

Shapere, D. [1966]: "Meaning and Scientific Change," reprinted in I. Hacking [1981], 28–59

Shapere, D. [1989]: "Evolution and Continuity in Scientific Change," *Philosophy of Science* **56**, 419–37

Sklar, L. [2000]: *Theory and Truth*. Oxford: Oxford University Press

Smolin, L. [1997]: *The Life of the Cosmos*. New York: Oxford University Press

Solla Price, D. J. de [1962]: "Contra-Copernicus," in M. Clagett *ed.*, *Critical Problems in the History of Science*. Madison: University of Wisconsin Press, 197–218

Swartz, N. [1985]: *The Concept of Physical Law*. Cambridge: Cambridge University Press

Toulmin, S. [1953]: *The Philosophy of Science*. London: Hutchinson

Watson, W. H. [1938]: *On Understanding Physics*. London: Cambridge University Press

Worrall, J. [1989]: "Structural Realism": The Best of Both Worlds?" *Dialectica* **43**, 99–124

Weinberg, S. [1977]: *The First Three Minutes*. London: Deutsch

Weinert, F. [1993]: "Laws of Nature: A Structural Approach," *Philosophia Naturalis* **30**, 147–71

Weinert, F. *ed.* [1995a]: *Laws of Nature – Essays on the Philosophical, Scientific and Historical Dimensions*. Berlin: de Gruyter

Weinert, F. [1995b]: "Laws of Nature – Laws of Science," in F. Weinert *ed.* [1995a], 3–64

Weinert, F. [1995c]: "The Duhem–Quine Problem Revisited," *International Studies in the Philosophy of Science* **9**, 147–156

Weinert, F. [1998]: "Fundamental Physical Constants, Null Experiments and the Duhem–Quine Thesis," *Philosophia Naturalis* **35**, 225–52

Weinert, F. [1999]: "Theories, Models and Constraints," *Studies in History and Philosophy of Science* **30**, 303–33

Weinert, F. [2004]: *The Scientist as Philosopher*. New York: Springer

Weinert, F. [2006]: "Einstein and the Representation of Reality," *Facta Philosophica* **8**, 229–52

Wittgenstein, L. [1921/⁸1971]: *Tractatus Logico-Philosophicus*. Frankfurt a./M.: Suhrkamp

Wendorff, R. [³1985]: *Zeit und Kultur*. Opladen: Westdeutscher Verlag

Wolff, M. [1978]: *Geschichte der Impetustheorie*. Frankfurt a./M.: Suhrkamp

Zeilik, M. [⁵1988/⁹2002]: *Astronomy*. New York: John Wiley & Sons

Essay Questions

1 Greek astronomy was concerned with '**saving the appearances**'. Explain what this means and what impact it had on the philosophical status of scientific theories.

2 Greek astronomy assumed a '**two-sphere universe**'. Explain what this means and what impact it had on the philosophical status of scientific theories.

3 What is the structure of **geocentrism** and why did the Greeks find it so compelling?

4 Explain how the replacement of the Aristotelian theory of motion by the **impetus theory** was a logical precondition for the development of Copernicanism.

5 Explain the major achievements of the **Copernican revolution**.

6 Explain why the Copernican worldview was only completed in the **Newtonian synthesis**.

7 What is the structure of **heliocentrism** and why did the Copernicans find it so compelling?

8 **Realism** is "the belief that a mere description of data is not all that should be required of a theory." [Bernard d'Espagnat] Discuss the significance of this statement, using appropriate examples.

9 There are two views on theories: **realism** and **instrumentalism**. Explain what they are. What arguments does the realist produce against the instrumentalist?

10 In which sense could you use *Copernicanism* to support, respectively, **instrumentalism** and **realism**?

11 Explain how the issue of **realism** versus **instrumentalism** arises from the Copernican turn and discuss some of the arguments in favor of realism and instrumentalism, respectively.

12 Explain, illustrate, and evaluate some of the typical arguments for and against **realism** and **instrumentalism**.

13 If **models** are ways of **representing** the natural and social world, how is this representation achieved?

14 Explain the role of **models** in science. What types of models are there and why are models important? Illustrate with respect to Copernicanism.

15 The DN model assumes the **symmetry** of **explanation** and **prediction**. Use examples from astronomy to evaluate the appropriateness of this assumption.

16 Explain and illustrate the role of **hypotheses** in the history of astronomy, from Ptolemy to Newton.

17 Explain the **underdetermination thesis** (Duhem–Quine thesis). What arguments can be advanced against it?

18 Explain and illustrate the **reversal** of **perspective** in Copernicanism.

19 What do we understand by a **scientific revolution**? Were *Copernicanism* and *Darwinism* scientific revolutions?

20 Explain why the Copernican heliocentric hypothesis was a **Copernican turn** rather than a scientific revolution.

21   Explain why it took **140 years** – from 1543 to 1687 – to complete the Copernican revolution.

22   If Copernicus initiated only a Copernican turn, why is Darwin's *Origin of Species* (1859) a scientific revolution and Copernicus's *Revolutionibus* (1543) not?

23   Explain what **advantages** the **Copernican model** had over the Ptolemaic model. Why do historians of science not regard Copernicus as a "true" revolutionary in science?

24   Critically discuss the applicability of **Kuhn's paradigm model** of scientific revolutions in the context of the Copernican model.

25   Explain for what reasons **J. Kepler** is regarded as the **true revolutionary** in the history of astronomy.

26   Critically analyze the role of **constraints** in science by reference to Copernicanism and Darwinism.

27   Explain the difference between the **laws of nature** and the **laws of science**. Why is this distinction important?

28   Critically evaluate the distinction between **theories** and **models**.

29   Critically discuss arguments in favor of and against the **structural view** of laws.

30   Critically discuss arguments in favor of and against the **necessitarian view** of laws.

31   Critically discuss arguments in favor of and against the **regularity view** of laws.

II

Charles Darwin: The Loss of Rational Design

Ape and man are similar forms, which developed in a dissimilar way from the same stock. [Rolle, Der Mensch *(²1870), 107; translated by the author]*

1 Darwin and Copernicus

Darwin seems to me to be the Copernicus of the organic world. [du Bois-Reymond, "Darwin and Copernicus" (1882–3), 557]

We have spoken, in Chapter I, about Copernicus. His heliocentric view of the cosmos led to a loss of physical centrality. This induced such a momentous shift in human perspectives that it is not surprising that Darwin and Copernicus should often have been compared. [Haeckel 1929, 207; Huxley 1860, 79] For Darwin, too, produced a change in perspective. He removed the idea of design. He placed all organic life, including humans, under the cosh of evolutionary thinking. As soon as Darwin had published *The Origin of Species*, on November 24, 1859, the tremendous impact of his theory was felt by his contemporaries. A. R. Wallace, co-discoverer of the principle of natural selection, recognized the revolutionary potential of Darwin's theory:

> Never, perhaps, in the whole history of science or philosophy has so great a revolution in thought and opinion been effected as in the twelve years from 1859 to 1871, the respective dates of publication of Mr. Darwin's *Origin of Species* and *Descent of Man*. [Wallace 1891, 419; Haeckel 1882, 534]

Almost 150 years after Darwin's publications, researchers in biology still agree on the upheaval in human thinking that Darwinism unleashed. Writing in 1974, Jacques Monod was quite confident

> that no other scientific theory has had such tremendous philosophical, ideological and political implications as has the theory of evolution. [Monod 1974, 389]

The distinguished evolutionary biologist Ernst Mayr is no less certain that the publication of the *Origin* "represents perhaps the greatest intellectual revolution experienced by mankind." [Mayr 2001, 9; Mayr 2000, 67–71] Reflecting the growing importance of the science of biology for the twenty-first century, Stephen Jay Gould accords Darwin a bigger impact effect than Copernicus.

> Of the two greatest revolutions in scientific thought, Darwin's surely trumps Copernicus in raw emotional impact ... [Gould 2002, 46; Büchner 1868, 267]

This chapter aims at giving a concise description of Darwinism, its principles and consequences. It will not be possible to survey the vast amount of literature about Darwin and Darwinism. Much of the written work on Darwinism deals with the principles of evolutionary biology. The focus of this book is on the impact of scientific thought on philosophical questions. We will therefore concentrate on relatively neglected aspects of Darwinism – the consequences of his evolutionary thinking for human self-images and philosophical issues. To understand the upheaval in human thinking, which we will later characterize as a **loss of rational design**, we need to canvass views of life before and after the *Origin*. To understand the impact of Darwinism on the question of human origins we need to look at work on the antiquity of mankind and the new idea of descent with modification. These strands run together in Darwin's *The Descent of Man* (1871). Through the study of this material we will gain a vantage point from which to tackle philosophical questions as they arise from Darwin's work. This will lead us into a study of the career of philosophical presuppositions in Darwinism, to questions of scientific method, of the philosophy of mind, of explanation and prediction.

2   Views of Organic Life

The scientific criticism of life "admits that all our interpretations of natural fact are more or less imperfect and symbolic, and bids the learner seek for truth not among words but among things." [Huxley, Science, Culture and Other Essays *(1888), 15]*

Let us start with some pre-Darwinian history. We consider the mixture of theoretical speculations about organic life and the emergence of empirical data that created the conceptual space in which Darwin's theory of evolution found a niche. It soon diminished the influence of its rivals and eventually established itself as the dominant view.

2.1   Teleology

If a scientific revolution is, in its primary sense, a rearrangement of conceptual networks, then Darwinism was a true revolution in science. Darwinism changed the view of organic life. The most striking change concerns the demise of teleological thinking. This is well illustrated in Kant's explanatory dualism. Kant applies different measures to the explanation of organic and nonorganic nature. In his *Theory of the Heavens* (1755) Kant proposed a purely mechanical view of the origin and evolution of the inorganic world. His cosmic history explained the emergence of order from an original chaos through the operation of purely mechanical causes. There is in Kant a very early anticipation of the Big Bang theory. At the beginning there is chaos. Today there is order. Most visible is the order of planetary systems. But planetary systems are grouped into galaxies. And galaxies are grouped into clusters. Kant was one of the first to conjecture that the specks of light in the night sky represent, in fact, other galaxies in the universe, which are structured in similar ways to our Milky Way. How did this order come about? Kant requires no more than Newton's laws to account for the cosmic order. The beauty of Kant's cosmic hypothesis is that it envisages the whole cosmos as a vast network of systems and subsystems, all locked into Newtonian regularities. Kant emphasizes that the establishment of today's cosmic order out of an original chaos can be explained solely by appeal to *mechanical* laws. Cosmic evolution, as conceived by Kant, lacks a purpose. The mechanical laws pursue no aim. Kant repudiates the postulation of final causes in the explanation of cosmic history. Inorganic nature follows no preexisting design.

But organic nature is strikingly different. Kant finds that organic nature cannot be explained without envisaging design. Design comes with a purpose in mind. A tool is designed to fulfill a purpose. Equally, organic nature cannot be explained without design. The employment of mere mechanical causes to explain a plant or an insect will fail at the earliest stage. To explain organic nature we need *final* causes. Kant predicts, in 1755, that it is more likely that a mechanical explanation of the formation of *all* planets, their orbits, and the origin of cosmic architecture will be achieved before "a single weed or caterpillar can be explained mechanically." [Kant 1755, Preface, 237] Kant's evolutionary account of cosmic history excludes the evolution of species. Exactly 104 years later, Darwin proposed a mechanism to explain the origin of weeds, caterpillars, and other creatures. Kant does not claim that purpose is observable in nature. But a mere mechanical description of the structure of, say, a bird would leave the impression of pure chance. Pure chance cannot explain the complexity of the organic world. In order to avoid explanations by blind mechanisms, purpose in nature must be stipulated as a regulative principle. This is the postulation of teleology. [Kant 1790, §§61, 80–4] The observable mechanisms must be interpreted *as if* they were operating according to a hidden blueprint (*Bauplan*). The purpose of the blueprint is to permit a plausible explanation of the intricate order of living things. Blind mechanisms are not sufficient, Kant declares, to explain how organic beings can be built. For in view of an organic creature it is always possible to ask: why is it there, what is its function? And mechanical explanations do not provide a satisfactory answer. Final causes are needed to complement mechanical explanations. Such final causes are also needed to make sense of human existence and morality. Humans before and after Kant regarded themselves as the final purpose on Earth.

Kant's explanatory dualism was understandable at the time. By the end of the eighteenth century, the physical sciences had eliminated teleological thinking.

Kant himself had produced a mechanical explanation of the evolution of the cosmos. But it was difficult to see, as Kant's dualism demonstrates, how a contemplation of the organic world could relinquish purpose and design.

At the time of Kant's death (1804), biology was far less developed than astronomy and physics. It lacked the fundamental laws and precise observations. It lacked an axiomatic structure. To Greek astronomers it seemed intuitively clear that the Earth is at rest, with the planets gyrating around its central position. To eighteenth-century biologists it was intuitively clear that the complexities of the organic world revealed order, built on purpose and design. Prior to Darwin several schemes reflected this belief in purpose-built nature. We will shortly look at the *Great Chain of Being* and Lamarck's evolutionary views. We must be aware that Darwin's revolution did not stamp out all of the older beliefs about organic nature. Teleology turned out to be a tenacious presupposition. It made a comeback in the 1880s and again the 1990s under the name of Intelligent Design. The resurgence of design scenarios is entirely due to the perceived inadequacy of Darwin's mechanism of natural selection. Even some of Darwin's staunchest supporters clung to teleological straws. Darwin could hardly have wished for a more enthusiastic proponent than August Weismann (1834–1914). Weismann made a tremendous discovery: that only *germ* cells transmit hereditary information. Weismann showed – against Lamarck's view of the transmission of acquired characteristics – that *somatic* cells cannot transmit characteristics from parents to offspring. Alfred Wallace illustrated the point:

> It has (…) never been alleged that the children of deaf-mutes exhibited any unusual difficulty in learning to speak, as they should do if the effects of disuse of the organs of speech in their parents were inherited. [*Nature* **48**, 1893, 267]

Yet Weismann, just like Wallace, found it difficult to jettison all teleological baggage. In his Darwinian exposition of a mechanical view of nature, he asks: "But how can we acknowledge a teleological principle without abandoning the purely mechanical conception of nature?" The essentially Kantian answer is: by the assumption of the "immanent teleology of the universe." Weismann, the scientist, embraces a mechanical conception of nature. Weismann, the philosopher, is frightened of crude materialism. He therefore attempts to marry a mechanical with a teleological conception of the universe. Like many of his contemporaries, Weismann fears that a renunciation of teleology will lead to a loss of culture and spirituality. [Weismann 1882, 710–18] In the eyes of Darwin's opponents, evolutionary theory had an unwelcome philosophical consequence: materialism. The evolution theory was more than just another piece of science. It threatened established philosophical convictions. The motivation behind the invocation of purpose and design in nature has not changed since Weismann's day. As we shall see, the intricate complexity of biological systems and functions still makes contemporary opponents of evolutionary theory infer intelligent design. In order to get a flavor of this thinking, let us first look at some typical teleological views of organic life at the beginning of the nineteenth century.

2.1.1 The Great Chain of Being

(…) we cannot understand how the trout north of the Alps can belong to the same stock as those south of the Alps, separated as they ever have been by insurmountable mountains (…). [Vogt, Lectures on Man (1864), Lecture VIII, 216]

One representative idea of the structure of organic nature is the so-called *Great Chain of Being*. It harks back to the Greeks. [Lovejoy 1936] The natural world was cast in the image of a ladder, ranking all created forms from the brightest angel to the humblest worm in a descending order. The Almighty embodied the highest form of perfection. Each creature occupied a rung on this hierarchical ladder. Shakespeare makes use of this scale of being in his history plays. [Tillyard 1943] As humans possess instincts, bodily sensations, and abstract faculties of reasoning, they stand between the animal world (ruled by instincts) and the realm of angels (governed by pure reason). The whole chain was seen as a *graded ladder of perfection*. It was complete, continuous, and harmonious. No chasms or gaps existed. Several features about this chain stand out:

- It does not allow evolutionary transformations. On the contrary, the chain of being in its pure form asserts the immutability of species. The scale is static. Each species, including humans, is consigned to a particular rung of the ladder. There is no room for descent with modification.
- The entire chain was believed to have been created in its present form. The resemblance between Man and Ape was not alarming to an eighteenth-century audience (as it would be for a nineteenth-century audience). This resemblance was just a reminder of the proximity of the rungs that apes and humans occupied, respectively. The continuity of the chain said nothing about descent. There was no place for evolution in the chain, since the world was believed to be only a few thousand years old.
- There was no room for the creation or extinction of species. The chain was complete.

The *Great Chain of Being* was an explanatory sketch. Like geocentrism it faced the verdict of evidence. At the end of the eighteenth century, the idea of a scale of being struggled with many difficulties.

First, geology came into its own. This new science discovered many fossil records, which documented the successive extinctions of species. In his *Dictionnaire Philosophique* (1764) Voltaire expressed his dissatisfaction with the fixity of the *Chaîne des Êtres*. [See Lepenies 1978, 47; Lovejoy 1936, 252] It was obvious, he observed, that there were extinct plants and animals. And humans had shown themselves capable of eradicating whole populations of animals, like the wolf in England. *Second*, it was obvious to Voltaire that breaks existed in the chain: "is there not visibly a gap between ape and man?" Scholars turned to the discontinuities and gaps in the animal world. For our purposes, the dissimilarities and more spectacularly the similarities between ape and human are of particular interest. [Gillispie 1959,

---

Box 2.1   The age of the Earth and the existence of life

- *Age of the Earth:* 4.6 billion years
- *Micro-organisms* in early fossil records: 3.5 billion years
- *Oldest animal fossils:* 700,000,000 years
- *First vertebrates:* 400,000,000 years
- *First mammals:* 200,000,000 years
- *Homo habilis:* 1,800,000 years
- *Homo erectus:* 500,000–1,000,000 years
- *Homo sapiens:* 25,000 years

17–18; Eiseley 1959, 5–10; Gould 1988, 6] *Third*, scholars like James Hutton and Charles Lyell began to doubt the traditional view that the Earth was just a few thousand years old. Tentative calculations extended the age of the Earth from thousands to millions of years. It began to dawn on scholars that the age of the Earth was much greater than had previously been conceived. This is often described as the discovery of *deep time*. In his *Theory of the Heavens* (1755), Kant had speculated that "millions of years and centuries" had flown past before the present order of the cosmos could be observed. During Darwin's lifetime, the physicist W. Thomson, later known as Lord Kelvin, estimated the age of the Earth to be around 98 million years. [Barrow/ Tipler 1986, 160–1; Gould 1988] [Box 2.1]

Faced with these difficulties, the *Chain of Being* ceased to serve as a description of the universe as it is and always has been. The *Great Chain of Being* underwent a conceptual revision. It became temporalized: the image of the scale of descent was converted into the ladder of ascent. [Crombie 1994, Vol. III, 1698; Lepenies 1978, 52–77] The *Chain of Being* was reinterpreted as a process of creation, occurring over extended periods of time. The German philosopher and mathematician Gottfried Wilhem Leibniz gave an eloquent expression to the temporalization of the *Chain of Being*:

Gottfried Wilhelm Leibniz
(1646–1716)

> Further, we realize that there is a perpetual and most free progress of the whole universe toward a consummation of the universal beauty and perfection of the works of God, so that it is always advancing toward a greater development. (...)

Although the *Chain of Being* had been transformed into a ladder of ascent, it retained some of its essential features. The universe evolves toward perfection; its progress is purposeful and follows a divine plan; the chain is no longer static but it still grows teleologically toward a goal.

To the objection that may perhaps be offered that if this were so the world would long ago have become a paradise, the answer is at hand: although many substances have already come to great perfection, yet owing to the infinite divisibility of what is continuous, there always remain in the abyss of things parts that are asleep, and these need to be awakened and to be driven forward into something greater and better – in a word, to a better state of development. Hence this progress does not ever come to an end. [Leibniz 1697]

With the temporalization of the Chain of Being a sense of development made itself felt in biological thinking. The *Chain of Being* did not stretch, statically, like a bridge from the beginning to the end of time. Rather, it sprouted, like a tree of life, from simple roots, at the moment of creation, to organic marvels in the fullness of time. Yet, "the notion that the organic creation was peculiarly a theater for the observation of Providence in action" remained in place. [Gillispie 1959, 19] Providence – the beneficial care of God for the inhabitants of Earth – was a central assumption prior to Darwin. There was a general conception that God was a kind of master workman who had personally supervised the creation of even the tiniest organisms in the natural world. Leibniz did not stray from this conception. He was as much an explanatory dualist as Kant. He held that mechanical laws were sufficient to explain the smooth running of the inorganic world. The laws of motion, governing all macroscopic objects, were so precise that there was no need for God to interfere in the mechanism. The universe was like a gigantic clockwork. It was a clockwork universe. The organic world was different. It required design, purpose, and teleological development. It would take hard conceptual efforts to overcome the belief in the progressive unfolding of organic complexity. In an attempt to construct a theory of the organic world at the beginning of the nineteenth century, the assumption of progressive evolution was an unquestioned building block. Lamarck made progressive evolution a central pillar of his theory. Lamarck needed design, like Leibniz. But a designer God played a negligible part in their accounts. In Leibniz and Boyle, God's role is confined to setting the clockwork universe in motion. Lamarck accepts at best that God lays down the basic design plan. Lamarck made giant steps toward a modern view of organic life. His was a lone voice. The dominant paradigm lay in design arguments.

2.1.2 Design arguments

According to Teleology, each organism is like a rifle bullet fired straight at a mark; according to Darwin, organisms are like grapeshot of which one hits something and the rest fall wide. [Huxley, "Criticisms" (1864), 84]

In discussions of the contrast between evolutionary theory and design arguments, it is customary to refer to Archdeacon Paley. William Paley was a well-known figure, who wrote a book entitled *Natural Theology, or Evidences of the Existence and Attributes of the Deity collected from the Appearances of Nature* (1802). The book presents the famous watchmaker argument: from the existence of a watch we infer a watchmaker; by analogy we ought to infer a Designer God from the order of nature. But Paley himself stands in a long tradition of teleological thinking. This design paradigm infers from the seeming order and perfection in nature a Designer God, a divinely gifted engineer. The Epicureans attempted to explain the beauty and symmetry of nature as a result of pure chance. Proponents of design arguments usually dismiss such attempts as highly implausible. It contradicts our sense experience. The ancient Greeks could not believe that the Earth turned around the sun. The evidence of the senses seemed to support geocentrism. If the Earth turned, there would be violent storms on Earth. And objects would not fall in a straight line. In a similar way design arguments rely on what seems manifest to human sense perception. Aristarchos's heliocentric view of the cosmos was a minority view; so was the Epicurean chance explanation. Geocentrism and design arguments constituted a majority view. They relied on plain reason and the testimony of the senses. Our senses tell us that the Earth resides stationary at the center of the universe. They also tell us that the order and symmetry in the organic world must be the work of an intelligent creator.

William Paley (1743–1805)

Robert Boyle felt provoked enough by the Epicureans of the seventeenth century to publish an extended proof of design: *A Disquisition about the Final Causes of Natural Things* (1688). He

Robert Boyle (1627–91)

first looks at the argument from design from the point of view of inanimate objects. His contemporaries indulged in the vulgar belief that the stars and the sun had been especially created to shed light and warmth on the inhabitants of the Earth. For instance, of the moon, a "secondary planet," Kepler says:

> this star has been assigned to the Earth as its private property, so that the moon might help with the growth of Earthly creatures and be observed by the speculative creature on the Earth, and that the observation of the stars might begin with it. [Kepler 1618–21, Bk. IV, Part I, §4 (43)]

This was too simplistic. The System of the Word was so complex, Boyle argued, that it was a mistake to infer from the regularity of planetary motions a particular purpose for human ends. Boyle believed, like Leibniz, in a clockwork universe. There was no need for corrective interferences on the part of the Artificer. This was a more radical mechanical view than that of Newton, who had assumed that God would have to make occasional adjustments to the clockwork universe. But in the domain of organic life, Boyle saw as little chance to avoid explanatory dualism as

Leibniz and Kant. Even so complex an object as a clock cannot match the complexity of animate objects in the biological world:

> (...) there is more admirable Contrivance in a *Mans Muscles* than in (...) the *Celestial Orbs*; and the Eye of a Fly is a (...) more curious piece of Workmanship, than the Body of the Sun. [Boyle 1688, 43–4]

The complexity and design of plants and animals was so intricate that an inference to particular ends was justified. It was obvious to Boyle that a divine engineer had created eyes to provide their lucky bearer with vision. This Artificer had even been wise enough to equip each animal with a *pair* of eyes, just in case one eye succumbs to disease. Humans cannot infer with certainty that a Deity created all these natural wonders for the sole purpose of human ends and delectation. Yet, they can be certain that this is one of their purposes. It would be the height of irrationality to infer from the beauty and perfection of nature that pure chance had been the author.

The French philosopher Pierre Louis Moreau de Maupertuis (1698–1759), however, refused to restrict the impact of the design argument to planetary order or organic structure. In his *Essai de Cosmologie* (1750) he argues that mathematics can establish the existence of God. The lawful regularity of the universe itself is based on a universal principle of least action. Maupertuis is referring to the existence in physics of least action principles, which prove that nature behaves in the most economical fashion. At the beginning of the seventeenth century Pierre Fermat proved that light rays move between any two points in such a manner that their paths take the least time. That may not be the shortest path but it is the path that takes less time than any other path. The same is true of particles, which move in accordance with Newton's laws of motion. Out of a number of possible paths, which a particle can take in space, the true path is the one for which a quantity called action, S, is the minimum.[1] So Maupertuis looks much deeper into the order of nature. He finds that the mathematical economy of nature, captured in the principle of least action, points to a divine Designer. The mathematical beauty, which reveals itself in the order of Nature, must be the work of a superior intelligence.

When Boyle and Maupertuis penned their respective arguments from design, the mechanical worldview was establishing its supremacy in the physical sciences. Newton had still believed that the clockwork universe needed the occasional interference of the Deity to keep it ticking with perfect regularity. But both Boyle and Leibniz saw in the need for occasional repair work a diminution of the power of the creator. Once the clockwork universe was set in motion by its divine author the universal laws kept it in reliable motion. However, in the order of the planets

---

[1] Roughly, the action, S, is defined by the integral $\int_{t_1}^{t_2} (KE - PE)\, dt$, e.g. the integral of the kinetic minus the potential energy over the period t_1 to t_2. Out of all possible paths, the true path is the one for which the action is the least.

and the galaxies no more than general design intentions could be detected. The organic world provided much more concrete evidence of divine artwork. Boyle concedes that pure chance may have made stones and metals, but not vegetables and animals.

> There is incomparably more Art expressed in the structure of a Doggs foot, than in that of the famous Clock at Strasbourg. [Boyle 1688, 47]

In Paley we find a similar preference for biology to establish the design argument. It is true, says Paley, that a clock reveals a considerable amount of design. But the organic world offers much more striking instances of design. We discover so much beauty, order, and regularity that only an intelligent designer could have created the world.[2]

> There cannot be design without a designer; contrivance without a contriver, order without choice. [Quoted in Gillispie 1959, 36]

With so much attention lavished on plants and animals, each species seemed to be purposefully crafted. Boyle lamented the strong tendency of his contemporaries to interpret all cosmic events as though specifically created for human purposes. This warning was understandable because design arguments emphasize the inclinations of humans to read into organic nature a Creator's intention. The organic world, though, seemed to offer a safer license to make inferences to a designer. Natural theology had the task of demonstrating "the final intention of the creator in respect to each structure." [Eiseley 1959, 178] Natural theology, though, is preoccupied with appearances. We found a similar anthropocentrism in the geocentric worldview. The appearances seem to suggest that the Earth sits at the center of the universe. Heliocentrism destroyed this belief. Now Natural Theology reaffirmed the centrality of human existence in the organic world.

Although the idea of a fixed chain of being faded away, the *search for design* continued. There had been suggestions in the eighteenth century that the age of the Earth was much older than the Bible's 6,000 years. Yet the evidence of the rocks was still too weak to dent the biblical story. To the mind of believers, geological evidence appeared to confirm it: a universal deluge had taken place and human existence could not be older than laid down in the Bible. These were central features of the biblical story of creation. They fitted in well with the general belief in teleology and progressive evolution. [Gillispie 1959, 107] In such situations, when the evidence cannot decisively credit one model and discredit another, philosophers speak of underdetermination. [See Chapter I, Section 6.4] The same evidence seems to be compatible with competing conceptions, for instance regarding the

---

[2] Discussions of design arguments are much older than Paley: they are to be found as much in Greek writers as in Hume's *Dialogues Concerning Natural Religion* (1779) and Kant's *Critique of Pure Reason* (1781, A620–631) – see Dawkins [1988], Ch. I; Barrow/Tipler [1986], Ch. 2; Dennett [1995], Ch. I, §4; Behe [1996], Ch. 10; Ruse [2003], Chs. 1–4.

age of the Earth. At the beginning of the sixteenth century, Copernicus clung to the equivalence of alternative hypotheses. At the beginning of the nineteenth century rival models for the origin of species made their appearance. They challenged the existing design models. But another 50 years would have to elapse before the evidence could weigh the competing models differentially. Philosophically speaking, the argument from design acts like a license for permissible inferences. The inferences take us from the existence of visible order in the natural world to design and from design to a benevolent creator. The argument from design works essentially by the power of analogies.

Every species seemed to be meticulously designed with a particular purpose in mind. A bird has wings in order to fly. Humans have brains in order to think. By common consent, humans occupy a superior place in the plan of creation. By implication, the animal and plant world was created for the purpose of Man. But we have to consider whether *permissible* inferences are also *admissible* inferences. Consider a debate at the very beginning of the nineteenth century between a Deist and an Evolutionist. The Deist relies on commonsense and the appearance of natural order to infer the existence of God. The Evolutionist relies on natural laws and mechanical explanations to infer regularity in nature. He shuns inferences to supernatural agents. For the Deist, function comes first; the organ is tailored to it. If the Creator wants a creature to see, He bestows eyes on it. It therefore makes little sense in this scheme to allow for change and variability. Creatures occupy their permanent places in the order of things. There is a ladder of ascent.

For the Evolutionists the organ rises first; it shapes its function. This reversal in perspective put into Huxley's hands a Darwinian rebuttal of the design argument:

> Suppose, however, that any one had been able to show that the watch had not been made directly and by any person, but that it was the result of the modification of another watch which kept time but poorly; and that this again had proceeded from a structure which could hardly be called a watch at all – seeing it had no figures on the dial and the hands were rudimentary, and that going back in time we came at last to a revolving barrel as the earliest traceable rudiment of the whole fabric. And image that it had been possible to show that all these changes had resulted, first, from a tendency of the structure to vary indefinitely; and secondly, from something in the surrounding world which helped all variations in the directions of an accurate time-keeper; and checked all those in other directions; then it is obvious that the force of Paley's argument would be gone. For it would be demonstrated that an apparatus thoroughly well adapted to a particular purpose might be the result of a method of trial and error by unintelligent agents, as well as of the direct application of the means appropriate to that end, by an intelligent agent. [Huxley 1864, 83]

Huxley benefited from Darwin's proposal of a mechanism by which functions might be causally explained. [See Section 6.7.3] Prior to Darwin there were often wild speculations as to how the organ could be shaped to perform its function. Lamarck made a brave attempt to provide a mechanistic explanation. Such mechanistic hypotheses had occasionally been made public and severely rebuked. In his *Système de la Nature* (1770) Paul Thiry d'Holbach declared humans to be products of nature, with respect to both their emergence and their moral and intellectual faculties. Plants and animals were adapted to the particular climatic conditions on Earth. If environmental conditions on Earth changed, he argues, "all products of the globe would also change." It is not unreasonable to suspect that all species are subject to change. Nature knows no permanent forms. And man is not the crown of creation. All these were vague speculations, a philosopher's fantasy running wild. Yet we should not underestimate the power of ideas. Ideas impose order on our observations. They make coherent what looks disparate. They create new models of reality. But when facts are scarce the ideas float too freely in a conceptual space. This was Lamarck's strength, which led to his ruin.

2.1.3   Jean Baptiste Lamarck

Unquestionably the greatest defect in Lamarck's work was the insufficiency of the stock of observations and experiments he brought forward in proof of his far-reaching principles. [Haeckel, Nature *26 (1882), 540]*

In many people's minds the name of Newton is associated with the falling apple and the law of gravity. This is a positive association for it is an indication of Newton's genius. The name of Jean Baptiste Lamarck invites a negative association. It brings to mind the giraffe, which stretches its neck to reach the leaves of tall trees; the neck grows longer and the giraffe passes its elongated neck on to the next generation of giraffes. That's why giraffes have long necks. And ducks have webbed feet because they spread their toes to propel themselves forward in water. Lamarck has been much ridiculed for his ideas of use-inheritance – the inheritance of acquired characteristics. The characteristics acquired by the hard-pressed individuals of one generation are miraculously passed on to their offspring. Darwin

Jean Baptiste Lamarck
(1744–1829)

held his achievements in low esteem. Yet there is much more to Lamarck than the "discovery" of soft inheritance. He had a successful career: he was professor of zoology at the Natural History Museum in Paris, member of the *Institut de France* and many other European Societies of Science. Yet his influence was minimal. He was little understood in his own time. He lived in the shadow of the powerful George Cuvier (1769–1832). And his ideas were eclipsed by Darwin's theory of evolution. The brilliance of Darwin's work should not let us forget Lamarck's achievements. Exactly 50 years before the publication of the *Origin of Species* Lamarck published his *Philosophie Zoologique* (1809), in which he presents his theory of transformism. This book contains his much-maligned doctrine of soft inheritance. But let us weigh his mistake against his achievement. Lamarck worked out a way of temporalizing the *Great Chain of Being*. In doing so, Lamarck abandoned the idea of the fixity of species and the theory of special creations. The immutability of species was a core element of the ladder of existence and design arguments. It lived on through the work of Cuvier until Darwin finally dismissed it. In place of the fixity of species Lamarck proposes the mechanism of *progressive modification*. Lamarck did not even assume an act of divine creation at the beginning of organic life. Life starts by spontaneous generation. Through this process very simple creatures are born. If it were left undisturbed, the complexity of life's scale would slowly evolve through progressive modifications. Would the simpler creatures not disappear? No, because spontaneous generation works continuously, always replenishing the world. Necessary inner propulsion would turn the simpler creatures into more complex organisms. Complications arise. Changes in the environment disrupt the mechanism of progressive modification and challenge the organisms to adapt themselves. Changing environmental conditions lead to changes in organisms. Some organs experience a reinforced use and others fall into desuetude. Strengthened organs are passed on, through use-inheritance. Organisms change within a few generations. The complexity of the organic world slowly unfolds. This is *linear* evolution, because the idea of hierarchy is retained. For Lamack man is a model of perfection and all other creatures are measured against man. Although Lamarck temporalizes the *Great Chain of Being* and abandons the idea of the fixity of species, he is still committed to teleology. That is, Lamarck thinks that the evolution of the ladder of existence follows some preset aim. The *telos* is to produce humans. Nature evolves from simpler to more complex organisms with the aim of bringing forth the most perfect being of them all: Homo sapiens. This has proved to be such a compelling image that even today evolutionary theory is often conflated with progressive modification; a confusion to which many contemporary documents testify. [Figure 2.1]

Lamarck was also a stern materialist. Materialism has many connotations. In the present context it expresses the belief that all the phenomena of organic life can be explained by appeal to mechanical causes. This includes the higher mental capacities of human beings. Kant was not a materialist. He even doubted that a caterpillar could be explained mechanically. But the doctrine had a long tradition in France (d'Holbach, Helvétius, La Mettrie). It led to the image of man-machine

Figure 2.1   The seven ages of an academic. An advertisement in *The Guardian* newspaper (23/4/2002), which draws attention to academic vacancies in the Tuesday editions. *Source*: *The Guardian*. © Guardian News & Media Ltd 2002

(*l'homme machine*). For the materialist, mental faculties and consciousness are nothing but expressions of brain states.

We shall see that materialism was one of the main philosophical objections against Darwin's theory. As the Darwinians were to find out, the biggest challenge to the materialist is to explain the phenomena of mind and consciousness.

It is an unfortunate effect of scientific revolutions that the ideas leading to them receive very little credit from the post-revolutionary generation. However, we learn from our mistakes, as Popper insisted. Mistakes even contribute to scientific progress. Lamarck's transformism put the fixity of species under a heavy cloud of suspicion. His materialism was a first attempt at overcoming Kant's explanatory dualism. Yet Lamarck's transformism turned out to be an inadequate model: soft inheritance does not exist; nor does progressive modification. Lamarck's trans-formism stood in need of empirical grounding. Like geocentrism, it saved the phe-nomena. Like geocentrism, it faced the question of reality. It happens in science that facts are in search of a theory. When Lamarck died there were still too many biological ideas in search of an anchor in evidence. In England, Eramus Darwin – Charles's grandfather – and Robert Chambers – the anonymous author of *Vestiges of Creation* (1844) – promulgated evolutionary ideas. In France, Isodore Geoffroy Saint-Hilaire anticipated the union of empiricism and rationalism in the study of organic nature, later espoused by Haeckel. He defended Lamarck's transformism against Cuvier's insistence on the fixity of species. But he corrected Lamarck's heavy emphasis on variability through the use of organs. The climate had to be taken into account. [Geoffroy Saint-Hilaire 1847; see Greene 1980, 300–1] As anatomy, embryology, geology, and paleontology uncovered new facts, these theoretical speculations began to strike roots in firmer ground.

3   Fossil Discoveries

When there are competing models vying for an explanation of the observable phe-nomena, each model tries to accommodate the available facts. The facts do not come with a ready-made assignment to the right model of explanation. Facts need interpretation. And the models serve to put them into a coherent order. When facts are few and far between, it is easy for the competing models to claim successful accommodation. This was the case between geocentrism and heliocentrism for a good 50 years. But eventually new facts which sit more comfortably with one model rather than another will emerge. The old and the new facts will then lend more credibility to one model than the other. That may not be the end of the dis-credited model, especially when new theoretical developments arrive. Yet for a while the victorious model occupies much of the attention. We have seen that geo-centrism did not die a sudden death. Brahe's observations, Kepler's laws, and Galileo's discoveries, however, made its life more uncomfortable. In evolutionary biology we observe a similar struggle between the old teleological view and the emerging mechanical view of life. The solar system cannot both be geocentric and heliocentric. Organic life cannot be the result of both design and "blind" chance.

Figure 2.2   Iguanodon. *Source: Nature* **55** [189], 463

If the planetary system is really heliocentric then a heliocentric model will be cognitively more adequate than a geocentric model. If the panoply of organic nature formed in the absence of design, then a mechanistic model will be cognitively more adequate than a teleological model. If the planets orbit the sun and complex organisms evolve then there should be facts that testify to these features. These facts will fit more elegantly into one explanatory scheme than the other. In the 50 years between Lamarck's *Philosophie Zoologique* (1809) and Darwin's *Origin of Species* (1859) new discoveries were emerging in diverse fields. In geology, for instance, the age of the Earth was gradually being extended from the miserly 6,000 years of the Bible to Kant's millions and millions of years of cosmic evolution. It was discovered that the rock layers contained a hidden arrow of time. The deeper the strata, the older the rocks. In paleontology, fossil finds added to the temporalization of the Earth. Fossils of extinct animals, of unknown life forms, revealed the fauna of ancient times. For instance, an ancient creature (*Archaeopteryx*), an intermediary between dinosaurs and birds, became the subject of intense discussions. This reptilia order was discovered early in the nineteenth century. But it only began to be properly understood in the 1860s. It was not until 1881 that Othniel Charles Marsh began to restore the skeletons of different dinosaur types, like the *Iguanodon*. The journal *Nature* printed impressive illustrations of these ancient life forms. [Figure 2.2] Voltaire's objections to the *Great Chain of Being* at last found support in concrete empirical data. Comparative anatomy demonstrated the close anatomical resemblance between humans and apes. Embryology revealed surprising similarities between embryos of different species in the earliest stages of development. [Figure 2.3]

As we are particularly interested in the philosophical consequences of scientific theories, we shall concentrate on the discoveries of human fossils and what they meant for the antiquity of humankind. Some of these findings were made well before Darwin's *Origin* was published. We should note that the question of the emergence of humans is independent of the question of man's descent from anthropoid apes. The debate about the antiquity of man is compatible with the view of

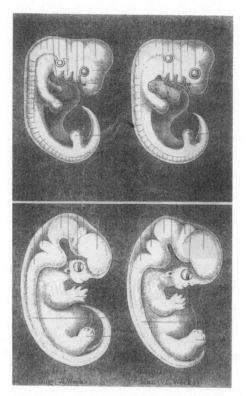

Figure 2.3   A comparison of embryos (left to right). Top: a dog (4 weeks old); a human (4 weeks). Bottom: a dog (6 weeks); a human (8 weeks). *Source*: Haeckel [1876], Vol. I, 306–7

special creation. The Bible's commitment to 6,000 years of human existence could be reinterpreted as merely metaphorical. But the view of man's descent from anthropoid apes is incompatible with the doctrine of man's special creation. When these two questions of antiquity and origin are combined, they form the explosive mixture that scandalized the public soon after the publication of the *Origin*. For the combination makes humans subject to natural selection in much the same way as Kant's weeds and caterpillars. Once Darwin's formula of evolution, driven by natural selection, became available as a solution to the species problem, it developed a logic of its own. Darwin's contemporaries were quick to extend Darwin's evolutionary model from the flora and fauna to the human realm.

3.1   Of bones and skeletons

Human fossils do not exist. [Cuvier, Ansichten von der Urwelt *(1822), 101, quoted in Rolle,* Der Mensch *(1870), 265]*

How do you establish human antiquity in the absence of precise dating techniques? You make inferences. A writer in the journal *Nature* expressed the severest doubts

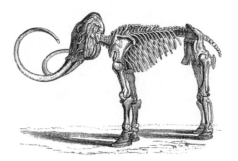

Figure 2.4   A drawing of a mammoth skeleton. *Source*: Rolle [1870], 286

that "we shall ever discover the exact cradle of our race." Still, the art of inferring can take us a long way. John Evans, president of the Anthropology Section of the British Association (1890), was confident

> that from time to time fresh discoveries may be made of objects of human art, under such circumstances and conditions that we may infer with certainty that at some given point in the world's history mankind existed (…). [*Nature* **42**, 1890, 507; cf. Rolle 1870, Ch. II, VI]

It may not have been possible, in the period before Rutherford's discovery of the radioactive decay law (1901), to establish the exact or even approximate age of skeletons. But comparative estimates could at least render certain inferences more likely than others. Human fossil records were found among fossils of extinct animals, like the cave bear, mammoth, reindeer, and the wild horse. [Figure 2.4] Also found were tools and weapons, made of flint stone. This suggested that hominoid creatures inhabited regions of Europe that were once home to now extinct animals. It also suggested that the Bronze and Iron Ages had not yet arrived. Human groups therefore must have lived in Europe during the Stone Age. This is a more likely inference than Cuvier's dictum that no human fossil remains can exist. It is more likely because it is more compatible with the facts. Several discoveries of human remains convinced many paleontologists that humans existed at a much earlier time than had hitherto been assumed. Philippe Charles Schmerling explored underground caves near the Belgian town of Liège. He found human skulls and bones of extinct animals. What is more, the human skulls had a somewhat different form from modern humans. This was still regarded as uncertain evidence. But in the 1840s Boucher de Perthes discovered primitive stone tools near Amiens in France among remains of extinct animals. Cuvier and others had already established independently that these animal fossils belonged to what they called the diluvial period or Ice Age. Today this period is referred to as the Pleistocene. [See Table 2.1] A human skeleton discovered in the Neander valley, near Düsseldorf (Germany), has gained celebrity as the Neanderthal Man (1856). No extinct animal remains were found. But the geological strata (clay) belong to the "diluvial" period. What struck

Figure 2.5   The skull of Neanderthal Man. *Source*: Rolle [1870], 309

contemporary paleontologists most then was the form of the skull. [Figure 2.5] It seemed to house a rather underdeveloped brain and showed strong similarities with the brains of anthropoid apes. (Note, however, that Neanderthal man is still under evaluation today.) All these discoveries were made before the publication of Darwin's *Origin*. Toward the turn of the century, another important discovery was that of Java Man (1891/2), a representative of *homo erectus*. Java Man served as a candidate for the missing link. The idea of a missing link made little sense prior to Darwin. In the old schemes humans were either especially created or the crowning moment of progressive evolution. Darwin inserted human lineages into the tree of life. Humans became subject to descent with modification. It therefore became imperative to investigate the existence of early anthropoid human forms. Researchers had better knowledge of human ancestry before Darwin than they had of the exist-ence of the dinosaurs.

3.2   The antiquity of man

The goal of humanity cannot lie in the end [Ende] but only in its highest speci-men. [Nietzsche, Thoughts out of Season *(1873/74), II, §9, quoted in Kaufmann (1974), 149)]*

It was clear before Darwin that mankind's pedigree had its roots in ancient times. This said nothing yet about human origins. But how long ago did these ancestral humans live? Researchers had no absolute measures of the ages of the Earth, only relative gradations. [Table 2.1] It was not clear how long ago the Ice Age had gripped the Earth. But there was a descending order into the past: from the Iron Age to the Bronze Age to the Stone Age; further, the Pleistocene, Pliocene, and Miocene. Again inferential practices played a major role. When human fossils were found with primitive stone tools and weapons, it was inferred that the early humans had to contend with such primitive instruments. For at later stages, bronze and iron tools replaced the stone tools. Hominoid finds in the presence of stone tools must therefore be older than hominids with more sophisticated gear. Most researchers placed the appearance of the earliest hominids in the Pleistocene period (Ice Age). In the 1860s writers were unclear about the extent of the geological ages in the history of the Earth, but human origins pointed well beyond the biblical limits.

Table 2.1 The evolution of *Hominidae*

Name	Period	Time	Characteristics
Australopithecus, gracile, robust (Southern Ape)	Late Miocene to Early Pliocene	4 to 2 million years ago	Cranial capacity: 430–450 cm³; upright posture
Homo habilis	Early Pleistocene	2.0–1.5 million years ago	Increased cranial capacity (640 cm³); smaller premolar and molar teeth; hand capable of fine manipulation; stone-working ability
Home erectus (homo heidelbergensis, homo ergaster, etc.)	Early to Middle Pleistocene	1.5 million–900,000 years ago	Cranial capacity 883–1043 cm³; mastery of fire; diminution in size of teeth; stone tools; Java Man; Peking Man
Homo neanderthalensis (Neanderthal Man)	Late Pleistocene (Stone Age)	200,000–15,000 years ago	Increased brain size; large long face, retaining browridges, projecting dentition and nose and absence of a full chin, large incisors and canines, primitive stone tools, first to intentionally bury dead; replaced by modern humans 50,000–30,000 years ago
Homo sapiens	Stone Age Early archaic Late archaic Modern	400,000–200,000 years ago 200,000–100,000 years ago 100,000–present	Mean cranial capacity: 1350 cm³, vertical forehead, rounding of cranial vault, reduction of masticatory complex; capable of making tools to make tools, true precision and power grip, derived from an opposable thumb

[1] Reconstructed from *Encyclopaedia Britannica* **18** (1991), 803–54 and *Scientific American* **13** (2003), with advice from Dr. Holger Schutkowski, Department of Archaeology, University of Bradford

[Rolle 1870, 159] The determination of the ages of the Earth through radioactive dating had not been developed. In 1868 Thomson (wrongly) estimated the age of the Earth as no greater than 100 million years. Only Darwin was seriously troubled by this estimate. Huxley and Wallace simply took the line that geology must determine the age of the Earth. And the biologist fits the tree of life into the available space. Global as it was, Thomson's estimate said nothing about the geological ages of the Earth and their extent. Friedrich Rolle reports the finding of a human skeleton in the Mississippi Delta (1852–3), whose age was estimated to be 57,000 years.

The antiquity of man was a debate of central importance well before Darwin formulated his theory of natural selection. The empirical establishment of the antiquity of man showed two things: (a) Humans were much older than the biblical 6,000 years. Hominids existed already during the Ice Age. There was also evidence of the migration of human groups into Europe. All this threw serious doubt on Cuvier's dictum that human fossil remains do not exist. It also suggested that species could become extinct, giving way to new species. (In 2003 what some regard as a new human species, nicknamed "Hobbits," was discovered on a remote island in Indonesia.) Structural similarities between humans, hominids, and apes became a serious scientific topic. All these findings challenged the view of the fixity of species; (b) they also undermined the idea of progressive modification, which required that certain organismic types must appear

Portrait of Charles Darwin (1809–82)

on Earth before more complex ones can make their entry. If early humans cohabited the Earth with now extinct animals, then progressive modification cannot be an adequate explanation. These considerations show clearly that Darwin's revolution occurred on the backdrop of available empirical discoveries and a number of competing theoretical accounts. They all tried to make sense of the observations and discoveries.

4   Darwin's Revolution

(...) Darwin converted evolution from speculation to doable science (...).
[Gould, Evolutionary Theory *(2002), 23]*

Darwin was *not* the father of evolutionary theory. Evolutionary attempts at explaining the diversity of life had already left their mark on the intellectual landscape of Europe. Lamarck's theory had not been influential but it had not been forgotten either. Darwin was aware of his predecessors. In later editions of his 1859 book, he added a historical sketch on early nineteenth-century views of the origin of species by way of a preface. He finds 34 authors who expressed their belief in the modification of species. Darwin also made his own important observations on the variability of species. During his journey on the *Beagle* he discovered puzzling varieties of finches and fossil bones from large extinct mammals in Argentina. When he came to write his *Descent of Man* (1871), he made extensive use of the empirical material regarding the antiquity of man. This material stretches from the 1830s to the 1870s. Darwin was perhaps less the Copernicus than the Kepler of biology. Kepler accepted the Copernican hypothesis but rejected the notion of uniform circular motion. He used Brahe's empirical data and his own to formulate mathematical laws of planetary motion. But Kepler could not properly explain why planets remained in their orbits. Darwin accepted the evolutionary viewpoint. But he rejected the Lamarckian mechanism of the inheritance of acquired characteristics. For Darwin the mechanism was natural selection. It operates on the fitness variations in individuals. But Darwin did not understand the genetic mechanism of heredity, which made the variability of individuals possible. The book in which he proposes his theory is called *The Origin of Species*. It was published on November 24, 1859. The full title of the book is quite revealing because it expresses in a nutshell the gist of Darwin's ideas: *On the Origin of Species by Means of Natural Selection, or the Preservation of Favoured Races in the Struggle for Life.*

As a historical footnote on multiple discoveries, it should be added that Darwin was rushed into publishing his work. He only meant it as an abstract to a fuller work on evolution to be written later. He was rushed into publication because he received a paper from a certain Alfred Wallace, in which the author presented ideas

on evolution very similar to Darwin's. The paper was entitled: "On the Tendency of Varieties to Depart Indefinitely from the original Type" and published in 1858. This alarmed Darwin because he had developed his theory of evolution already by 1838, after a five-year journey on the *Beagle*, but had never published it. Nobody likes to be scooped. So although Darwin was the first to propose a testable mechanism of evolution, Wallace arrived at the same ideas independently, but 20 years later. Already in 1855 Wallace had published an article in which he related the "gradual extinction and creation of species" to gradual changes in the environment. [Wallace 1855] He even argued that geological barriers, as in the Galàpagos Islands, would lead to different species. Distancing himself

Alfred Wallace (1823–1913)

from Lamarck, Wallace warned that he was not propounding a theory of progression but one of gradual change. This earlier paper presented a theory of evolution, but offered no mechanism by which evolution would be propelled. It was his 1858 paper that repeated many of Darwin's most cherished ideas. In this later paper Wallace insisted on the importance of the struggle for existence as a "powerful check on the dramatic increase of a species." The abundance or rarity of species, he held, was dependent upon the more or less "perfect adaptation to conditions of existence." Species face a triple challenge: food supplies, natural enemies, and climatic conditions. In this struggle, useful variations will tend to increase the number of individuals who possess them, while useless variations will tend to diminish them. If there was a difference between Darwin and Wallace it lay in differences of emphasis. Wallace explicitly stated that no useful inference could be made from domestic to wild animals. The inference from artificial to natural selection was one of Darwin's most important extrapolations. Wallace was also more concerned with species than individuals, which preoccupied Darwin. Finally, Wallace excluded humans from the impact of natural selection.

4.1   The Darwinian view of life

The basic clue in the discovery of natural selection was the realization that bio-logical groups may form populations or units of interaction in nature. [Ghiselin, The Triumph of the Darwinian Method *(1969), 56]*

A very rough and erroneous idea of the theory of evolution states that evolution is the survival of the fittest. It is interesting to note that Darwin, in the first edition of *Origin*, uses neither the term "evolution" nor the term "survival of the fittest." It was the philosopher Herbert Spencer who associated evolution with survival of the fittest. Darwin had influential opponents whose interests were served by the association of evolution with some form of neces-sary progress toward higher forms of life. This is a misunderstanding of Darwinism. There is no neces-sary progress. But it was a certainty before Darwin. It was the certainty of teleology. It was the convic-tion of design. It is this conviction that Darwin attacked. Yet it is rare for a scientific revolution to eliminate established ideas completely in one fell swoop. Even revolutionary ideas do not lead to an enthusiastic conversion, a sudden *gestalt* switch. Take for instance one of the protégés of George Cuvier and Alexander von Humboldt: Louis Agassiz (1807–73). Agassiz was born in Switzerland but emigrated to Boston in 1846. As professor of zoo-

Portrait of Louis Agassiz (1807–73)

logy at Harvard University, he had a formidable influence on biological thinking in America and a worldwide reputation as a scholar. He was regarded as the von Humboldt of America. He published important studies on glaciers and was one of the discoverers of the Ice Age. Agassiz was also an inveterate opponent of the theory of descent.

> The species, he insisted, were "categories of thought embodied in individual living forms," and natural history was ultimately "the analysis of the thoughts of the Creator of the Universe, as manifested in the animal and vegetable kingdom."[3]

---

[3] Quoted in Menand [2001], 128. For more on Agassiz, see Browne [2003]; Gould [2002]; and the obituary published in *Nature* **19** [1879], 573–6.

Agassiz was eager to demolish the "development" hypothesis, which he saw as incompatible with "the great truths of morality and revealed religion." He welcomed the publication of Hugh Miller's *Footprints of the Creator* (1849, 1861). He praised it for "its successful combination of Christian doctrines with pure scientific truths." [Miller 1861, 1–35; Ellegård 1958, 336–7] Agassiz wanted to restore design against Darwin's mechanical conception of nature. Darwin's contemporaries found the idea of natural selection so little convincing that Lamarckian ideas of acquired inheritance resurfaced. Even Darwin entertained his doubts. In *Descent of Man* (1871) he expressed some regret for his earlier wholehearted rejection of Lamarck's soft inheritance. He was clearly shaken by the lack of enthusiasm for his principle of natural selection.

Why does Darwin avoid the term evolution? As we have already seen, this term had strong associations with Lamarck's progressive modification. It meant a progress of development from a rudimentary to a mature or complete state. The term also had a rather technical meaning in embryology. Albrecht von Haller, professor of medicine and botany, had taught in the middle of the eighteenth century that embryos grew from preformed homunculi enclosed in the egg or sperm.

Darwin speaks of descent with modification. The driving force of descent with modification is natural selection. By this he means the preservation of favorable variations in individual organisms. If a variation is favorable for survival in a particular environment, there is a tendency for it to be preserved. Darwin observes that Spencer expressed the idea of the preservation of favorable variations by the term "survival of the fittest." But he defends the use of his own term, although it seems to imply conscious choice, because it brings together the production of domestic races by man's power of selection, and the natural preservation of varieties and species in a state of nature. [See Crombie 1994, Vol. III, 1751]

Darwin did not think that the evolution of organic life could be depicted as a necessary progression from "lower" to "higher" forms. For Darwin, evolution just meant a better adaptation of organisms to their natural environment. During this process an evolution to forms of higher complexity could occur. There are worms and there are wombats. This could be explained by the operation of natural selection. Or so Darwin hoped. Evolution pursues no plans. It is not goal-directed. Yet, it does not proceed by blind chance. It must preserve the favorable characteristics. It operates by cumulative selection. In the felicitous phrase of François Jacob, evolution is a tinkerer, not a gifted engineer.[4] Predictably, the Darwinians faced their biggest challenge when they applied the idea of natural selection to the emergence of humans, creatures capable of higher forms of consciousness. The objection was, as we shall see, that natural selection could not explain the higher mental functions in human beings. The Darwinians never contested the superior place of humans in the scale of organic life. But it makes a fundamental difference whether this is understood in a spatial sense, as in the *Great Chain of Being*, or in

---

[4] Jacob [1977], 1161. Using the German expression "Handlanger," Fritz Müller employed Jacob's expression already in his "Für Darwin" [1864], 259.

a temporal sense, as in the theory of evolution. [Wendorff 1985, 403] The theory of evolution therefore has significant implications for the philosophy of mind: how to explain the emergence of higher mental functions out of the neural activities of the brain.

4.1.1 Principles of evolution

Evolution proceeds like a tinkerer who for millions of years reshapes his work (...). [Jacob, Le Jeu des Possibles (1981), 66; translated by the author]

Before Darwin tackled the descent of man, he addressed the origin of species. Darwin's title *The Origin of Species* is misleading. It is not the book's intention to explain the origin of life at the dawn of time but how new species originate. His theory attempts a naturalistic explanation of the preservation and modification of organic beings of the recent past and present time. Instead of saying, like Lamarck, that the series of species tended toward some kind of perfection – culminating in the emergence of humankind – Darwin's idea was that organic change leads to local adaptations in response to a changing environment. Adaptive responses to changing environmental conditions lead to diversity. This process may lead to evolution within a lineage (*anagenesis*) or to the splitting of a lineage (*cladogenesis*). Some creatures become more complex than others. The cranial capacity of humans increased dramatically as they evolved from *homo habilis* to *homo sapiens*. But this is not a necessary march toward perfection. An organism can lose the function of its eyesight, if it adapts to living conditions in dark caves. So the fittest are not the best in an absolute sense. They are simply organisms with the most adequate adaptation to a local environment. Biologists cite many examples of the modification of body parts: from the loss of external and internal ears to the loss of forelimbs in snakes, from the simplification of eyes in snakes (loss of eyelids) to the loss of eyesight in cave fishes. [Raff 1996, 207–9; Gould 2002, 203–4, 218–9]

To understand the main ideas of evolution, consider three levels: the level of species, of individual organisms, and of genes.

On the **species level**, Lamarck had offered a *linear*, progressive view of evolution. Nature brings forth an increasingly complex series of life forms. The series terminates in humans, the crown of creation. The Darwinians responded with a *branching* view of evolution. Survival needs push organisms into ecological niches. They impose constraints on their morphology and structure. The adaptive responses produce the colorful panoply of life.

Wallace's focus was on the waxing and waning of *species*. This inspired him, already in 1855, to a new metaphor: the branching tree of organic life. From a bird's-eye view, evolution for species means local adaptations to changing environmental conditions. The result is a great diversity of species. The tree of life sprouts many branches and twigs. T. H. Huxley, an early master of the popularization of science, captured the essence of the change in a metaphor:

> Instead of regarding living things as being arranged like steps on a ladder, modern investigations compel us to dispose them as if they were the twigs and branches

of a tree. The end of the twigs represents individuals, the smallest groups of twigs species, larger groups genera; the main branch is represented by a common plan of structure. [Huxley 1888, 300]

The more species diversify, the more they exploit what local ecological niches offer in terms of resources. The environment acts as a constraint. The Kuala bear is an extreme form of diversification, as it only relies on one food source. This study of the diversification of species is called *macroevolution*. On this level, evolution has two essential properties: (a) *speciation*: the splitting of lineages and the evolution of new species;[5] (b) the morphological divergence of lineages: species change their appearance and internal structure until they are unable to interbreed. It is this process that Darwin dubbed *descent with modification*. It is the subject of the only tree diagram in his *Origin*. Following these leads, one of Darwin's staunchest German allies, Ernst Haeckel, introduced genealogical trees, to represent descent with modification. [Figure 2.6]

But let us leave the bird's-eye view and descend to the **level of the individual organism**. How does evolution look from this point of view? On this level, a struggle for life rages. To explain this struggle for life, Darwin introduces some explanatory principles. First, many more individuals are born than can possibly survive.[6] They differ from their parents by slight variations. This is the **principle of hereditary, isotropic variation**. Among the offspring some will be born with favorable variations, others with injurious variations. Any given environment can support only a limited number of individuals of a particular species. This has the effect that a constant struggle for survival ensues. In a given environment, individuals with a slight advantage over others have the better chance of surviving and of procreating their kind. Individuals with variations in the least injurious would tend to be eliminated. This is the **principle of natural selection**: *preservation of favorable variations and rejection of injurious variations*. Most adaptations follow from the struggle for life: not only the survival of the individual but also success at leaving progeny. "You cannot get adaptive complexity without natural selection." [Ruse 2003, 333]

We have so far descended from the evolution of species down to the survival of individuals. If it is true that some individuals are born with favorable and others with less favorable characteristics, it is natural to ask: "What causes these variations in individuals?" The modern answer is: random genetic mutations. This allows the identification of a third level, **the level of genes**.

---

[5] See Raff [1996], 59; Mayr [2001], Part III; Ridley [1997], Part E. There are two main types of speciation: allopatric speciation (due to geographical isolation) and sympatric speciation (due to non-random mating or mate preference), see also Williams [1973].

[6] Modern biologists make a distinction between reproductive success – producing offspring – and genetic success (genetic survival beyond any individual's lifetime): success of getting one's genes into future generations. The overall ability of an individual to get genes into future generations is termed *inclusive fitness*. The evolutionary process that maximizes the ability to treat others according to their genetic similarity to oneself is termed *kin selection*. See Williams [1996], 43–7, 47–51 for this terminology and a special example referring to bees.

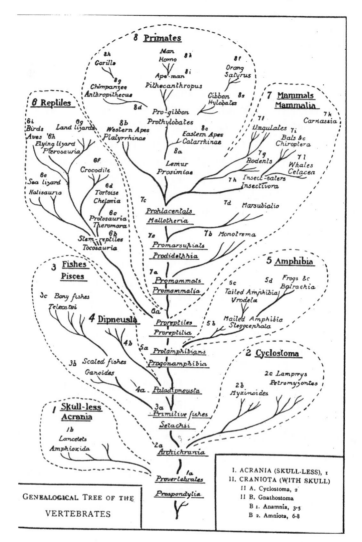

Figure 2.6 A genealogical tree of life. *Source*: E. Haeckel, *Last Words on Evolution* [London 1910], 32

   These three levels provide the units of evolution, selection, and variation. [Figure 2.7]

- Species are the unit of evolution.
- Individual organisms are the unit of selection.
- Genes are the unit of variation.

Darwin could offer only speculations regarding the unit of variation. He often speaks of "our" ignorance of the cause of each particular variation (in individuals). Darwin had no knowledge of Mendel's laws but he suspected that "disturbances in

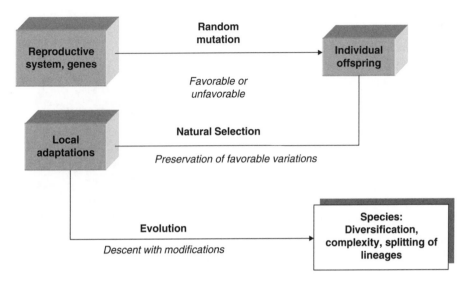

Figure 2.7 Three levels of evolution

the reproductive system," as he put it, chiefly contribute to the varying or plastic condition of the offspring. [Darwin 1859, 173]

According to Darwin, the process of evolution is an immeasurably slow affair, which leads to minute, imperceptible modifications. This is often called *gradualism*. (Gould [2002] discusses and criticizes Darwin's commitment to gradualism.)

> I cannot see any difficulty in natural selection producing the most exquisite structure, if such structure can be arrived at *by gradation*. [Darwin 1859, 435, 153]

Natural selection can also prevent the slow modification of species over thousands of generations. Natural selection then "culls departures from the currently optimum development" of the organisms' features. For instance, Herman Bumpus found in 1899 that sparrows whose wings deviated from the norm were more abundantly killed in a storm than those with average wingspans. The advantage of having intermediate character development (wing length, coloration) is called *normalizing* or *stabilizing selection*. [Williams 1996, 32–4]

4.2   The descent of man

Man is more like a gorilla than a gorilla is like a lemur. [Huxley, Collected Essays, II: Darwiniana (1907), Essay II, 61]

We have already mentioned the discovery of human fossils. This was in connection with the question of human antiquity. Darwin's work replaced the age-old need for teleology with the mechanism of natural selection. There is only one sentence in the *Origin* where Darwin mentions humans: "Light will be thrown on the origin of man and his history." [Darwin 1859, 458] Twelve years after the publication of the *Origin* he finally tackled the question of man's origin in a book called *The Descent of Man* (1871). But Darwin's contemporaries did not wait that long. As soon as Darwin had published his original idea, which coupled the idea of evolution with the mechanism of natural selection, his contemporaries were quick to draw inferences to the origin of humans. During the 1860s a number of important texts spelled out what in the popular press was traded as the ape theory. The year 1863 saw the publication of Thomas S. Huxley's *Evidence as to Man's Place in Nature* and Charles Lyell's *The Geological Evidences of the Antiquity of Man*. It was followed by Edward Tyler's *Researches into the Early History of Mankind and the Development of Civilization* (1865), and later his *Primitive Culture* (1871) and *Anthropology* (1881). On the Continent, Friedrich Rolle published *Der Mensch, seine Abstammung und Gesittung*

Julia Pastrana, the "gorilla woman," was a curiosity of the 1860s. Darwin did not regard her as proof of the ape ancestry of humans

im Lichte der Darwin'schen Lehre (1865, ²1870). The book presented the series of human fossil discoveries and defended the Darwinian view of the descent of humankind from earlier anthropoid forms. Karl Vogt published his *Vorlesungen über den Menschen* (1863), translated as *Lectures on Man* [1864]. Vogt explained the origin of races by relating their descent to different anthropoid ancestors. Most Darwinians rejected this polygenetic view. They saw all races as originating from the same common ancestor. [Browne 2003; *Scientific American* **289**, 2003, 50–7; Menand, 2001] Ernst Haeckel published his two-volume *Generelle Morphologie der Organismen* (1866), followed by *Natürliche Schöpfungsgeschichte* (1868). Haeckel was one of the most determined and ardent defenders of the Darwinian faith. He combined scientific acumen with philosophical insight. He turned Darwinism into a worldview. What is important about this flood of publications on the question of human origins is its anchorage in the Darwinian theory. As we have already observed, the questioning of the biblical chronology did not pose a real threat to the orthodox view. Six thousand years did not matter, literally. It may be a metaphorical time scale. What mattered were fundamental presuppositions, inherited from the *Great Chain of Being* and the design argument. Humans were distinctly, manifestly different from other animals. Their origin could not have the same roots. But Darwin's descent theory opened a conceptual space in which the question of human origins could be linked to the mechanism of natural selection.

The question of human origins was extremely sensitive, since there had been a long tradition of assuming man's uniqueness. This assumption was not tied to a non-evolutionary model. It was based on teleology. It was already difficult to imagine that the human brain should have evolved in the same manner as other "organs." The existence of the human mind and the phenomenon of consciousness seemed forever removed from the reach of naturalistic explanations. Just as it had appeared natural to the Greeks to assume a unique position of the Earth, so it appeared natural to most pre-Darwinian biologists to assume a unique position of humans in the story of creation. There was a great unwillingness to accept continuity between all creatures. Pre-Darwinian thinkers were as much preoccupied with saving the phenomena as Greek astronomers. What was needed to break this spell was a switch of perspective. Copernicus had told his incredulous contemporaries to contemplate the dance of planets and stars from the perspective of a spinning, not a stationary Earth. From an evolutionary perspective, the Darwinians recommended a consideration of the human brain as a natural organ, the seat of the human mind.[7] Do not look at human intelligence from a divine perspective, envisage it as a product of evolution.

It is natural to object that the human brain is far superior to any other "organ" and therefore cannot be the result of evolution. It is not adapted in the sense other "organs" or "organisms" are adapted. Just compare the human hand and the human brain, Wallace admonished his readers. Wallace adopted a dualist viewpoint, reminiscent of Tycho Brahe. Brahe could not accept the loss of physical centrality of the Earth, which follows from Copernicanism. He kept the Earth in its central Aristotelian position and made the moon and the sun orbit the central Earth. The other planets, however, spin in orbits around the sun. Wallace submitted the human body to the workings of evolutionary principles. For the creation of the human mind he stipulated a divine source. The Darwinians wanted to jettison such a two-sphere view as much as the Copernicans rejected the two-sphere universe. If humans must trace their origins to earlier animal forms of life, then the emergence of both brains and minds must yield to evolutionary explanations. This was again a question of inference.

It was of course easier to begin with the human bodily frame. And this is where the Darwinians started. An earlier generation had already inferred the antiquity of human life on Earth. The challenge now was to establish continuity between the bodily frames of humans and anthropoid apes. Once this was established an explanation could be attempted, which treated the human mind as emergent from the human brain. Thus the Darwinians moved from biology to philosophy, from evolution to the philosophy of mind.

The general strategy was to stress the continuity between the animal and the human world. At that time no genetic studies were available to inform us that humans and chimpanzees share 95 percent of their genetic material. For the Darwinians, embryology, anatomy, and paleontology carried the same message. The study of embryos of different species revealed that at least during the first phase of their development it was hardly possible to distinguish a human from a chicken. [Figure 2.3] Anatomy informed the Darwinians about the similarities in bodily structure between humans and other primates. From head to toe, the

---

[7] Tylor [1881], Chapter II characterizes the brain as the organ of mind. See also *Nature* **16** [1877], **21** [1879–80], **22** [1880], **29** [1883–84].

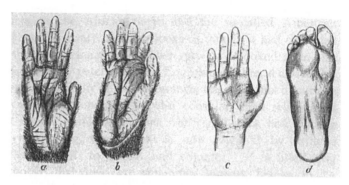

Figure 2.8 A comparison of the foot and hand of man and chimpanzee. *Source*: Tyler, *Anthropology* [1881], p. 42

skeletons of humans and monkeys revealed surprising similarities. The patterns of convolutions in the brains of humans and anthropoid apes are the same, so are muscle dispositions in the foot. [Figure 2.8] The disposition of abdominal organs is similar. There is a close resemblance in molar teeth.

Research on the question of human antiquity had already revealed fossils like the Neanderthal Man, which differed from the appearance of modern humans. Darwinism now put a heavy emphasis on historical narratives. As organisms, accord-ing to Darwin's evolutionary view, are related by descent, it was not a far-fetched conclusion to infer from anatomical similarities to common descent. The study of life in the geological past revealed descent with modifications. In his *Man's Place in Nature* (1863), Huxley drove this point unmistak-ably home, when he illustrated the anatomical rela-tionship between various primates. His sketch compares the skeletons of gibbon, orang, chimpan-zee, gorilla, and man. [Figure 2.9] The point about such detailed studies was to find evidence in favor of the descent theory. It was also a way of bringing the creation scenario into disrepute. The descent theory claimed that it could account for all these

Portrait of Thomas Huxley (1825–95)

phenomena with one explanatory principle. The principle of natural selection brought about unification. The Darwinians were keen to stress that the cerebral capacity of humans was not so very different from that of animals. It was a differ-ence of degree, not a difference of kind. At the end of the nineteenth century many books addressed the question of animal intelligence. Lauder Lindsay published *Mind in the Lower Animals* (1879). George Romanes published his *Mental Evolution in Animals* (1883). R. C. Lloyd Morgan offered his views in *Animal Life and Intelligence* (1890–1). The journal *Nature* opened its pages to amusing anec-dotes, apparently demonstrating animal cleverness. [*Nature* 29 [1883–84], 336; see Box 2.2]

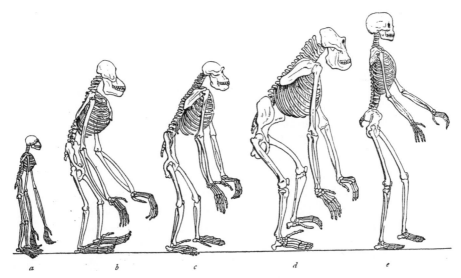

Figure 2.9 The anatomical relationship between gibbon, orang, chimpanzee, gorilla, and man according to T. H. Huxley. *Source*: Tyler, *Anthropology* [1881], p. 39

It was one of the recurring charges of the Darwinians that the creation theory needed to appeal to numerous acts of special creations to account for the diversity of life. "I disbelieve in innumerable acts of creation," Darwin declared. Organisms have many features some of which are not adaptations and do not promote survival directly. He postulates another type of selection – sexual selection – which does not depend on a struggle for existence but on a struggle for possession of the females. [Darwin 1859, 136; Darwin 1871, Ch. 8] Darwin explained the origin of features that appeared to be irrelevant or harmful in the struggle for existence – antlers of deer, feathers of peacocks – as devices for increasing success in mating:

Box 2.2 Animal intelligence

A certain F. J. Faraday, of Manchester, offered the following anecdotes to readers of *Nature* as evidence of animal intelligence.

A fish, unable to seize a morsel of food lying in the angle formed by the glass front and bottom of the tank, raised himself into a slanting posture, the head inclined upwards and the under surface of the body towards the food, and, by waving his fins, caused a current in the water which lifted the food straight to his mouth.

A bun was thrown into a pond, and fell at the angle beyond the reach of the bear. The animal thereupon commenced stirring the water with its paw, so that it established a sort of rotatory current, which eventually brought the bun within reach.

When males and females of animals have the same general habits of life, but differ in structure, color or ornament, such differences have been mainly caused by sexual selection. [Darwin 1859, 137; Gould 1987, 46]

Although Darwin thought that sexual selection was a process separate from natural selection, many modern biologists regard it as a special category of selection for social status. It is a kind of natural selection. [Williams 1996, 28–31] Darwin also holds that natural selection is the main but not the exclusive means of modification. [Darwin 1859, 69] This observation shows its true significance in Darwinian explanations of mental matters.

Thus we see the Darwinians simultaneously engaged in science and philosophy, in matters of biology and matters of philosophy. For it is not enough for a scientist to collect empirical data. The data have to be put into a coherent order. As the Darwinians were fully aware, their evolutionary view of life faced rivals. Lamarck's linear view of evolution with its insistence on inheritance of acquired characteristics had never held much sway. True, it rejected the constancy of species. Yet, it was committed to teleology, which it shared with creation scenarios. Creation theories had no truck with the variability of species. Even though the idea of evolution became generally acceptable, Darwin's principle of natural selection remained controversial. In 1859 Darwin's book entered a conceptual landscape which sustained several rival conceptions of the origin of species, all tottering on a rather thin empirical base. Darwinism needed both empirical and philosophical support. Let us turn to the philosophical matters to show that Darwinism inherited from the philosophical tradition certain presuppositions. This earned it accusations of disrespect for established scientific methods, of materialism and immorality.

5   Philosophical Matters

All true science is philosophy and all true philosophy is science. In this sense all true science is natural philosophy. [Lamarck, Zoological Philosophy *(1963),* Chapter II, 33; translated by the author]

Einstein once said that science without philosophy is muddled, and philosophy without science is an empty scheme. This may at first glance be a surprising statement. We are used to thinking of science and philosophy as two separate domains. It is true that much of what we call science today was once called natural philosophy. The term "science" came into current usage at around the time of Darwin's revolution. Since then the increasing specialization in all areas of knowledge has led to the erection of boundaries between the disciplines. Philosophers, especially in recent times, have sought a rapprochement with the sciences and created disciplines like the philosophy of science, the philosophy of physics, the philosophy of biology. But the fact is that scientists have often been interested in philosophy. This is especially true of great scientists whose innovations have led them to a consideration of the philosophical consequences of scientific discoveries. There have been so many that we may rightfully suspect that a scientist's philosophical *penchant* is more than a hobby. There must be some dimension in the very activity of science that tempts scientists to think philosophically. The general answer, which is spelt out in the three chapters of the book, is that the sciences do not only operate with specialized mathematical techniques but also have recourse to quite fundamental concepts, which help to build a consistent, coherent theory of a particular realm of phenomena. These may take the form of unspoken assumptions about the nature of reality or the nature of theories. But often scientists make their philosophical presuppositions quite explicit. What is more, these presuppositions are so general that they have a transdisciplinary status. It is therefore possible to find philosophical issues like chance and necessity, materialism and mechanism, being and becoming, time and space, in the discussions of physicists, biologists, and social scientists alike. These issues are typically also of great concern to the philosopher. In attempts to clarify such fundamental notions, the philosopher meets the scientist on common ground. As we shall see in the following sections, fundamental philosophical presuppositions were made in the realm of evolutionary thinking. A consequence of Darwin's new view of organic life was that such philosophical presuppositions dropped sometimes from the thin air of philosophical abstraction onto the hard ground of empirical science. Often a revolution in thought leads to a questioning of philosophical presuppositions. This is what Einstein meant. Science is muddled if it remains unclear about its philosophical commitments toward realism and the scientific method. Philosophy needs science to throw an anchor into the real world. Philosophy needs to embrace the results of science to prevent her from turning on empty wheels. [Weinert 2004]

5.1 Philosophical presuppositions: Mechanical worldview, determinism, materialism

For my own part I would as soon be descended from that heroic little monkey, who braved his dreaded enemy in order to save the life of his keeper, or from that old baboon, who, descending from the mountains, carried away in triumph his young comrade from a crowd of astonished dogs – as from a savage who delights to torture his enemies, offers up bloody sacrifices, practices infanticide without remorse, treats his wives like slaves, knows no decency, and is haunted by the grossest superstitions. [Darwin, Descent of Man (1871), Pt. II, Ch. XXI, 404–5]

We saw that Darwin's evolutionary theory entered an empirical space in which the fossil records told a revised story of human antiquity. To demonstrate human antiquity and common descent, researchers relied on evidence from embryology, comparative anatomy, and paleontology. During his voyage on the *Beagle* Darwin added a fair amount of observations about the geological distribution and the diversity of species. The factual material was still weak enough to permit the coexistence of several conceptual schemes. There were the creation scenarios with their belief in the fixity of species (Cuvier, Agassiz).

Box 2.3   The puzzle of genetic mutations

One of the central insights of Darwinism, as has repeatedly been pointed out by Ernst Mayr, is *population thinking*. Species are not to be thought of as Platonic types but populations of individuals. [See Mayr 2001; Ghiselin 1969, 56; Büchner 1868, 1. Vorlesung.] These individuals differ slightly from each other and the parents differ from their offspring. This enables some of the off-spring to be better equipped than others. There is much talk in the literature of Gratio Kelleia, a Maltese who was born with six fingers and six toes. [See T. H. Huxley, *Collected Essays II* (1907), 37, 406; F. Rolle ²1870, 82–4, 102–3] He married a woman with the ordinary number of fingers and toes. The couple had four children. Their first child was born with six toes and six fingers, like his father. The third child was normal, like the mother. The other children had five toes and fingers but their limbs were slightly deformed. When these children became adults and married, they again produced children, some of whom had six toes and fingers and others only the normal five.

   Darwin [*Nature* **24**, (1881), 257] discusses the case of an American gentle-man who began to turn gray at the age of 20. When he married he had four daughters, two of whom also began to turn gray at the same age as their father, while the others kept their mother's dark hair.

There were also several evolutionary scripts (Lamarck, Saint-Hilaire, *Vestiges of Creation*), whose most progressive element was the variability of species. They struggled, however, to propose a credible mechanism that could explain the diversity of species.

With his proposal of natural selection, Darwin made an important step in the right direction. But the cause of hereditary variation remained a puzzle. Spontaneous mutations attracted much attention in the literature. [See Box 2.3]

These various empirical and conceptual strategies occurred on the background of a philosophical canvas that had been painted by the Enlightenment. So it is no surprise to find the Victorian age imbued with three philosophical commitments: the first is the **mechanical worldview**; the second is **determinism**; and the third is **materialism**. There is no need for us to search for subterranean channels that would link Victorian science with Victorian philosophy. The philosophical presuppositions are embedded in the writings of Victorian scientists. They inherited them from the philosophical tradition.

1.   In 1829 Thomas Carlyle dubbed the nineteenth century the Age of Machinery. [Carlyle, *Signs of the Times* (1829), quoted in Bachmann 1995, 11] This epithet struck a chord for at the end of the century Haeckel opposed "our *naturalistic* century" to the earlier "anthropistic centuries." Steam engines and electricity had given the nineteenth century a "machinist stamp." [Haeckel 1929, 279, 307; italics in original] The commitment to the *mechanical worldview* had been inherited from the Scientific Revolution and the Enlightenment. According to the founding fathers of

Ernst Haeckel (1834–1919)

science, the physical world consists of matter and motion. Robert Boyle called this the corpuscular worldview. This accounted for all natural phenomena by reference to the laws of motion, which operated between the smallest units of matter, atoms or corpuscles. [See Ch. I. 5]

2.   The commitment to *determinism* also stemmed from the Scientific Revolution. In the philosophical literature, determinism is often characterized as a belief in the strict predictability of natural phenomena from the knowledge of deterministic laws and initial conditions. Its enduring image is the Laplacean demon, whose gaze over cosmic history reveals the dynamic state of all events, past and future, as if they were beads on an infinite string. Closer inspection shows that the Laplacean demon also embraces an ontological notion of determinism. The Laplacean demon can predict and retrodict cosmic history, because cosmic events are locked into a unique chain of prior causes and posterior effects. No chance events happen. Yesterday's causal events lead to today's effects, which become the cause of tomorrow's effects. [Weinert 2004, Ch. 5] This commitment to ontological determinism is reflected in the writings of physicists and medical practitioners as much as in the pages of biologists. According to Ernst Haeckel,

> the science of evolution made it clear that the same eternal iron laws that rule in the inorganic world are valid, too, in the organic and moral world. [Haeckel 1929, 285; 1866, Ch. IV; 1878, 509]

This was a view that Huxley shared.[8]

We should note in this connection that Darwin shared the commitment to materialism but not to strict determinism. Evolution is a stochastic process, in which contingency rather than necessity is the order of the day. The principle of natural selection acts on isotropic variation. It tends to select favorable variations and to weed out unfavorable variations. This should not be misread as blind, random chance. Evolution must preserve and accumulate favorable variations. But an organism's fate is always at the mercy of random genetic mutations and environmental constraints. Evolution acts in a strictly local context, which is shaped by a changing environment. Darwin left his readers in little doubt about the contingency of evolution:

> I believe in no fixed law of development (...). The variability of each species is quite independent of that of all others. Whether such variability be taken advantage of by

---

[8] Huxley [1862], 288; in another passage, Huxley claimed that "the fundamental axiom of scientific thought is that there is not, never has been and never will be, any disorder in nature," quoted in Ellegård [1958], 183.

natural selection, and whether the variations be accumulated to a greater or lesser amount (...) depends on many complex contingencies ... [Darwin 1859, 318; Dennett 1995, Ch. 10]

3. The Darwinians were thus committed to *materialism*, again a heritage of the Scientific Revolution. This commitment became particularly trenchant when Darwin's evolutionary model was applied to the appearance of humans on Earth. Darwin's contemporaries found it especially difficult to accept the earthly, common descent of man. It was held that Darwin's transmutation theory was based on "supposititious" facts. Worse still, it committed a heresy against Christianity and the immortality of the soul. Darwin's work was dangerous because it eliminated the need for special design.[9] Why did they feel such revulsion when materialism had been a venerable theme among the Enlightenment philosophers? [LaMettrie 1747; Helvétius 1758]

Lamarck, too, was a materialist. In his *Philosophie Zoologique* he argues that

all the faculties without exception are purely physical, i.e. each of them is essentially due to activities of the organization, from humblest instinct to intellectual faculties.

As Lamarck admits, materialism faces the objection that the connection between brain and mind cannot be understood.

What is the mind? It is a mere invention for the purpose of resolving the difficulties that follow from inadequate knowledge of the laws of nature. Physical and moral have a common origin; ideas, thought, imagination are only natural phenomena. [Lamarck 1809, Part II, Introduction; Lange 1873, Part III, I]

The French materialists went much further than Descartes. In his *Traité de l'Homme* (1664), Descartes had already examined the human body as an earthly machine. But this seemed to belittle the superiority of the human mind. Descartes's solution was substance dualism. The body was an extended, physical substance. The mind was an immaterial substance, connected to the human body through the pineal gland. Blood flows in the veins. When it reaches the brain, it turns into animal spirits. The animal spirits are able to move the limbs. Mind and body inhabit separate worlds. They are connected through mysterious fluids. Descartes's philosophical heirs were not content with his dualist solution. Mental phenomena also had to yield to a mechanistic explanation. The French materialists did their best to account for all mental processes as manifestations of physical processes.

The biggest challenge to the materialist is to explain the phenomena of mind and consciousness. Are mind and brain identical? Is the mind a mere epiphenomenon? Do mental processes emerge from brain processes? Such philosophical questions

---

[9] See Mrs. Miller's Preface in H. Miller's *Footprints* [1861]. This was not an isolated view. The literature at the time speaks of the general impression among the public that Darwin's view was an affront to morality and religion; see Jaeger [1869], Chs. I, IV; Braubach [1869], 17–18; *Nature* **28** [1883].

are a direct consequence of the commitment to materialism, which most Darwinians endorsed. Once they had accepted the mechanism of natural selection, which unified many diverse phenomena, they were loath to reintroduce a dualistic solution for the purpose of solving the puzzle of the mind. The emergence of the human body could be traced to the mechanism of natural selection. If the mind were exempted from the realm of evolutionary explanations, design and teleology would be reintroduced. The Darwinians were too committed to materialism to leave a loophole for mind–body dualism. In embracing a materialist theory of the mind, they were obliged to move from evolutionary biology to the philosophy of mind.

5.2   From biology to the philosophy of mind

Man still bears in his bodily frame the indelible stamp of his lowly origin. [Darwin, Descent of Man *(1871), Pt. II, Ch. XXI, 405]*

With the publication of Darwin's *Origin* the issue of the origin and nature of humanity suddenly acquired a new theoretical framework. Darwin's famous promise that light would be thrown on human origins[10] was immediately taken up by a number of researchers who attempted to provide an answer in terms of natural selection. When Darwin finally published *The Descent of Man* (1871) he was not entering an uncharted terrain. As we have already seen, Haeckel, Huxley, Lyell, Rolle, Tyler, and Vogt had already spelled out the lessons of applying natural selection to the appearance of humans on Earth. Any application of the theory of natural selection to the origin of mankind has to grapple at least with two issues. First, there is the *empirical* challenge of placing the origin of humans within the realm of the organic order. Second, there is the *philosophical* challenge of accommodating man's superior mental and moral faculties within the evolutionary framework. Until Darwin's *Descent* these attempts were undertaken without a major change in the fundamental philosophical presuppositions of determinism, materialism, and mechanism.

5.2.1   Empiricism   The most important *empirical* argument in favor of placing human origins in the animal world lay in structural similarities. In the 1860s the argument from similarity had to rely on embryology, comparative anatomy, and paleontology. Researchers undertook painstaking anatomical comparisons, especially between the skeletons of old-world monkeys and humans. Friedrich Johann Blumenbach (1752–1840) named monkeys *Quadrumana*: four-handed creatures,

---

[10] Darwin [1859], 458. The evolutionist, materialist explanation was fiercely resisted in some quarters. But the literature of the period from 1859 to 1871 also testifies to numerous cases of rapid endorsement of the Darwinian cause. In order of national impact, Darwinism had its greatest effect in England and Germany, its least in France. And some of the most advanced men of science, like E. Haeckel and T. H. Huxley, lent their intellectual support to the evolutionary theory. This observation throws some welcomed doubt on Max Planck's often-quoted comment that a revolutionary idea gains acceptance in science not because of the conviction but the death of its critics. See Planck [1933], 275. It was T. S. Kuhn who made this quote famous. [Kuhn 1970, 151] It is as little true of the early reception of Darwinism as it is of the early reception of the Special theory of relativity.

and humans *Bimana*: two-handed creatures. This distinction was popular until Huxley showed it to be inadequate. But the resemblances between their body plans, extremities, and skulls were plain to see. [See Figures 2.8–2.9]

The antiquity of man and his close structural similarity with the ape was established. Once the principle of natural selection became available, the hypothesis of man's descent from the ape was a natural corollary. But which form did this descent take? Lamarck had already offered his theory of progressive modification, which defined man as the crowning moment of linear evolution. Man therefore appears as the most complex and highest form of life. Lamarck then proceeds to compare the perfection of humans with the graded imperfection of lower organic forms. However, the fossil records did not speak in favor of progressive modification. Although Huxley and Haeckel rejected the Lamarckian theory of progressive modification through use-inheritance, they both considered some form of direct descent of man from lower organic forms. According to Huxley, it is plausible that man may have originated by the gradual modification of a man-like ape:

> But if Man be separated by no greater structural barrier from the brutes than they are from one another – then it seems to follow that if any process of physical causation can be discovered by which the genera and families of ordinary animals have been produced, that process of causation is amply sufficient to account for the origin of Man.[11]

In an address to the French Association of Science (1878), Haeckel dismissed the popular misunderstanding, which stemmed from the association of Darwinism with the "ape theory": the belief of man's direct descent from existing anthropoid apes. The more accurate picture was that "man and the apes of the Old and New World are descended from a common ancestor." [Haeckel 1878, 509; see Figure 2.9] But in a popularization of his philosophy – Monism – Haeckel advanced a more radical claim. It is an "incontestable historical fact"

> that man descends immediately from the ape, and secondarily from a long series of lower vertebrates. [Haeckel 1929, 69]

In Huxley we find a similar vacillation. On a more cautious note he considered that

> man might have originated (…) as a ramification of the same primitive stock as those apes. [Huxley 1863a, 125]

The sobering thought that man may have descended directly from the apes was grist to the mill of Darwin's friends and foes. Many heated exchanges at the time fed on the mistaken assumption of *anagenesis*: man had evolved out of existing apes. Haeckel uses it to shock his audiences into humility. Even today a picture of a massive gorilla menacingly staring out of a poster, bearing the rhetorical title: "Is this

---

[11] Huxley [1863], 125; Huxley [1864], 151 also states: "There is nothing in man's physical structure to interfere with his having been evolved from an ape."

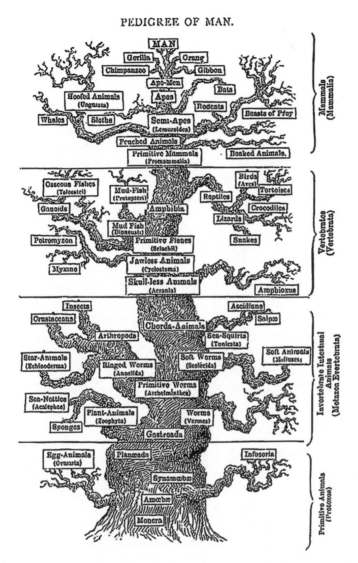

Figure 2.10 Haeckel's Pedigree of Man (1874) illustrates the idea of "branching" evolution. *Source*: Wikimedia

your granddad?" exploits the linearity to ridicule the Darwinian view of descent. [See also Figure 2.1]

In his more radical mood, Haeckel attributes to Darwin and Huxley the view of man's direct descent from the apes. But Darwin was as circumspect as Huxley. In his *Descent of Man* (1871) he staggers toward the image of commonality of descent. Within the span of a few pages in this long book, he speaks of man as *descending* from some lower form (1871, Pt. I, VI, 185), as the *co-descendant* with other mammals of some unknown or lower form (1871, 186), as having *diverged* from Old World monkeys (1871, 199, 201). Then, finally he suggests the image of **branching**. [Figure 2.10]

The Simiadae then branched off into two great stems, the New World and Old World monkeys; and from the latter, at a remote period, Man, the wonder and glory of the universe, proceeded.[12]

But even if we accept the branching idea – not that humans descended directly from the ape but that both share a common ancestor – a formidable stumbling block remained. Even the staunchest materialist cannot deny the intellectual gulf between men and brutes. Man's mental and moral faculties are far superior to the apes'. So man cannot have descended from the ape. It became a major challenge to Darwin's supporters to explain the appearance of mental and moral abilities within the limits of natural selection and materialism.

5.2.2   Philosophy of mind
Survival of the fittest led to survival of the brainiest. [Nozick, Invariances *(2001), 296]*

Many pre-Darwinian materialists had faced up to this *philosophical* challenge. They rejected Cartesian dualism. Just like Lamarck, they accepted a thoroughgoing materialism. Following a long tradition in French philosophy, Paul Thiry d'Holbach treated man as a machine. From this assumption he derived the conclusion that the intellectual capacities and moral properties of human beings must be derived from the same material causes that also affect the human body. [D'Holbach 1770/1978, 110] D'Holbach fully embraced the consequences of his materialism. Humans have no reasons to consider themselves privileged creatures in the order of nature. [1770, 81] Man is not the final purpose of creation. [1770, 452] Others could not bring themselves to accept such a radical conclusion. Alfred Wallace, the co-discoverer of the principle of natural selection, became increasingly dissatisfied with the ability of this principle to account for the essence of humanity. He was tempted by design arguments. He accepted as an established fact the antiquity of man. He saw the human body as having emerged from some primitive ancestor. But he could not accept that the human mind was the result of natural selection. In human con-sciousness and moral probity, Wallace detected evidence of design. Soon he added man's naked skin, his hands, feet, and voice as further manifestation of intelligent design. In some of his last writings he claimed that "life on Earth culminates in man" and "his existence was the purpose of the universe." [Wallace 1903; 1891] Wallace reverts to the Cartesian mind–body dualism. In his apelike state, man was subject to the forces of natural selection. But then, in the distant past, a revolution occurred. It brought forth a creature with a mind and blessed with consciousness.

---

[12] Darwin [1871], Pt. I, 213. But this terminology was so little fixed that the idea of branching also appears in Lamarck 1809, 37: "I do not mean that existing animals form a very simple series, regularly graded throughout; but I do mean that they form a branching series (*série rameuse*), irregularly graded and free from discontinuity, or at least once free from it." For Lamarck this branching is the unavoidable result of environmental interferences with the ideal, necessary progressive evolution of the tree of life. For Darwin, branching is a contingent event.

Man's bodily development ceased. His ascent was henceforth focused on the blossoming of his mental capacities. Natural selection shifted to cultural selection, in which superior races will replace inferior races. The mental abilities of man far exceed what would have been useful for his survival.

> Natural selection could only have endowed savage man with a brain a few degrees superior to that of an ape, whereas he actually possesses one very little inferior to that of a philosopher. [Wallace 1870, 202, cf. 1891, 474]

Wallace therefore concludes that a superior being must have guided the "development of man in a definite direction." [Wallace 1870, 204]

But a materialist explanation of mental functions cannot reach for supernatural causes. They would constitute a breach of determinism, which for almost all nineteenth-century scientists presented a unique chain of physical events stretching from the past to the future. They would also constitute a breach of materialism, for they run counter to the reduction of the mental to the physical. On the basis of these presuppositions the materialist must seek a plausible explanation of mental functions, without falling outside the scope of the set limits. The materialists were keen to avoid dualism. The pages of *Nature* reflect again the progressive thinking of the day. George Henry Lewes published *The Physical Basis of Mind* (1877), in which he proposed the view that the mental and neural processes are simply different aspects of the "same" reality. But the Darwinian materialists had lost some of the confidence of their French predecessors. The emphasis was on Mind and Matter, that is, the dependence of consciousness on the nervous organization. Huxley hypothesized that "molecular change in the nervous system causes states of consciousness." Yet it remained a puzzle, he continued, "how consciousness and material organism are related." [Huxley 1874, 365; cf. *Nature* **15**, 1876–7, 78–9] How should this puzzle be approached? First, the similarities between man and lower organisms, as established through the study of embryology, comparative anatomy, and paleontology, carry much of the evidential weight. The structural similarities suggest the emergence of humans from lower forms of life, from an apelike progenitor. The intellectual gulf between man and the brutes, however, cannot be denied. The materialist must find a bridge: accepting the gap but explaining it from below. Huxley once more reaches for the watch analogy. All faculties depend on structures. The organ shapes the function. But a certain variation in structure does not correspond to a linear variation in function. The function is an expression of molecular arrangements. A small change in structure can lead to a vast change in function. The smallest grain of sand will adversely affect the workings of a clock mechanism. In a similar vein, variations in the human body structure – upright position, the freeing of the hands, and increasing brain capacity – have led to vast functional differences between humans and apes: the emergence of mental functions, the development of language and other symbolic forms. [See Darwin 1871, Part I, Ch. 2; Huxley 1863a, 112–25; 1863b, 470–5; 1909, 51; Dennett 1995, 117, 287] In his *Descent* Darwin argues that there is no fundamental difference in kind between the mental functions of humans and higher organisms.

Complex animals feel emotions and manifest signs of intelligence. According to Darwin, the principle of natural selection is able to explain the evolution of mental and moral faculties. In the idiom of the day, the brain is the organ of mind. But the human intellect is such a sophisticated and complex organ that it is implausible that each of its functions would have been especially selected. For Darwin, natural selection was not the exclusive means of modification. [Darwin 1859, 136] We have already mentioned that he introduced the principle of sexual selection to explain the origin of features, like antlers in deer, that seem injurious to the survival prospects of an individual. Darwin argues that the mental and moral faculties have, most probably, been "perfected through natural selection," adding, "either directly, or more commonly indirectly." [Darwin 1871, Part I, Ch. 2, 80] A Darwinian can account for nonadaptive modifications by advancing two considerations.

1. "Organisms are integrated systems and adaptive change in one part can lead to nonadaptive modifications of other features." [Gould 1987, 45] Darwin calls this the *Principle of the Correlation of Growth.* [Darwin 1859, 182; 1871, Part 1, Ch. 2] Slight differences in human body structure lead to parallel changes elsewhere in the system. This could explain the doubling of the cranial capacity of the human brain. [See Table 2.1]
2. "An organ built under the influence of selection for a specific role may be able, as a consequence of its structure, to perform many other, unselected functions as well." [Gould 1987, 50] This could explain the evolution of mental capacities. Our large brain may have originated "for" some set of necessary skills in gathering food, socializing, etc., but these skills do not exhaust the limits of what such a complex organ can achieve. [Darwin 1859, Ch. V; 1871, Part I, Ch. V; Ayala 1987; Crombie 1994, Vol. III, 1759] "Natural selection makes the human brain big, but most of our mental properties and potentials may be spandrels – that is, nonadaptive side consequences of building a device with such structural complexity." [Gould 2001, 104]

Hence humans for Darwin are as much a product of evolution as any other organism. The Darwinians vigorously rejected an appeal to "new creative forces" to explain human brains, and hence followed the principle of unification that the same phenomena must be explained by the same principles.

5.2.3   Emergent minds

What's matter? Never mind. What's mind? Doesn't matter. [Gould, The Structure of Evolutionary Theory *(2002), 97]*

The materialist hopes to explain how the brain can cause the mind. At Darwin's time this was more of a research plan than a detailed program. Even today it remains one of the great mysteries, the focus of intensive research. Philosophies of mind and cognitive science together have produced a number of possible answers, all still thoroughly underdetermined by evidence. We cannot review these models here.

[See Blakemore/Greenfield 1987; Lyons 2001; Blackmore 2003; Searle 2004] However, we can ask to which modern-day position the Darwinian reflections on the mind may be most aligned.

On the Darwinian view, mental states are *emergent* properties of the brain. An **emergent property** is not reducible to the base from which it emerged. It constitutes a qualitatively novel phenomenon, which requires new levels of descriptions and possibly new physical laws. Mix flour, butter, eggs, and sugar, put the mixture in the oven, and you are rewarded with a cake as an emergent phenomenon. If this is true of the cake, it may be true of the mind. Subjective mental processes – conscious planning, reflection, problem-solving – constitute novel phenomena. The brain is a self-organizing system, which brings forth novel, higher-order mental processes. Mental processes are the emergent properties of interacting neuronal networks. As such, mental functions are distributed across the brain. Emergent properties belong to the whole system rather than its constituents. The Darwinians applied the principle of the correlation of growth to the brain. They argued that the brain may take on mental functions for which no direct selection had taken place. This may indeed lead to a snowball effect. [Bunge 1977; 1980; Sperry 1983; Chalmers 1996; Clark 1997; Damasio 1999; Edelman 1992; Humphreys 1997; Seabright 1987; *Scientific American* 2004] Huxley had already argued that small changes in the brain may lead to vast changes in the mind. Small increases in the neuronal wirings, from which humans benefited in their evolutionary past, may first lead to the invention of novel cultural tools, like symbolic language. Once these tools are available, their rapid exploitation in the interaction with the material world gives humans a major evolutionary head-start. The objective manifestations of subjective mental processes – culture, symbolic language, and science – outlive their original creators. Emergence implies the reality of the mental. Emergent properties are furthermore wrapped in causal loops, as illustrated in the phenomenon of group behavior.

A group can affect how its individual members behave. As we shall see in Chapter III, Durkheim argued for an emergent holism with respect to the existence of social groups and society. Freud, in his work on group psychology, also attributed to crowds causal powers which could not be reduced to the sum of the individuals that made up the crowd. According to this view, the interaction of individuals leads to higher-order phenomena, like social relations. If the mental is an emergent phenomenon of a higher order than the brain states from which it emerges, we should be able to find such causal loops in mental affairs. Mental processes are heavily dependent on symbolic means, both on the personal and on the societal level. We talk to ourselves and to others in a public language. Social life is propped up by institutional frames. Money, for instance, is a symbolic means, which allows it to be exchanged for goods. If the mind is able to produce personal intentions and cultural products, which in turn can affect individual and group behavior, the mind cannot be a mere epiphenomenon. There must be mental causation. Mental disturbances, as Freud insisted, result in troubled individuals. A mental decision to perform a physical action usually causes the action. Equally revealing are the effects that objective ideas have had on our worldviews and our social actions. Humans no longer behave, unquestionably, as if they were the center of the universe, the crown of creation. The ideas embodied in Copernicanism and Darwinism have channeled our practical attitudes toward the natural and social world into new directions. They have changed philosophical assumptions. Public institutions (education, justice, politics, and the market) are also symbolic entities, which heavily affect the behavior of the group that has adopted them. The behavior of individual economic agents differs strikingly, depending on whether they operate within the institutional framework of a free or a directed market.

If mental causation is real, it must be captured in any model of the human mind. Emergent properties seem to satisfy this requirement. They open up causal loops. The emergent properties can causally interfere with the elements from which they have arisen. Mental processes can affect other mental processes, but also brain processes and physical processes. A good idea may engender other good ideas. Emotional stress will affect the physiological states of the individual. Depression drives some to suicide. It is an important idea in the social sciences that ideas and institutions can channel social action. [Weber 1948, 280]

Although our Darwinian materialists did not have the notions of emergence and embodied mind at their disposal, in their discussions they characterized the mind as what we would call an emergent property. They considered the mind to be an important force in nature and society. The mind can develop new functions, like symbolic language, which will give any organism a huge advantage in the struggle for existence. The mind also creates, say, moral ideas by which social relations are regulated. In this way, the cruel force of natural selection has been tempered by cultural evolution.

Yet even today, many of the old questions remain unsolved. Let mental states "arise" from neuronal activity. How can empirical science explain the rise of conscious minds? Neuroscience has discovered neuronal networks in the brain. They are correlated with mental functions. The correlation can be localized. Speech, for instance, seems to be located in Broca's area. Other cortex areas are correlated with motor movement and vision. Darwinism has inspired a research program. Cartesian substance dualism has been laid to rest. The brain is seen as a biological organ, the seat of the mind. The focus of Darwinism encourages questions about the growth of new mental functions and the evolutionary advantage of consciousness. [Barlow 1987; Nozick 2001] (In Chapter III, Section 5 we will look at the development of evolutionary psychology – an attempt to explain mental functions through the application of evolutionary principles to the mind.) Darwinism has also had reverberations on human self-images: it has brought about a loss of rational design.

5.3   The loss of rational design

Looking into the future of the human race is more satisfying for our pride than looking into the past. [Büchner 1868, 256; translated by the author]

The Darwinian, naturalistic explanation of the emergence of humans constitutes a complete loss of rational design. No supernatural intervention sets up the cradle of humanity. Materialist Darwinians reject Wallace's compromise position. If natural selection can explain the emergence of the human body, it must also explain the emergence of the human mind. Humans, in a *literal* sense, are not the crown of creation. The Darwinians reject the Lamarckian solution of presenting humans as the target of progressive evolution. As a mechanism, progressive evolution does not exist. With the appearance of human beings the evolutionary process does not reach its pinnacle. Humans do not represent the peak of perfection, to which all other creatures are unfavorably compared. On the models of design and linear evolution, humans were relatively late additions to the panoply of nature. The ground needed to be laid for their complexity. Fauna and flora were therefore harnessed to the ultimate purpose of the creation of complex humans. Hence, humans could consider plants and animals as subservient to their needs. It was geocentrism in the biological realm. Both Copernicus and Darwin chipped away at such blatant biocentrism. Rational design was lost. Material forces took their place. The problem is, as Karl Vogt observed, that

> the whole inherent pride of human nature revolts at the idea that the lord of creation is to be treated like any other natural object. [Vogt 1864, Lecture I, 10]

Could humans figure at least as the crown of creation in a *metaphorical* sense? Even amongst materialists, reactions to the loss of rational design differed widely. Ernst Haeckel was the most pessimistic. He equated the loss of rational design with the cosmic insignificance of man. Haeckel refuses to grant humans any comfort, even in a metaphorical sense. On the contrary, the discovery of man's lowly origin sees man topple from the preeminent rung in the scale of being:

> Our own "human nature," which exalted itself into an image of God in its anthropistic illusion, sinks to the level of a placental mammal, which has no more value for the universe at large than the ant, the fly of a summer's day, the mircroscopic infusorium or the smallest bacillus. [Haeckel 1929, 199–200]

Not all Darwinians, by contrast, read cosmic inconsequence from man's lowly origin. Huxley rejects a crude version of materialism and regards the mind as a third force in nature. It is often stated that Huxley is an epiphenomenalist for whom consciousness is causally inefficacious. [Honderich 1987; Chalmers 1996; Blackmore 2003] Yet, if conscious states are a subclass of mental states, then Huxley was no epiphenomenalist with respect to mental states. Huxley emphasizes that mental states produce ideas, which can change the world. [Huxley 1874; 1886; 1893] The development of civilization consists in the gradual deflection of the forces of natural selection. Social life in civilized societies is suspended from the cosmic process of nature, in which the brutish struggle for existence prevails. Civilized life resembles the horticultural process, in which the struggle for existence is largely eliminated. Like the good gardener, humans modify the conditions of existence to suit their needs. Social Darwinism is a false extension of evolution to the human sphere.

> It is as if nature herself had foreseen the arrogance of man, and with Roman severity had provided that his intellect, by its very triumphs, should call into prominence the slaves, admonishing the conqueror that he is but dust. [Huxley 1863a, 125]

Darwin concurred with these views. He did not see human descent from anthropoid ancestors as a humiliating revelation. Where the anti-Darwinians sought to maximize the gulf between ape and man, the Darwinians were eager to minimize it. They only saw qualitative differences. Animals had intelligence and sensibility. It was not demeaning to count them as our ancestors.

The more optimistic Darwinians turned their attention from descent to ascent. Mental states emerge from cerebral states. This leads to symbolic language, to culture, and to values. The argument is open to the materialist that human values, even if their roots lie in social instincts, take on a life and dynamic of their own. Values channel social life in a direction opposite to the struggle for existence. Darwin saw the cosmic significance of humans in their ascent from humble beginnings to the summit of the organic scale, through cultural evolution.

In its temporalized form, the *Great Chain of Being* has a built-in dynamics, propelling evolution toward the highest perfection, which is invested in the human race. Humans therefore literally become the *crown of creation*. Darwin and the Darwinians reject the teleology, which is central to progressive modification. The emergence of human beings does not constitute the pinnacle of progressive modification. Humans are an offshoot of evolutionary branching. The tree of life might have grown without sprouting a human branch. Rather than necessity, it is contingency that builds the cradle of humanity. The assessment of the place of humans in the cosmos must necessarily change. It need not be as somber as Haeckel's. As many Darwinians stressed, the dignity of humans derives from their cultural evolution:

> Our reverence for the nobility of manhood will not be lessened by the knowledge that Man is, in substance and in structure, one of the brutes. [Huxley 1863a, 132]

> Important as the struggle for existence has been and even still is, yet as far as the highest part of man's nature is concerned there are other agencies more important. For the moral qualities are advanced (...) much more through the effects of habit, the reasoning powers, instruction, religion etc. than through natural selection. [Darwin 1871, Pt. III, Ch. XXI, 688–9]

The Darwinians agreed with Copernicus's rational centrality. For their more scientifically minded contemporaries, Darwin's revolution in science had led to the loss of rational design. The old design arguments failed to cope with the explanatory task of giving a coherent account of the diversity of the tree of life. Biblical-style creationist accounts are simply too ad hoc to be able to claim much credibility. Consider, for instance, the age of the Earth. There is independent evidence that the Earth was formed some 4.5 billion years ago. It will not do to claim that the Earth is really only 6,000 years old, as the Bible states. Clinging to this dogma requires a deceitful Creator who surreptitiously plants the geological evidence to make it appear *as if* the Earth were much older. Orthodox foot-stamping is of little avail when a serious scientific theory is the challenger of orthodoxy.

Even during Darwin's reign, some of his opponents moved away from biblical creation scenarios. They took science seriously. From scientific findings they inferred design, rather than a designer. Intelligent design scenarios were upheld as competitive rivals to Darwin's insistence on natural selection. We should not forget that natural selection became highly questionable during the 1870s. Even T. H. Huxley entertained some doubts about the explanatory power of natural selection. Looking at man's naked skin, his hands and feet, his voice, let alone his intellect, Wallace suspected the work of intelligent design. Miller's anti-Darwinian tract *Footprints of the Creator* (1861) argued that indications of harmonious design and eternal Providence could be gleaned from the geological facts. The anti-Darwinians agreed that the evidence supported natural selection insufficiently. It could not be a true physical cause. And it failed to account appropriately for the evidence of complexity. Opponents of Darwinism, like the Duke of Argyll, H. Miller, and St. George Mivart, contended that so much coordination is required in the growth of organisms that it could not be attributed to the work of "blind" chance. The rejection of natural selection, though not evolution, was also a stab at the mechanical philosophy.[13] But if natural selection and the mechanical philosophy fail, then the appeal to design regains some purchase. At the root of organic complexity there must lie an intelligent plan. The whole of organic evolution must have directionality built into it.

Biblical creation scenarios do arise from time to time. But a credible challenge to Darwinism must take scientific methods and facts seriously. In the 1990s a modern theory of intelligent design emerged. Its aim is to demonstrate scientifically that natural selection cannot do the explanatory work it claims to do. From a philosophical point of view intelligent design theory is interesting because the arguments are cast as questions about permissible inferences. We have already seen that inferences play a major part in Darwinian histories.

5.4   Intelligent design (ID)

Each revolution in the natural sciences has generated new problems for philosophy. [Rosenberg, Philosophy of Social Science *(1995), 212]*

The old design arguments inferred from visible order to design, from seeming perfection to a divine designer. Darwin confirmed that there was order in Nature, but not perfection. The new design arguments infer from organic complexity and improbability to intelligent design. Paley's further inference from design to a designer is deemed unnecessary. The observation of complexity is the result of scientific investigations, not pious contemplation. Intelligent design theory, like its nineteenth-century precursors, takes science seriously. In particular, it relies on

---

[13] See *Nature* **34** [1886], 335–6, *Nature* **54** [1896], 246–7; Browne [2003].

scientific discoveries from molecular biology and mathematical results from probability theory. It takes the fight to the evolutionists. It has as little patience with orthodox creationism as the Darwinians. It accepts that the Earth is approximately 4.5 billion years old. It accepts that natural selection plays some role in evolution. It sees no need to invoke a designer. It treats intelligent design as a hypothesis of "unevolved intelligence" (Dembski/Ruse 2004, 3) or even as an inference to the best explanation (Menuge 2004, 32; Meyer 2004, 371–2). From a logical point of view, it seems to employ inferential practices just like the Darwinians.

To simplify, let us say that there are two explanatory accounts – the familiar (neo-)Darwinian story and an intelligent design account. But today evolutionary biologists no longer rely on anatomy, embryology, and paleontology; the range of evidence has extended into the genetic and molecular realm. In particular, modern design theorists focus on (a) the complexity on the molecular level, (b) the transition from inorganic to organic life and the emergence of novel animal body plans, and (c) mathematical probability in order to argue in favor of ID. Let us briefly characterize these arguments and then look at some of the criticism of modern ID theories.

Ad (a)   *The argument from irreducible complexity.* [Behe 1996; 2004] According to Behe modern molecular biology has established an irreducible complexity of systems within the cell. Behe's real-life scientific examples refer to the clotting of blood, the biochemistry of vision, and, as part of cell chemistry, the so-called bacterial flagellum (e.g., an ion-powered rotary motor which, according to Miller [2004, 82], is "anchored in the membranes surrounding the bacterial cell"). Irreducible complexity means that "the removal of any of the interacting parts causes the system to cease functioning." [Behe 2004, 353] The point about irreducible complexity is, according to Behe, that it cannot be explained by a Darwinian process of gradual modification. If natural selection and gradualism fail to explain the complexity, it seems reasonable to Behe to infer intelligent design to explain this type of molecular complexity. The complexity reveals intelligent design. From a philosophical point of view, it is remarkable that ID theorists, like Behe, are willing to jump from the question of admissible inferences, on which all parties agree, to the question of permissible inferences. Thus Behe declares that "the design process may have contravened no natural laws at all." [Behe 2004, 357]

Ad (b)   *The argument from the emergence of life and novel body plans.* Biologists have tried to explain the improbable transition from inorganic to organic life. The problem is that the first molecules were already quite complex, and the origin of this complexity needs explaining. According to the ID theorists, appeals to chance and natural laws are insufficient. The "origin of life seems to be the ultimate example of irreducible complexity" and the "most compelling example of Intelligent Design in nature." [Bradley 2004, 350] Note again the form that the argument takes: traditional accounts "fail" so that the inference to ID seems to impose itself. In a further development of the argument, ID is proposed not only as an adequate solution to the "origin of life" question but also the emergence of novel animal body plans in the period of the Cambrian explosion. [Meyer 2004, 372] ID theorists accepts that "the Darwinian mechanism can explain micro-evolutionary adaptation," like the variety of beaks in Darwin's finches. What they contest is that evolution can explain "the appearance of complex specified information in living organisms." [Meyer 2004, 386; but contrast Gould 1991; 2001 for an evolutionary account] Again we observe the willingness to proceed from admissible to permissible inferences:

Determining which among a set of competing possible explanations constitutes the best one depends upon judgements of causal adequacy or "causal powers" of competing explanatory entities. (...) Intelligent agents have causal powers sufficient to produce increases in CSI, either in the form of sequence-specific lines of code or hierarchically arranged systems of parts. [Meyer 2004, 386–7]

Meyer appeals not just to an inference to the best explanation. It is relatively easy to claim that one's favorite theory is the best explanation of some available evidence. Meyer refers to inference to the best *contrastive* explanation. In this type of inference, the evidence is taken to lend empirical support to one of a number of competing hypotheses. It credits one hypothesis and discredits its rivals.

Ad (c)   *The argument from specified complexity.* The arguments from irreducible complexity are only "a special case of complex specified information (CSI)." [Menuge 2004, 47] CSI is characterized as information with a low probability value and the existence of an independent pattern, which is capable of eliminating chance. [See Dembski 1998, Ch. 2; 2004] Dembski sees specified complexity as a statistical criterion which helps to identify the effects of intelligence. To illustrate, imagine you travel abroad with a friend about whose linguistic abilities you know nothing. When you arrive in your country of destination you notice that your friend is fluent in the language of your host country. Clearly the probability that your friend learned the foreign language on the day of your arrival is close to nil. The only reasonable hypothesis is that your friend was a fluent speaker of that country's language prior to your arrival. In this example we have a specified pattern – the ability to speak a foreign language, *L* – which explains your friend's fluency much better than the hypothesis that he learned it by chance on the day of your arrival, *A*. Your friend's unexpected mastery of the foreign language, *L*, is detached, that is, independent of the specified event *A*, although *A* of course instantiates the pattern or design.

Specifications are those patterns that in combination with small probability warrant the elimination of chance. [Dembski 1998, 136]

Once chance and regularity have been ruled out as possible explanations of an event that displays all the features of an *as-if* design, inference to design seems to remain the only option. This idea of specified complexity is then applied to evolutionary biology. Molecular systems, like the flagellum, are so improbable that

Darwinian mechanisms fail to explain their emergence. If cumulative selection fails, random selection is even more unlikely, given the probabilities involved. Dembski appeals to the procedure of eliminative induction [Dembski 2004, §7]: where rival mechanical explanations fail to account for the evidence, *E*, this same evidence is then taken to support the account, which seems to explain it.

There is a vast amount of genetic and molecular evidence. Does the evidence favor the Darwinian explanation or does it support intelligent design? This boils down to a question of *admissible*, not just permissible, inferences. Both camps agree that the evidence reveals order and regularity. Is natural selection sufficient to account for complexity? The Darwinian answer is in the affirmative. As a theory, natural selection is jointly necessary and sufficient to explain the rich tapestry of life. [Dawkins 1988; Dennett 1995; Ridley 1997; Jones 1999; Mayr 2001; Gould 2002; Ayala 2004] Intelligent design theorists disagree.

Who is right? Should we opt for a design or evolution inference? Empirical science must establish the nature of the complexity of organic systems (e.g., the clotting of blood, cell chemistry, and the biochemistry of vision). [See Miller 2004; Kauffman 2004; Weber/Depew 2004] But this is not just an empirical question. The preference for a certain type of inference also reflects a philosophical attitude toward science. Modern evolutionists will argue that the inference to intelligent design, while permissible, returns us to supernatural explanations. Darwinism has always aimed at unification and a naturalistic explanation of biological phenomena. Even if natural selection cannot give a complete account of evolutionary phenomena, a naturalistic approach is still better than inference to design. [Pennock 2004] For design lies beyond the reach of testability.

Testability is a contrastive notion. We use the same evidence to assess the respective values of two competing theories. [See Sober 1999; 2002; 2004] In the present case, the two competing theories are Intelligent Design and Darwinism. The question of testability becomes the question of whether the weight of evidence supports the hypotheses of ID and evolution *differentially*.

The insistence on the original complexity of biological systems ignores, in the eyes of modern Darwinists, a Nietzschean insight. We cannot infer, Nietzsche held, from the nature of the present function of an organ to the nature of its past function. There is a difference between the "origin or emergence of a thing and its ultimate usefulness." [Nietzsche 1887, Sect. XII; Dennett 1995, 470; Gould 2002, 1216] But evolutionary biologists will insist that even incipient organs, like rudimentary wings or light-sensitive cells, will have offered some evolutionary advantage to the organism in the past. They also find in molecular biology support for evolutionary explanations. The study of molecular evolution reveals valuable, often quantifiable information about the rate and range of evolutionary changes. The evolutionary trees of genes and species can be constructed from the precise analysis of molecular sequences.

These are not just empirical matters. The empirical data require understanding. How can they be incorporated into a coherent scientific theory? At this level, philosophical questions of method arise. Among the permissible inferences, which are the admissible ones?

6   A Question of Method

We are true moles in the path of Nature. [LaMettrie, L'Homme Machine *(1747), 79; translated by the author]*

We have noted that the Darwinians operated in a conceptual space, in which various views of organic life competed against a backdrop of empirical discoveries and philosophical presuppositions. We have also seen that the Darwinian revolution had significant consequences for the place of humans in organic nature: the loss of rational design. The popular opposition to Darwinism partly stemmed from a facile association of the descent theory with the ape theory. It was claimed that materialism would lead to immorality. Scholarly opposition to Darwinism was more concerned with the logic of the Darwinian argument. St. George Mivart and Wallace, for instance, questioned whether the mechanism of natural selection was sufficient to explain the complexity of human beings. Their doubts may have been motivated by religious allegiances. But the motivations had to be converted into counterarguments for the purpose of a scientific debate. There is hardly a more effective attack on a scientific result than questioning the scientific method, which established it. Bishop Wilberforce epitomizes the worry of the anti-Darwinian camp, when he accused the Darwinians of disloyalty to Baconian induction. [Cohen 1985b; Ellegård 1958, Ch. 9] Today, proponents of intelligent design accuse the Darwinians of drawing the wrong inferences from molecular complexity.

Our aim is to explore the philosophical consequences of scientific theories. It therefore behooves us to investigate the question of method. Which scientific methods did the Darwinians employ? Did they deviate from the true path of the Baconian method? What is Baconian induction? How does Baconian induction compare with Popper's testability?

6.1   Darwinian inferences

While asking general questions led to limited answers, asking limited questions turned out to provide more and more general answers. [Jacob, "Evolution and Tinkering" (1977), 1162]

We have already seen a number of inferences at work. From the coexistence of fossil records of extinct species, human bones, and artifacts, researchers in the 1840s inferred the antiquity of man. Darwin made similar inferences. Emphasizing the well-known power of artificial selection among breeders, he inferred a principle of natural selection. It could explain, he held, the diversity of species. The principle of natural selection is an explanatory principle. It can be broken down into a number of sub-principles:

- the principle of variation (random mutations);
- the principle of variation in fitness (un/favorable modifications);
- the principle of the struggle for existence;
- the strong principle of inheritance. [Kitcher 1993, 19]

Thanks to those principles, Darwin is able to unify diverse areas of inquiry, which had hitherto stood on their own. **Unification** is a powerful methodological principle in scientific research. We have already encountered it in the transition form geocentrism to heliocentrism. The disappearance of the two-sphere universe and its fundamental presupposition, the Aristotelian theory of motion, enabled a unification of terrestrial and celestial physics. Darwin's principles provide powerful techniques of biological explanation. They permit us to infer histories of descent for numerous biological phenomena.[14] Consider three such phenomena:

1.	Darwin was struck by the biogeographical distribution of species and their varieties on the Galàpagos Islands. How was it possible that the same species of finches could show slightly different characteristics from island to island? Prior to Darwin the predominant view had been the theory of special creations. Darwin's answer was to treat the modifications of each current group "G" of organisms from its ancestors as the organisms' response to changes in the environment. This could be extended to explain the observed differences between, say, Neanderthal skulls and modern human skulls.

2.	The origin of man was established by investigations in comparative anatomy. As Huxley had stressed, there were only minor anatomical differences between humans and apes. If humans and apes, belonging to different species, share common properties in the bone structure of limbs, how is this to be explained?

	In order to tackle such questions, the Darwinians distinguished **homologies** from **analogies**. A *homology* is the presence of the same organ in different organisms in a variety of forms and functions, like avian wings and human arms. Homologies are due to descent with modification. The presence of a homology in different organisms can be inferred through histories of descent from a common ancestor. An *analogy* is the presence of similar organs, with similar functions, in unrelated animals, like avian and insect wings. Analogies are due to convergent evolution, i.e., the result of similar environmental pressures. Their presence can be explained by tracing the history of the emergence of the analogy along the different lineages of different species. They have no recent common evolutionary origin. Although this distinction is clear in theory, it can be difficult to establish in practice. [Darwin 1859; Rolle 1870, 151–2; Crombie 1994, Vol. II, 1291; Dennett 1995, 357; Raff 1996, 34–6; de Beer 1997, 213–21]

3.	Finally there are questions of adaptation: why organisms living in certain environments display certain properties, which contribute to their fitness in this particular environment. These questions, too, can be answered by rehearsing the historical process through which the adaptation emerged. Such answers

---

[14] Kitcher treats a Darwinian history as a narrative which traces the successive modifications of a lineage of organisms from generation to generation in terms of various factors, most notably natural selection. [Kitcher 1993, 20, 26; see also Lloyd 1983]

replace the old design argument. They also dismiss the insistence on absolute perfection. As Darwin stresses, adaptation is a question of degree:

> Natural selection tends only to make each organic being as perfect as, or slightly more perfect than, the other inhabitants of the same country with which it has to struggle for existence. [Darwin 1859, 229]

The mechanism of descent with modification produces the relative perfection of the adaptation of organic beings to their physical environment. But other organisms, with which they enter into competition, also inhabit their physical environment. There is a tug-of-war for food and shelter, for eschewing injury and death. The struggle for existence has a "corollary of the highest importance": the structure of every organism is fundamentally interrelated with "that of all other organic beings." [Darwin 1859, 127]

It was already clear to Darwin's contemporaries that the theory of descent with modification was about drawing acceptable inferences. The real question is not simply: What inferences are *permissible*? But what inferences are *admissible* in the light of the available evidence? This accentuation is reflected in the difference between possible and actual solutions. For Darwin's friends and foes it all boiled down to a simple alternative: does the evidence of exquisite adaptation justify an inference to a supernatural Designer or to the operation of natural selection? [See *Nature* 27 (1882–83), 362–4, 528–9] For the Darwinians the inference underlying the design argument – from the order of Nature to a Designer God – is not admissible, since it does not belong to the realm of natural science. But as Intelligent Design theory illustrates, there is, logically speaking, no obstacle to blocking the inference at least to design. Its proponents agree with the Darwinians that it boils down to a question of support. Which of the two hypotheses – design or natural selection – receives better support from the evidence? The Darwinians never claimed that the hypothesis of natural selection is dead right and that of design is dead wrong. Many inferences are logically permissible, but are they admissible? Admissible inferences are a matter of evidential support. We shall soon draw a distinction between *positive* and *supportive* evidence. Here it is sufficient to characterize supportive evidence as evidence that credits one hypothesis at the expense of the other. Given the evidence it is admissible to infer the plausibility of one model of explanation at the expense of its rivals. This is how the Darwinians argued. In some cases the evidence is such that it lends support to one hypothesis *and* discredits its rival. Darwin did not argue from the rather vague notion of perfect design. He looked at the biogeographic distribution of species, their variations, similarities, and adaptations. He argued that this body of evidence could be accounted for by appeal to natural selection. If there are anatomical similarities between apes and humans, it is more likely that they derive from a common ancestor than that they were created separately. [Darwin 1859, Ch. XIII, XIV] If the fossil records reveal the antiquity of man, it is more rational to infer that humankind is older than biblical chronology allows. Note that this makes explanatory attempts on the part of the old-style creationists more difficult, not impossible. The creationist can always argue that a Deity planted the fossil records approximately 6,000 years ago.

The creationist is forced into such ad hoc maneuvers because the age of the fossil record is prima facie incompatible with the evidence. Intelligent design theorists avoid such ad hoc maneuvers. They prefer inferences to design. Natural selection refers to physical causes; design arguments rely on supernatural agency. Orthodox design arguments name the agency, modern design arguments leave it unnamed. Modern design theorists agree with the Darwinians on the need for the availability of supportive evidence. There is logically nothing wrong with the move from evidence to design. But it requires a break in the chain of natural explanation. We are asked to make a leap of faith from admissible to permissible inferences.

Darwinism entered a crowded conceptual space of competing models of explanation. The Darwinians were always engaged in contrastive explanations. [See Box 2.4] Given Weismann's negative results on the inheritance of acquired characteristics, it became less likely that Lamarck's explanation would gain much support from the evidence. Natural selection, by contrast, worked on the phenotypic manifestations of random genetic variations. If only genetic cells are transmitted, this is compatible with Darwin's belief in isotropic variation. Given the lack of perfect adaptations and harmonious order in nature, the force of the design argument was diminished. The Darwinian theory itself struggled with various difficulties. There was the "missing link" argument, which Darwin, in an ad hoc fashion, blamed on the poverty of the fossil record. Then there were Thomson's calculations of the age of the Earth. Ninety million years were not enough to perform the invisibly slow gradualist process of evolution that Darwin envisaged. And there were the doubts about the explanatory weight of natural selection itself. Although the Darwinians acknowledged the difficulties, they still considered the Darwinian model superior to the theory of special creations.

Darwin's detractors accused him of a shameful neglect of Baconian principles. The true method, they held, was to observe nature and generalize from the observation of some cases to all cases. Darwin indulged in wild speculations. Much "pseudo-scientific cant" has been written about the Baconian philosophy, replied

---

Box 2.4   Inference to the best explanation (IBE) or abduction

If inference to the best explanation just means that we infer from the available evidence to an account which we regard as the best explanation of this evidence, then *IBE* is not to be confused with eliminative induction. We shall see in Chapter III that Freud inferred from observational phenomena, like neurotic behavior and dream reports, an explanation in terms of the dynamic Unconscious. However, this cannot provide an adequate explanation until we have contrasted it with a number of rival explanations. [Lipton [2]2004] We need an inference to the best *contrastive* explanation, such that the evidence can assign probability weights to the contrasting explanations. It is precisely on this point that Freud's procedure is woefully defective. It takes the meager evidence to confirm the model of the Unconscious, without weighing the likelihood of rival explanations.

Huxley. [Huxley 1907, 364–5; Haeckel 1866, Ch. 4; Greene 1980, 314] Both Huxley and Haeckel became aware of the importance of affirming the credibility of Darwinian methods. It required, both agreed, a combination of critical induction and deduction. They credited John Stuart Mill with the discovery of this synthesis.

Darwin claimed to have arrived, inductively, at the principle of natural selection. But this does not establish it as a valid theory. Darwin made a number of observations, which suggested to him that the diversity of species and their adaptation to the natural environment may be explained by the physical mechanism of natural selection. But such generalizations from a relatively small sample space may be seriously misleading. Europeans habitually inferred from their observation of white swans that all swans were white. This inductive generalization was dramatically refuted when settlers discovered black swans on their arrival in Australia. Already Francis Bacon, on whose authority Darwin's critics attacked his "shaky" method, called

the induction, which proceeds by simple enumeration (…) childish; its conclusions are precarious, and exposed to peril from contradictory instance; and it generally decides on too small a number of facts, and on those only which are at hand. [Bacon 1620, Book I, §105]

We shall see in the next section that both Francis Bacon and John Stuart Mill proposed a much more sophisticated method than induction by enumeration. They call it **induction by elimination**.

6.2   Philosophical empiricism

But let us first complete the steps by which Haeckel and Huxley defended Darwin's method. We may arrive at a model, like natural selection or heliocentrism, by an inductive step from a certain number of observations, aided by some theoretical principle. But they remain mere conjectures or hypotheses. Once such models are available they become subject to tests. Here the deductive part sets in. According to standard wisdom, we derive from the model certain predictions, which must be tested. This demand could be satisfied in Copernicanism but not in Darwinism. Darwinism was not able to offer precise numerical predictions about the future behavior of biological organisms. Darwin was able to accommodate hitherto known facts, which had remained puzzling under the rival theories. For instance, the great similarity in the anatomy of apes and humans, the existence of rudimentary organs, found a natural explanation on the Darwinian hypotheses. They remained puzzling on the theory of special creations. Evolutionary theory provided a framework, which could be used to derive the accommodated facts. There is more: once the theoretical framework is available, it can be used to infer deductive or inductive consequences. The most spectacular fact, which the Darwinians derived from the descent theory, was the probable emergence of modern humans from earlier hominids and ultimately from anthropoid apes through branching. Darwinism established a new link between the antiquity of man and the trajectory of descent. This deduction cannot be treated as a novel prediction, since it incorporates an independently known fact into the new theory. The German biologist Fritz Müller focused on the developmental history of Crustacea. Many of the physiological details, which he described in great detail, find a natural explanation in the theory of descent but stretch credulity on the theory of special creations. In the complicated physiological minutiae of Crustacea Müller found little evidence for Agassiz's "fixed plan of the Creator." He pointed out that anatomical details of crustaceans could be derived from Darwin's position. Such inductive consequences constitute supportive evidence for Darwinism. [Müller 1864, §§III, IV] Darwin was so impressed with Müller's work – calling it an "admirable demonstration" of his doctrine – that he arranged for an English translation, which was published in 1869. [Browne 2003, 259–60]

Haeckel calls this interaction of induction and deduction *philosophical empiricism*. It is an interplay of experience and reason, a synthesis of empiricism and rationalism. But as many scientists at that time also observed, there is no good observation without some prior theory. As we have seen, Copernicus and Darwin operated against the background of rival accounts. We should add an important further requirement in this interaction. Any good theory, which gains accreditation either by the confirmation of novel prediction or by successful accommodation, must discredit some rival theory. The true Baconian or Millian methods are methods of elimination. As Mill observed:

> Most people hold their conclusions with a degree of assurance proportioned to the mere *mass* of the experience on which they appear to rest; not considering that by the addition of instances to instances, all of the same kind (…) nothing whatever is added to the evidence of the conclusion. A single instance eliminating some antecedent, which existed in all the other cases, is of more value than the greatest multitude of instances. [Mill 1843, Book III, Ch. 10.2]

For instance, 20 people enjoy a meal at a restaurant but eight of them come down with food poisoning. On investigation we find that these eight had the chicken dish, while the remaining 12 had various other dishes – all 20 having eaten different desserts. We are entitled to conclude that, probably, the chicken dish is responsible. We eliminate the other dishes, because none of the 12 people fell ill. This is Mill's **method of agreement**. The unfortunate victims all had one thing in common: they ate the chicken dish. Mill proposes a number of methods of elimination, including the **method of difference**. The true "Baconian" principles, as Haeckel, Huxley, and Mill recognized, work by elimination. Darwin's critics were mistaken. Let us investigate the power of elimination.

Francis Bacon (1561–1626)

6.3   Some principles of elimination

Did Darwin heap fact upon fact, or was it fact upon theory? [Ghiselin, The Triumph of the Darwinian Method *(1969), 232]*

In his *Autobiography* Darwin claimed that he had carried out his work on true inductive principles. "My mind seems to have become a kind of machine for grinding general laws out of large collections of facts (…)." [Darwin 1958, 54; Huxley 1907, Ch. X, 284] In reality he had followed true "Baconian" principles. Neither for Bacon nor for Mill did induction mean simple induction by enumeration. Induction is a two-edged affair. While it lends credit to some model, it simultaneously discredits a rival model. This is called induction by elimination. [Norton 1994, 1995; Weinert 2000] Some phenomenon P is at hand. Often in the history of science competing models of explanation offer their services. Which model is correct? Let P be the occurrence of a vicious crime, like in Lee Harper's novel *To Kill a Mockingbird.* Before detectives gather evidence about P, any human being alive on Earth could logically have been the perpetrator. But soon the evidence begins to eliminate vast numbers of potential suspects. P happened at a particular location, L_1, and at a particular time, T_1. The elementary facts exclude the vast majority of potential perpetrators. At T_1 most of them were at L_2. The evidence gets more sophisticated. Before long, only a small list of names remains. Let us say that two suspects, S_1 and S_2, remain on the list. Detectives will want to make the evidence so precise that it definitely points to either S_1 or S_2. Let us say that DNA evidence links suspect S_1 to the crime. This makes S_1 the prime suspect and reduces the probability that S_2 committed the crime. The scientist's procedure is not much different from the detective's work. The scientist is like a detective who will want to link the available evidence to preferably just one of a number of explanatory models. The weight of the evidence must decide on the probability of the explanation. We can infer the credibility of an explanatory model from how strongly the evidence points to it.

Where does the evidence come from? It may come from observations, as in the case of Copernicus and Darwin. It may come from laboratory experiments. Physical principles may also provide evidence. Physics tells us that there can be no *perpetuum mobile* and that no material process can travel faster than light. Where do these models come from? They may have been arrived at through simple inductive steps, through hypothesis or conjecture. Once the models are available, we require evidence that has the ability to weigh more in favor of one model than its rivals. Bacon was fully aware of such procedures of elimination. Having dismissed enumerative induction as "childish," he proceeds to specify a more positive account:

> But the induction, which is to be available for the discovery and demonstration of sciences and arts, must analyse nature by proper rejections and exclusions; and then, after a sufficient number of negatives, come to a conclusion on the affirmative instances … [Bacon 1620, Book I, §105]

Mill reminds us that we must repeat our experiments and observations to exclude errors of observation or measurement. Our experimental and observational results must be reliable. But once this is done,

> the multiplication of instances, which do not exclude any more circumstances, is entirely useless ... [Mill 1843, Book III, Ch. 10.2]

But mere rejection and elimination will not do. We are interested in some working model, which can adequately cope with the evidence. If one model of explanation gets discredited, the same evidence must credit its rival model. What does it mean to say that a model can "cope" with the evidence?

We must discuss four essential features of eliminative induction: (1) the distinction between positive and supportive evidence; (2) the exploration of the space of possibilities; (3) the distinction between alternative and rival models; and (4) the accommodation of known facts as against the prediction of novel facts.

→ 6.4   Essential features of eliminative induction

1.   Modern discussions often draw the distinction between **positive** and **supportive** evidence or instances. *Positive* evidence is a deductive consequence of some competing explanatory models. Positive instances follow deductively from the principles of a theory. *Supportive* evidence, by contrast, comes ideally in the form of novel predictions, which are compatible with only one of the competing models. Supportive instances constitute tests for competing theories. It may also take the form of successful accommodation of known facts into the framework of one model, while it resists such integration into the other. Bacon appealed to crucial experiments, which stand at the crossroads between two opposing explanations. [Bacon 1620, Book II, §36] They have the power to sway the argument in favor of one account at the expense of the other. Bacon considers the Copernican hypothesis – whether the rotation of the Earth is real or apparent. He suggests various testable empirical consequences, which could decide between, say, geocentrism and heliocentrism. Consider some examples:

Positive instances of both geocentrism and heliocentrism are the regular motion of planets and the behavior of objects on Earth. Positive instances of both natural theology and Darwinism are the diversity of species and certain fossil records. Positive instances are compatible with the logical framework of competing models. They do not distribute probability values unevenly over these models. This creates a serious explanatory difficulty. If, despite the available evidence, there are still rival models which claim to explain it, we have no explanation at all. If all terrestrial phenomena can be explained on geocentrism or heliocentrism alike, if all biological phenomena can be explained equally on the theory of special creation and the principle of natural selection, we really have no explanation to hand. For the solar system cannot be both geocentric and heliocentric; and species cannot be both created and evolved. We know for certain that the solar system is heliocentric. This makes geocentrism incompatible with the facts. Most biologists are fairly certain that species have evolved. This makes creationism implausible. Just like the detective, the scientist needs evidence which points in the direction of one model and away from its rivals.

Supportive evidence plays the dual role of confirmation and disconfirmation. As Bacon already emphasized, crucial experiments can increase the credibility of one model at the expense of its rival model. The observation of new facts can play a similar role. In the history of heliocentrism such supportive evidence weighed heavily in favor of the Copernican hypothesis: Brahe's observations of comets and the supernovae, Galileo's observation of the Jupiter moons, and the joint prediction by Adams and Leverrier of the existence of Neptune (1846) were compatible with Copernicanism but not with geocentrism. Supportive evidence also favored Darwin's theory of natural selection against rival explanations: the discovery of extinct human and animal species in the fossil record, anatomic homologies between anthropoids and hominids, similarities in embryonic development, the diversity of species in their biogeographic distribution.

It is important to realize that the surviving model is the most adequate model available in relation to the available evidence. It need not be the true model. Whatever truth means, it is very difficult for scientific theories to avoid abstractions and idealizations. It is not possible to arrive at complicated scientific theories via simple enumerative induction. All theories are conjectural. From a logical point of view, an infinite number of theories may exist. We must always be aware of a space of possibilities – that is, the space of possible and actual models, which may equally well account for the evidence. In the history of science, on the other hand, only a few competing models are available at any one time. Heliocentrism faced geocentrism and some variants. Darwinism faced design models, and some alternative evolutionary models. How can we explain the paucity of models in practice?

2.   Let us envisage a **space of possibilities**. Such a space of possibilities constitutes a logical or constraint space, for it harbors both actualized and non-actualized models. Think of a constraint space as an abstract grid, with cells in rows and columns. The color charts from a paints manufacturer can be thought of as logical spaces. There are, for instance, many cells of yellow hue. These may not be all the hues of yellow that can be manufactured. Indeed with each season new hues are offered. The yellow chart has room for many shades of yellow but it excludes all shades of other colors. They belong to a different logical space. Certain elements are allowed to enter a constraint space but by no means all. A large board with triangular and rectangular holes will let pass triangles and rectangles smaller than the respective openings. It will bar all other shapes above a certain size. We can think of *constraints* as delineating these logical spaces. Think of constraints again as restrictive conditions of an empirical and theoretical kind. They act as "goal-keepers." Constraints allow certain parameters to enter the domain while excluding others. A simple selection procedure on a fairground illustrates the power of

constraints. If you can pass below a certain bar of height, h, you are too small to go on the fun ride. This is an empirical constraint. There are also theoretical constraints. If you cannot solve differential equations, your career as a physicist will be short-lived. Think of constraints as throwing boundaries around constraint spaces. [See Chapter I, Section 6.4.2]

3.   We can use constraints to make a needed distinction in the constraint space of models. So far we have freely spoken of competing models, without specifying whether they were **alternatives** or **rivals**. In a conceptual domain some models will be mere alternatives, others will be genuine rivals. *Alternative models* are competitors but logically compatible. Alternative models are compatible because they share similar constraints. *Rival accounts* mostly satisfy divergent constraints and share only certain overlapping presuppositions. For instance, all heliocentric models are rivals to geocentric models. All non-teleological models of evolution are rivals to teleological accounts of evolution. They belong to different constraint spaces. The constraints, which define these spaces, erect different criteria of membership. Thinking of the situation in terms of rivals has an immediate advantage. A certain set of constraints not only eliminates one explanatory account but a whole class of rival accounts in one stroke. For instance, the confirmation of heliocentrism dispenses with all geocentric accounts. The *algebraic–topologic* structure of heliocentrism is incompatible with that of geocentrism. In other words, if the planetary system is sun-centered, governed by Newton's laws, then no geocentric account can satisfy the constraints, which are imposed by a sun-centered planetary system. Darwinian accounts of evolution dispense with all teleological theories, whether they come in the shape of design arguments or progressive evolution. The *algebraic–topologic* structure of Darwinian histories is incompatible with that of teleological accounts. The topologic structure resides in the model of branching evolution, their algebraic structure in the mechanism of natural selection. That is, if evolution is driven by natural selection, this cannot be squared with any idea of design or progressive evolution. Rivals are competitors but logically incompatible. Rival models are incompatible because they fall into distinct constraint spaces.

   We said: depending on which constraints are operative, they form distinct constraint spaces. We can use Copernicanism and Darwinism to provide illustrations of constraint spaces and of rival and alternative models. Lamarckism and Darwinism agree on the non-fixity of species but disagree on the teleological nature of evolution and the mechanism of selection. Kepler and Copernicus agree on the central position of the sun but disagree about the particular shape of planetary orbits. Stripped of its metaphysical trappings, the circle can be regarded as a mathematical idealization of the ellipsis. Copernican models are *alternative* models. Mature Copernican models are radically incompatible with the Ptolemaic model, since they do not share the same algebraic–topologic structure, and therefore not the same constraints. They are rivals, not alternatives. Lamarck's evolutionary model is an *alternative* competitor to Darwin's evolutionary model, since they both agree on the variability of species. But evolutionary models are radically different from teleological design models. They are rivals, not alternatives.[15]

A useful criterion to decide whether two models are alternatives or rivals resides in the degree of permissible change. Both the Lamarckian system and the Copernican models could be modified. What was modified in both cases was the algebraic structure: from soft inheritance to natural selection, from circular to elliptical orbits. In the Copernican case this change did not affect the topologic, sun-centered structure. But it meant transforming the circle from a metaphysical to a mathematical device. In the Lamarckian case, the change in algebraic structure led to a change in topologic structure, from linear to branching evolution. But the design and geocentric models cannot be modified in this way: both their topologic and algebraic structures militate against the adoption of natural selection and elliptical orbits respectively, despite the evidence.

When we think about the space of possibility, logic compels us to regard it as inexhaustible. We can always make changes to existing models to generate new potential models. In practice, science always works with a limited pool of available models. When we engage in eliminative induction, we use empirical data, physico-mathematical results, metaphysical presuppositions, and methodological beliefs as constraints to guide our choice. We have seen how the inference to design or evolution models faces us with a philosophical choice. It is to put our trust in either evolutionary theory or design models. To trust in evolutionary theory means to adopt naturalism. This expression of trust includes the hope that outstanding problems will be solved in the future. To trust in design models means to declare natural selection incapable of accounting for complexity and improbable structures. The hope is that the inference from order to design will become scientifically respectable. As we exclude one model in favor of its rival model, let us say the *Great Chain of Being* in favor of descent with modification, we exclude a whole class of rival models, say all those that postulate the fixity of species. We exclude a class, not just individual rival models, because all models in this class (potential and actual) fall foul of the constraints. This leaves us with a class of alternative models. But they are not particularly threatening because they share a large number of constraints. Often they can be sorted on a scale of plausibility with the discovery of new evidence.

4.   One vexing problem is that of **accommodation**. Philosophers are often agreed that novel predictions constitute the purest kinds of tests, to which explanatory models can be subjected. Accommodation is often regarded with suspicion. It is

---

[15] See Kepler [1618–21], BK. V, 2nd book, Part I, 127; Rosen [1959], 49 writes that elliptical orbits can be approximated by circles because of their small eccentricity. Darwin conceded the similarities of some approaches when he wrote: "Whether the naturalist believes in the views given by Lamarck, by Geoffroy St. Hilaire, by the author of the 'Vestiges,' by Mr. Wallace or by myself, signifies extremely little in comparison with the admission that species have descended from other species, and have not been created immutable ..." [Quoted in Greene 1980, 315]

feared that the models will be constructed with the known facts in mind, exposing them to the weakness of *adhocness*. [Psillos 1999, Ch. 5; Howson/Urbach 1993, 147–60] For instance, when Darwin was confronted with the missing link argument, he responded with the claim that the fossil record was very poor. In Darwin's time this was an ad hoc defense, which protected Darwin's model against a serious objection. Luckily for Darwin the fossil record improved, removing the sting of the missing-link argument. Still, accommodating models to existing, known evidence always presents particular problems, as our analysis of Freudianism will soon illustrate. Nevertheless, there are famous examples of accommodation in the history of science. The perihelion advance of Mercury was known to classical physics but could not be explained on the Newtonian model. The perihelion is the closest point of a planet's orbit to the sun. In the case of Mercury, this point advances over a period of time. Instead of the perihelion always returning to the same point after an annual orbit, Mercury's orbit forms a rosette pattern. Einstein was able to accommodate the problem in his general theory of relativity (1916). When Niels Bohr constructed his famous atom model of the hydrogen (1913), he built it to accommodate various atomic phenomena known at the time: the scattering experiments, the discreteness of atomic spectra. Copernicus did not make a single novel prediction, nor did Darwin. Their work consisted in accommodating the amount of observational data. They provided a coherent framework, within which the data would make sense. Such cases of accommodation are unproblematic for two reasons. (1) In all these cases the new models were proposed on the background of rival accounts. If a model can account for a known fact, which threatens the coherence of a rival model, accommodation poses no serious danger. We have encountered several examples. Copernicanism naturally accounts for retrograde motion, which forces geocentrism to adopt an unphysical solution. And Darwinism elegantly accounts for the diversity of species, which forces design arguments to adopt a multiplicity of creative acts. (2) These theories had testable consequences, even if they fell short of novel predictions. Once a theory offers a coherent framework to known facts, deductive and inductive consequences follow. They do not have the impact of novel predictions but may nevertheless strengthen the empirical support of the theory. Copernicanism can explain the seasons, as a deductive consequence of its structure. And the antiquity of man naturally follows from descent with modification.

We have distinguished four essential features of eliminative induction. Both the Copernican and the Darwinian models fare well on this account. The Copernican account can, among other things, provide an accurate estimate of the relative order of the planets from the sun. The Darwinian account can, among other things, explain the adaptation of species to their local environments. These deductive and inductive consequences enhance unification. They should count as supportive evidence. Both Copernicus and Darwin explored their rival models. But they saw the space of possibilities as exhausted by their respective rivals. Copernicus sought to discredit geocentrism; Darwin was mostly preoccupied with exploring the weakness of the theory of separate creations. Looking at the algebraic–topologic features of heliocentrism and geocentrism, it is fairly easy to distinguish rival models from mere alternatives. Looking at the algebraic process of natural selection in Darwinism, as against divine selection in natural theology, it is again easy to distinguish the rivals from the alternatives. We proposed, as a criterion, the degree of acceptable change in a model in the light of evidence. The Copernicans and Darwinians were much more in the business of accommodating known facts than making novel predictions.

6.5 Falsifiability or testability?

Popper fully accepted Hume's criticism of simple induction, induction by enumeration. There is no logical proof that we are justified to infer from the observation of some "p" to the universal statement that "all p." Nor is there an empirical proof. We cannot infer from the success of our inductive practices in the past to their success in the future because this presupposes the very principle of induction, which we aim to prove. Popper concluded from this state of affairs that it was impossible to establish the truth of a universal scientific theory on the basis of a limited number of observations. Universal theories in science cannot be established by inductive reasoning. Popper stood this situation on its head. We cannot verify the truth of a universal theory by observation or experiment. But we can show at least that a theory is false. If a universal theory is cast in such a form that it may contradict some observation statements, from which they are derived, then we have reason to believe that the theory is false. This is the asymmetry of verification and falsification. A theory in science cannot be verified but it can be falsified. It is falsified when one of its deductive consequences contradicts our empirical discoveries about the world. [Popper 1959; 1963; 1973] Kepler's first law of elliptical planetary orbits is a discovery that falsifies the presupposition of circular orbits in geocentrism and Copernican heliocentrism. At the same time it constitutes a rejection of geocentrism and a modification of heliocentrism because the elliptical orbits take the sun as a focal point. And the paleontological discoveries of the nineteenth century contradict the fixity of species in non-evolutionary accounts. At the same time they constitute a rejection of the *Great Chain of Being* and a modification of earlier evolutionary accounts insofar as they were still committed to teleology.

Popper claimed that he had disposed of inductive practices in science for good. Popper's towering influence over twentieth-century philosophy of science has created the impression that inductive practices are unimportant events in the history of science, while deductive practices are the norm. Popper was much impressed with Einstein's theory of relativity. Here was a good conjecture, freely invented, which resulted in three precise testable predictions (the perihelion advance of Mercury, the redshift of light in gravitational fields, and the bending of light near gravitational bodies). But our study of Copernicanism and Darwinism has convinced us that inferential practices are widespread in science. They take the form of eliminative rather than enumerative induction. The physicist Steven Weinberg has reacted to the dominance of falsificationism in his own way: he finds no single instance of falsificationism "in the past one hundred years." [Weinberg 1993, 102] In fairness to Popper we should recall that he requires the refutability of theories rather than their actual refutation.

Should we read Popper's criterion as falsifiability or testability? If we take Popper too literally – insisting that theories are scientific only if they are falsifiable – we are led to inconsistent conclusions. Perfectly respectable scientific statements about the impossibility of perpetual motion machines cannot be falsified. The cosmo-logical search for the existence of dark matter cannot be falsified. The criterion of falsifiability excludes the search for the positive existence of some entity or prop-erty. It also renders theories unscientific, which the history of science regards as scientific. Ptolemaic astronomy wished to save the appearances by ingenious geo-metric constructions. There was no commitment to the reality of the geometric devices, no novel predictions. Ptolemaic astronomy fails as a scientific theory in Popper's sense. Still, it ranks as a serious contender among Greek models of astron-omy. It predicted and ordered the known phenomena; it only became testable in the long run. And what do we think of theories, like astrology and Freudian psycho-analysis, which could have falsifiable consequences? If astrology and Freud's theory are falsifiable in principle, by making predictions that turn out to be wrong, should we regard them as scientific? And what do we make of paradigmatic theories, like Newton's mechanics, which face anomalies, like the perihelion advance of Mercury? The perihelion advance of Mercury fell into the domain of Newton's theory but could not be explained. But Newton's theory was not rejected because of this failure.

Popper imposes some stringent constraints on new theories, T_2, which replace older theories, T_1. The new theory, T_2, must make more precise assertions than T_1; it must explain more facts than T_1 and it must explain them in more detail; T_2 must pass tests which T_1 has failed, and it must pass new tests; T_2 must also unify hitherto unrelated phenomena. [Popper 1963, Ch. 10] In particular, a theory must define the parameters between which it establishes relations. Both the parameters and the relations must be quantifiable. As we have seen, such quantifiable relations between parameters often form the basis of the laws of science. Astrology stumbles over such a requirement. The positions of the stars in the sky can be stated with numer-ical precision but the same is not true of character traits. This drawback makes it difficult to quantify any supposed correlation between character traits and the position of the stars.

We can summarize these criteria under the term *testability*. We can cast testability as a method of elimination or a *contrastive* activity. [See Sober 1999, §1] After all, Popper's criteria are based on a contrast between T_1 and T_2. The easiest situation occurs when T_1 makes prediction P_1 and T_2 makes a different prediction P_2. If, on the Popperian scheme, P_2 does not occur, T_2 is regarded as falsified and T_1 as cor-roborated, if P_1 is observed. What happens, however, if T_1 and T_2 make the *same* prediction, say, regarding the motion of Mars, and T_1 and T_2 are based on different principles, respectively? According to eliminative inductivism, these confirmations do count as positive but not as supportive evidence; we need to wait for an instance of supportive evidence that differentiates between the two theories. Popper's evolutionary model of the rational growth of scientific knowledge populates the space of possibilities with a number of alternative and rival accounts: a number of tentative solutions to a particular problem are allowed to compete, until error elimination weeds out some tentative solutions, leaving a number of more sophisticated tentative solutions now tackling a more refined problem. In his criticism of falsificationism Kuhn, too, stresses the need for a number of rivals, among which a choice must be made according to certain criteria. [Hoyningen-Huene 1993, §7.4] The method of eliminative induction tells us which models we should regard as the best candidates in terms of explanatory power and in the light of the available evidence. We expect the surviving model not only to describe but also to explain the phenomena. It should allow us to derive new phenomena or accommodate known facts. Once the method of eliminative induction has selected a model as the best candidate to cope with the available evidence, the question arises: How does it explain the evidence? What is scientific explanation? We can best answer this question by studying some models of explanation.

6.6   Explanation and prediction

A phænomenon is explained when it is shown to be a case of some general law of Nature (...). [Huxley, "Origin of Species" (1860), 57]

Are you a patient observer? Imagine you live by the sea and become inquisitive about the pattern of high and low tides. You decide to construct tables, in which you enter the times of the receding and swelling waters. After a while you will no doubt discern the pattern and make predictions about the next high tide. Your predictions may not be accurate to the minute and the second. But you will be able to predict the tides with sufficient accuracy not to be out by hours. Emboldened by your success, you turn your attention to the moon. Soon you observe the phases of the moon, making sketches as best as you can. You notice that it takes approximately 28 days between two full moons. Before long, you will be able to predict the phases of the moon to within acceptable margins of error.

You can predict the tidal and lunar phases. Does this mean that you can explain them? Most certainly not! To be able to predict is not to be able to explain. To be able to explain is not necessarily to be able to predict.

Your tabular results of the phases of the moon and the tides permitted you to make rather accurate predictions. You may have no idea how these natural phenomena come about. You are in good company. The ancient Greeks predicted the motions of the planets with an accuracy of within 5 percent of modern values. The Greeks had an "explanatory" model, on which their predictions were based. But their model – geocentrism – was mistaken. The physical world is not geocentric. The Greeks were able to make good predictions but they had a completely false explanation. With heliocentrism prediction and explanation moved closer to each other. Copernican heliocentrism is a good approximation of the topological structure of the planetary system. But the algebraic structure is mistaken. Copernicus still worked within the Greek paradigm of perfect circular motion. Still, the Copernican model satisfied some of the constraints of the logical space it inhabited. Within this constraint space, it experienced a number of important improvements, associated with the names of Kepler, Galileo, and Newton. Through these giant steps the explanatory structure of heliocentrism improved markedly. The Newtonian model of the solar system displays an algebraic and topologic structure which fits the structure of the physical system rather well. Therefore the Newtonian model provides us with a good explanation of the planetary system. This improved the accuracy of predictions of this sophisticated Copernican model. This success culminated in the prediction of the existence of a new planet, Neptune. John C. Adams and J. J. Leverrier (1811–77) calculated from Newtonian mechanics that observable perturbations in the orbit of Uranus

Portrait of John C. Adams
(1819–32)

must be due to the existence of another planet. Johann Galle (1812–1910) in Berlin went to look for the new object and found Neptune in 1846. But the perihelion advance of Mercury eluded a Newtonian explanation.

One is tempted to think that the explanatory adequacy of a theory will give rise to predictability. But the example of astronomy can lead us astray. When the world is very complex the availability of good explanatory principles does not guarantee the derivation of good predictions. If you are a good observer of lunar and tidal phases, perhaps you are a good observer of human moods. A colleague whose friendliness is as blameless as his politeness ignores you as you cross his path on campus. You later learn from a friend that your colleague suffered a bout of bad news on that particular day. You construct an explanation: he was in such a foul mood that he did not wish to look at anyone, let alone speak to anyone. Understandable! You attribute your colleague's unusual unfriendliness to his considerable irritation on that day. You can sympathize. And you can explain. Does this mean that you can predict that your colleague will be unfriendly once more, if bad news arrives again? No way! Humans are too complex to make such easy inferences. Next time your colleague, eager to forget the bad news, may be particularly friendly when your paths cross again. So it is possible to explain without being able to predict.

In a certain way Darwinian evolutionary theory is in this situation. Evolution in the Darwinian sense is a statistical theory. The theory of natural selection speaks of the *tendency* of favorable characteristics to be preserved and unfavorable characteristics to be eliminated. The success of a species – a population of individuals – does not just depend on its favorable characteristics. What is favorable is relative to a particular type of environment. The fitness of a species is a function of the environment in which it tries to survive and reproduce. It is therefore impossible to predict with astronomical accuracy the trajectory of a species through the phase space of life. But as evolutionary theory is a statistical theory it is able to make statistical predictions. Statistical predictions refer to the future behavior of collections of individuals rather than to the individual members of the collection. In a certain sense it is correct to say that evolutionary theory can explain but not predict. But then we apply a notion of strict predictability. It derives from astronomy, where the orbit of individual objects can be predicted with great accuracy. Astronomy can predict into the future and retrodict into the past. But we should not forget that statistical predictions are a valuable source of information. Economists make predictions about the economic performance of whole countries over limited periods. Sociologists make predictions about the effect of social conditions on health standards or educational performances of certain age groups. Criminologists predict rises in criminal offenses at least over a limited period of time.

As population thinking is the predominant paradigm in evolutionary biology today, it is only to be expected that it will give rise to statistical predictions. Consider the Hardy–Weinberg ratio: $p^2 : 2pq : q^2$, where p and q represent two alleles of a gene in a population. This proportion will be maintained in the population from generation to generation, if no interfering factors occur. The loss of existing genes or the arrival of new genes may constitute such interferences. If this happens the proportion will be disturbed and evolution may occur. [See Mayr 2001, 96–7; Fisher 1930, 22; Williams 1973; Rice/Hostert 1993] Consider, for instance, the predictable effect of the introduction of gray squirrels on a population of red squirrels. The effect of natural selection will be such that the red squirrel population is likely to decline. Similar changes in population numbers will occur if there are artificial or natural interferences with the food chain of a species. Other predictions concern the speciation process. Why do populations diverge and develop along different lineages? One influential model predicts that geographic isolation of a population into two subgroups can lead to the splitting of lineages. A predictive implication of allopatric speciation is that some transitional forms will be found. The existence of transitional forms follows from the logic of Darwin's argument. As we mentioned earlier, researchers in the 1830s and 1840s found hominoid remains, which differed slightly from modern humans. Such transitional, homologous forms lend support to the hypothesis of descent with modification, understood here as allopatric speciation. The discovery of the fossil remains of primitive birds like *Archaeopteryx* and *Hesperornis Regalis* revealed more transitional forms. They resembled reptiles in one important respect – they possessed strong teeth. [Huxley 1876, 94–113; Mayr 2001, 14, 67] [Figures 2.11a, b]

Our discussion leads to the conclusion that there exists a certain asymmetry between prediction and explanation. Ideally our explanatory principles allow us to make precise predictions. But there are many areas of human knowledge where this ideal case does not obtain. We cannot always predict even though we can explain. And we cannot always explain even though we can predict.

→ 6.7 Some models of scientific explanation

Some of the most influential models of scientific explanation are reflected in the Copernican, Darwinian, and Freudian traditions.

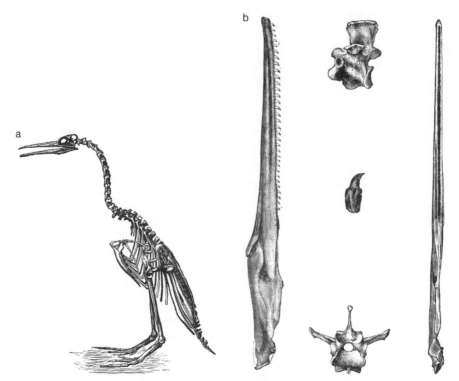

Figure 2.11   (a) *Hesperornis Regalis* measured five and six feet in length and resembled reptiles in sporting teeth. *Source*: T. H. Huxley, *Collected Essays*, Vol. IV [1898], 96; (b) *Hesperornis Regalis*, showing side and upper views of half the lower jaw. Also views of a vertebra and a separate tooth. *Source*: T. H. Huxley, *Collected Essays*, Vol. IV [1898], 97

→ *6.7.1   Hempel's models*   What is a scientific explanation? Carl Hempel is famous for his two models of explanation. He proposed the deductive-nomological model (*DN* model) and the inductive-statistical model (*IS* model). [Hempel 1965] According to Hempel the explanation of a phenomenon, *P*, is its subsumption under the regularity, *R*. More precisely, Hempel's general idea is that we provide an explanation for some event *E*, if we can show it to follow either deductively (*DN*) or with high probability (*IS*) from general laws and initial conditions. Any explanation scheme has two components. There is the phenomenon to be explained (the *explanandum*). Then there are the principles, which provide the explanation (the *explanans*). The explanans itself consists of two parts: general laws and initial conditions. The general laws are expressed in general symbolic forms (e.g., $F = \mathbf{ma}$). To derive the explanandum we need to insert particular values into the general form. The particular values come from the initial conditions (for instance, a value for mass and acceleration). The explanandum then follows either as a logical consequence from the explanans (*DN* model) or with high probability (*IS* model). Let *P* be the period of Mercury around the sun. *P* is the explanandum. According to Hempel's *DN* model we explain *P* by showing that it falls under the explanans. In this instance, the explanans consists of Kepler's third law, $A^3 \propto P^2$ (*A* is the average distance of a planet from the sun, *P* is its period, which is the explanandum) and the particular initial conditions (the average distance of Mercury from the sun). From the explanans, the value of *P* follows. Let *R* be the recovery of a particular person from an infection after treatment with penicillin. We can explain this recovery (explanandum) by considering (a) that 95 percent of patients with infections respond well to penicillin; (b) our particular patient suffers from a penicillin-treatable infection {(a) and (b) make up the explanans}. On the *IS* model, as stated in the scheme, the patient was likely to recover. This is a statistical explanation of why the patient got better. It is helpful if we think of both the physical parameters and the general laws, which enter the explanatory schemes, as quantifiable. Laws are (strict or statistical) structural regularities about the natural world. Laws of science encode structural information about the natural world. Hempel became persuaded that, under certain conditions, explanation and prediction are symmetric. Once general laws are available, the symmetry thesis between explanation and prediction becomes tempting. We *explain* an existent phenomenon by showing that its present occurrence falls under the law. We *predict* a not-yet-existent phenomenon by deriving its future occurrence from the law. We *retrodict* an already-gone phenomenon by deriving its *past* occurrence from the law.

Hempel's models have been criticized on a number of particular issues, like the symmetry between explanation and prediction [Achinstein 1985, 169–81], the assumption that high probability is both necessary and sufficient for statistical explanation, and the general idea that subsumption of a phenomenon under a law explains the phenomenon. [See Kitcher/Salmon 1989]

Scientific explanations can be given of *particular* events, like the appearance of Halley's Comet in 1759 or the mutation of a particular gene. More often science attempts to explain *general* features of the natural world, like the nearly elliptical orbits of planets or the "origin of species." The more general explanations indicate that science is not really concerned with the deduction or inference of particular events from general laws, as the Hempel model seems to suggest. As we can see from Copernicanism and Darwinism, the task of science is to identify the most general structures of the physical world: the structure of planetary systems, the structure of the organic world. Laws play an important part in the identification of such structures. Physical systems are manifestations of structures. They consist of components, like the planets of the solar system, and their relations. The way the components are related is captured in the laws of science, like Kepler's planetary laws. Darwin showed an appreciation of the importance of the structure of natural systems when he declared that "the structure of every organic being is related (...) to that of all other organic beings." [Darwin 1859, 127; Weinert 2004, Ch. 2.5.1]

As explanation in the natural and social sciences involves different types of phenomena, they may require different types of explanation. Not only are there different models of explanation; the same phenomenon can be explained from different explanatory perspectives.

→ *6.7.2 Functional models* Organisms are so beautifully adapted to their local environments that deliberate design seemed to be the only possible explanation.

Teleological models, whether based on design or progressive evolution, assume that final purposes operate in nature. The function precedes the structure. The structure of an organ derives from the function it must satisfy. The Artificer desired blood to circulate in creatures. The function of the heart is to circulate blood. Therefore the Artificer created the heart. The Almighty Creator desired creatures to see. Therefore He created eyes. Lamarck is not committed to creative design. Still the use of the organ determines the structure. [Lamarck 1809, 113; Gould 2002, 177] One difficulty with this teleological reasoning is that eyes are analogous structures, which developed independently 40 times in the history of life. This makes it less likely that function comes first. The evolutionists reverse this chain of reasoning. The function is the effect, not the cause, of an organ. The organ and its structure develop before the function. All differences of function, says Huxley [1863a, 115–23], stem from a difference of structure. Then the organ may adapt to new situations. It may even develop functions for which it was not originally selected. This is the Darwinian explanation of the emergence of conscious minds. Darwin's evolutionary theory explains the adaptedness and diversity of organisms through the theory of descent with modification. Darwinism is therefore often seen as a functionalist theory, "leading to local adaptation as the environment proposes and natural selection disposes." [Gould 2002, 31; Dennett 1995, 228; Ruse 2003, 264–70] The fitness of organisms is a function of their morphological characteristics and the environmental conditions, not of preordained design.

With Darwinian inferences it becomes possible to descend from the dizzy heights of pre-given functions to the mundane level of evolved functions. In the teleological picture a preexisting function needs an organ. In the evolutionary picture the organ acquires a function. For instance, the ability to walk upright was a selective advantage for early hominids who emerged from the forests to conquer the savannah. The selection of favorable characteristics is a response of the organism to environmental conditions. Functions are responses to selective pressures. Function can be explained causally, not teleologically. There is a set of causal factors, including the differential fitness of organisms and the structure of the environment, which make the effect probable: the ability to fly (the function of wings), to see (the function of eyes), to breathe (the function of lungs), to think (the function of the human mind). In order to avoid explanations in terms of final purposes, the biologist turns to Darwinian inferences. The Darwinian inferences expose a set of causal conditions, which are likely to have produced the observed effect. The Darwinian evolutionist does not claim that a given set of conditions will determine a resultant effect. Evolutionary theory is not Newtonian astronomy. The future development of a species cannot be determined with astronomical precision from past and present conditions of its lineage. Darwinian inferences revisit the past. The diversity of life, the adaptedness of organisms to their respective environments can be adequately explained by relating these observable effects to some prior causal conditions. This prior set is unlikely to include a traceable causal link. The prior causal conditions are to be located on the trajectory of the lineage to which the individual species belong. Functions are not mysterious. They receive a thoroughly naturalistic, causal explanation. [Ayala 1995] In fact, they can be reduced to causal explanations.

→ *6.7.3   Causal models*   On entering the house, you switch on the light. Flipping the switch, so it seems, causes the light bulb to glow. You cannot see how the electric current makes the bulb glow. But it is always possible to construct a circuit in a laboratory, which demonstrates how the light switch closes the electric circuit, allows the electrons to flow around it and to meet a resistor in a filament, which begins to glow. The good thing about a laboratory circuit is that the experimenter can control all the parameters that enter the physical situation. This control makes it very unlikely, although not logically impossible, that any other parameters are responsible for the observable phenomenon.

Causality is an important issue in human affairs because of its practical implications. Causality is the glue that holds events together. It matters in natural and social affairs whether we understand why some event happened. It allows us to tackle the medical and social ills that beset us. It helps us control and manipulate the environment. Understandably, philosophers have been interested in developing some conceptual models, which help to explain the issue. As with models of the mind, we will limit our attention to models of causality, which are of interest in the problem situations encountered in Copernicanism, Darwinism, and later Freudianism. [See Psillos [2002] for a discussion of various philosophical accounts of causation]

Hume's insight was that causation was a matter of the regular succession of events: an effect, which regularly follows a cause. The cause was always prior to the effect and cause and effect were in spatial proximity. Such a characterization is obviously insufficient, for several reasons: (1) it is not the case that every *E*, which regularly follows an event, *C*, can be regarded as the effect of the prior event. The day regularly follows the night, low tide regularly follows high tide, and yet the night is not the cause of the day and high tide is not the cause of low tide; (2) some effects follow their cause only with a certain statistical frequency, which may be quite low. Consider the condition called *ptosis*. The drooping-eyelid condition affects older people but with a low frequency. While it is a statistical fact that ptosis is the effect of old age, it is not the case that old age is the regular cause of ptosis.

Hume's original insight has led to modifications – as in Mackie's *INUS* account (Mackie 1980) – or alternatives – as in Lewis's counterfactual analysis (Lewis 1986). We will first consider a counterfactual approach, which is based not on Lewis's notion of possible worlds but on Woodward's idea of hypothetical interventions. Then we generalize Mackie's *INUS* account to a conditional view of causality, for reasons which have to do with the applicability of causality to the cases at hand.

6.7.3.1   A counterfactual-interventionist account   Woodward [2003] rightly stresses that causal models must be practical in the sense that they must reflect our causal practices. We encounter causation in accidents when the causal situation is beyond our control. But in human attempts to learn from the accidents and prevent them in the future, a central feature of causation emerges. This central feature is that we analyze causal situations in counterfactual terms. Causal explanations answer "what-if-things-had-been-different-questions." [Woodward 2003, 12] According to Woodward, we envisage hypothetical experiments or interventions, by way of which we answer counterfactual questions: (1) "What would have happened to the plane *if* its engines had not caught fire?" (2) "What would have happened to the asteroid *if* its orbit had been more elliptical?" (3) "What would have happened to the red squirrel population in Britain *if* the gray squirrel had not been introduced?" Finally, a hypothetical question, which Voltaire could have asked: (4) "What would have happened to the wild life in Britain *if* the wolf had not been exterminated in that country?" On counterfactual views of causation we will have grasped the causal conditions of an actual situation, if the envisaged change in the selected parameters, due to a hypothetical intervention, shows us *how and by how much* the actual causal situation would have differed from its actual occurrence. With respect to the first two counterfactual questions: if the plane engines had not caught fire and if the asteroid had had a more elliptical orbit, the plane would not have crashed into the mountain and the asteroid would not have crashed into Jupiter. Astronomers can calculate how much the hypothetical orbit of the asteroid would have had to differ from the actual orbit if its collision with Jupiter was to be avoided. The third question can be answered by looking at Continental countries: the red squirrel population in Britain would not have been decimated if the gray squirrel had not been introduced, all other conditions remaining equal. Sometimes biologists can calculate the effect of the introduction of a predator into a new environment, if controlled experiments can be carried out.[16] The evaluation of the fourth counterfactual question involves the whole ecosystem. The multifarious relations between the inhabitants of the ecosystem and the statistical nature of the principle of natural selection make it very difficult to answer this counterfactual question.

   The counterfactual account then amounts to the view that a given condition is the causal factor, C, of a certain effect, E, if a specifiable hypothetical variation in the causal factor shows that without it the effect would not have occurred or would have occurred differently. This variation must be sufficiently large and invariant to have an impact on the counterfactual situation. A minute change, say in environmental conditions, may not threaten the livelihood of a species. Even an unusual climatic event, like the rare eruption of a volcano, may not wipe out a species. On Woodward's account, causation is a matter of counterfactual dependence between antecedent and consequent conditions. Yet the scientist cannot ask "what-if-things-had-been-different-questions" before reasonable answers to actual causal questions are available. As long as no robust regularities are known, answers to counterfactual questions remain mere speculations. The scientist therefore needs to ask what the actual world is like before s/he can extrapolate from what is known about the actual world to a hypothetical world. Actual regularities imply counterfactual situations.

   How, for instance, do we evaluate counterfactual questions in the social sciences? Could Freud have answered the question "What would have happened to my

---

[16] For instance, the coloration of guppies, a fresh water fish, is correlated to the number of predators in their environment: basically, the more numerous and aggressive predators are, the less colorful ("drabber") male guppies will become. This pressure, however, is counterbalanced by sexual selection: female guppies seem to prefer colorful males.

patient if s/he had not experienced this particular neurosis-provoking event in her/his childhood?" Freud once said:

> We find the most different reactions in different individuals, and in the same individual the contrary attitudes exist side by side. [Freud 1931, 233]

This admission did not prevent Freud from claiming that psychoanalysis was a science, capable of general statements about human nature. If that is the case, how, say, will historians evaluate counterfactual questions about historical situations? "Would World War II have occurred if Hitler had been killed in a march on November 9, 1924?" As long as there are no regularities it is difficult to evaluate such questions with some degree of confidence. In the next chapter we will argue that the social sciences can rely on regularity patterns in the social world. But from a philosophical point of view, these patterns have the status of *trends* rather than laws. Nevertheless, as long as there exist relatively reliable patterns of behavior in the social world, at least some qualitative evaluation of counterfactual questions becomes possible. Given patterns of human behavior and evidence about historical events, it will be possible for historians to assess counterfactual situations. But this assessment will differ from the hypothetical situations that Woodward has in mind.

The preoccupation of the natural and social scientist with actual causal situations suggests a different approach to the topic of causality. What we will call a conditional view of causality is a generalization and modification of Mackie's *INUS* account of causation. [Mackie 1980]

6.7.3.2 A conditional model of causation   Let us start with the causal–mechanical model, according to which there must be a traceable link between cause and effect. It is the most satisfying to our sense of explanation. Hume's idea of regular succession between cause and effect is replaced by the idea of a traceable causal link between a cause, like the activation of a light switch, and the consequent effect, the burning of a lamp. It is the most satisfying but also the most restrictive model. [Salmon 1984; 1998; Dowe 2000; cf. Woodward 2003, Ch. 8.1–8.5] Similarly to Hume's account, it imposes three conditions on a causal situation: that the cause, C, be temporally prior to the effect, E; that the cause, C, be in spatial proximity to effect, E; and that there be a traceable mechanism, which links cause, C, and effect, E. Accident investigations show that these conditions are often satisfied. But there are also many real-life situations where a traceable link cannot be established. As you lie on a hot summer beach, do you see how the ultraviolet radiation attacks the cells in your skin? Do biologists see how lineages split? Can neuroscience explain how consciousness emerges from brain processes? Did Newton see how a combination of the First Law and the law of gravitation kept the planets in their orbit? The answer is "no," yet in all these situations we are comfortable to speak of a causal situation. Why? We are able to identify a cluster of actual causal conditions, which can be regarded as jointly sufficient to produce the effect, at least with an acceptable degree of probability. That is, given the effect, the likelihood that these conditions are not causally responsible for the effect is very small. In laboratory situations in the natural sciences, any external influences can often be excluded as practically negligible. The potential effect of an excluded parameter can also be calculated. For instance, if the nucleus of an atom is fired at another atom, the observed "deflexions" will not be due to the presence of electrons, because of the energies involved. In the natural environment, the influence of certain conditions on an observed effect may also be extremely unlikely. It is extremely unlikely that your morning coffee will contribute to your sunburn. This can easily be tested. Drink coffee in the morning and stay out of the sun. You will not get sunburn. By contrast, the exposure to ultraviolet radiation is very likely to be the cause of sunburn.

Any situation, which may be the effect of some prior state of affairs, is embedded in a great number of conditions. Of these, not all are causally relevant. The background conditions allow the normal running of things. Birds fly, plants grow, rivers flow. These processes happen according to regular patterns. We do not normally ask, "Why?" But planes crash, plants wilt, and rivers burst their banks. These events interfere with the normal running of things. We normally ask why the disturbance happened. Out of the normal background conditions, we select a cluster of (necessary and sufficient) causal conditions.

A *conditional* model of causation is a generalization of Mackie's model in terms of necessary and sufficient conditions.[17] [Weinert 2004, §§5.35–5.36; 2007] It has three main features. Firstly, it claims that causal relations can exist between some antecedent and some consequent conditions even where we lack a traceable causal mechanism that would allow us to travel all the way from the cause to the effect and back. We may also lack the advantage of a regular succession, where regular succession means a law of nature. [See Chapter I, Section 6.5.3] As we shall see in Chapter III, causal relations between social events can be established in the social sciences, even though there are no social "laws" and only sometimes traceable mechanisms. Secondly, a conditional model is concerned with the actual causal conditions, which obtain in causal situations, rather than counterfactual situations. As we argued above, counterfactual scenarios are projections from lawful regularities. The natural and social sciences seem mostly to be concerned with *actual* causal conditions. Thirdly, the conditional model sees causal relations as questions about *conditional* dependences between antecedent and consequent conditions.

---

[17] If X is a *necessary* condition for Y, then in the *absence* of X, Y cannot occur. In Mackie's words, "whenever an event of type Y occurs, an event of type X also occurs." [Mackie 1980, 62] Thus the atmosphere around the Earth (X) is a necessary condition for life on Earth (Y). If X did not exist, life on Earth would be impossible; e.g., in the absence of X, Y would not have occurred. And in the absence of natural selection, on Darwin's account, the diversification of species would not have occurred. Recall, however, a difference between Lamarck's and Darwin's view of evolution. On Lamarck's linear view of evolution the emergence of humans is a necessary consequence of the evolutionary process. But on Darwin's branching view of evolution, the emergence of the human species is only a contingent event in the history of life.

If X is a *sufficient* condition for Y, then in the *presence* of X, Y will occur. But if X is not present, Y may still occur through some alternative process. Thus our sun is a sufficient condition for the solar system, which makes up the familiar nine planets. But there are other stars in the Milky Way, which are at the center of different solar systems. According to the evolutionists, certain environmental conditions (X) are propitious to the differential survival of a species (Y). But if the environmental conditions X change, the survival of the species Y may not be threatened if its members can adapt to the new conditions.

In the absence of a traceable mechanism and regular succession, we may still be able to determine conditional dependences by focusing on a cluster of (necessary and sufficient) conditions, which are jointly sufficient to explain the effect. We can call the cause the *antecedent* (or prior) condition and the effect the *consequent* (or posterior) condition. We will consider some examples from the realm of astronomy and biology; Chapter III, Section 4 will analyze examples from the social world.

1. *Planetary motion.* We would be hard pressed to call the Copernican model a causal explanation. Copernicus offers no dynamic theory of planetary orbits. The Copernican model remained a kinematic description until Newton offered a dynamic theory of planetary motion. This can, to a certain extent, be regarded as a causal explanation. In a sense, planetary orbits are explained by taking the vector product of Newton's law of inertia and the law of gravitation. Newton argued that in a world without gravitational forces, planets would move in constant rectilinear motion. The effect of gravitational pulls makes planets fall continuously toward the sun. The combination of rectilinear motion and the fall toward the sun produces the elliptical orbits of planets. But gravitation is not a causal mechanism that would allow us to trace the gravitational effects of the sun on the planet. Newton regarded gravitation as a force but he himself was mystified as to how the sun could exercise an effect on a distant planet. It would have to "reach" over immense distances in empty space. Although gravitation falls short of a causal, traceable mechanism, the combined effect of the law of inertia and the law of gravitation goes some way toward explaining why planets move in elliptical orbits around the sun. [Figure 2.12] Yet this explanation establishes a *conditional dependence* of the consequent conditions

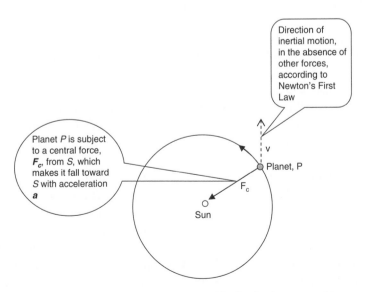

Figure 2.12   A classic case of causal explanation – idealized, planetary orbits

on the antecedent, causal conditions. The causal conditions in this example can be further differentiated into necessary and sufficient conditions. *Sufficient* conditions are gravitational bodies – not necessarily planets, since satellites also fall under the Newtonian explanation. The *necessary* conditions are found in the lawful regularities, which govern the respective phenomena.

2. Can Darwinian histories, which we presented as inferences from the present state of an organism to its trajectory of descent, also be cast in the language of conditional dependence? There may be a problem, which will make its presence felt very strongly in the social sciences. It may not be possible to define a closed set of such prior causal conditions. Darwinian evolution is based on a statistical principle, giving rise to stochastic histories. Necessary conditions for adaptation and diversity are the presence of oxygen and food resources for carbon-based life. Such necessary conditions seem rather trivial, but it is sometimes possible to be more specific. For instance, is geographic isolation a necessary condition for the splitting of lineages, as Darwin thought? If sympatric speciation (by mate preferences) is observed to occur, geographic isolation cannot be a necessary condition for speciation. Is gradualism a necessary condition for evolution to occur? The true significance of gradualism is still controversial today. Although it is part and parcel of orthodox Darwinism, its controversial status in evolutionary theory prevents it from being simply postulated as a necessary condition. Such conditions pose challenges to future research in evolutionary biology. What about the sufficient conditions? The Darwinians explain that in the presence of a number of environmental and organismic conditions, the adaptation of an organism is a likely outcome. Knowledge of such sufficient conditions is of great importance where humans want to interfere in the ecosystem of a species. The decline of a species, like the otter, has been halted because the conditions for their differential reproduction – the modern view of natural selection – are sufficiently known.

In terms of the availability of a cluster of necessary and sufficient conditions, the Darwinian is in a situation similar to that of a sociologist or a historian of political history. As we shall see, the historian can name a cluster of social and political conditions which is *likely* to have caused a war in a country. But it is very difficult for the social scientist to delineate a cluster of necessary and sufficient conditions which can be regarded as jointly necessary and sufficient for the occurrence of some historical or social event. Nevertheless, causal accounts exist in the social sciences, as we shall see. To explain them we require a philosophical model of causation which is adapted to social relations. Max Weber developed an account of what he called "adequate causation." This account is a version of the conditional model, developed in this section. It is adequate for the social sciences because it does without any idea of regular succession or causal links between two social events. It shows that the job of the social scientist is to construct causal models out of the existing social data such that a cluster of prior conditions can objectively be regarded as the most likely conditions to have brought about the effect in the social world.

Like some social science accounts, Darwinian histories are explanatory. They may lack a causal, traceable mechanism in a strict sense. Under laboratory conditions the mutation of a virus, like the AIDS virus, is directly observable. But outside of the laboratory, the work of natural selection can only be observed indirectly, as for instance in the case of industrial melanism (the blackening of moths in heavily polluted areas in nineteenth-century Britain). This lack of "direct" observation is due to the time scales involved in evolutionary changes. Most Darwinian trajectories fail to specify a closed set of necessary and sufficient conditions which jointly determine the occurrence of some evolutionary event. Nevertheless, Darwinian inferences are rightly regarded as explanatory. Which type of explanatory model do Darwinian histories satisfy? Let us say that Darwinians regard them as conditional causal models, since the cluster of necessary and sufficient conditions spell out an adequate explanation of some evolutionary event.

But heliocentric and evolutionary explanations can also be viewed from the point of view of *structural models* of explanation, which will be treated here as a form of unification.

→ *6.7.4 Structural explanations*

In scientific investigation, it is permitted to invent any hypothesis and, if it explains various large and independent classes of facts, it rises to the rank of a well-grounded theory. [Darwin, The Variations of Animal and Plants Under Domestication, *quoted in Smolin,* The Life of the Cosmos *(1997), 161]*

There is no privileged account of explanation. Vast areas of scientific research will have to make do without the identification of causal mechanisms. As we approach the more historical sciences, we lose the ability to delineate a closed cluster of necessary and sufficient conditions. Enhanced unification is, however, a feature of the hard sciences, like astronomy, physics, and biology. Unification occurs when apparently unrelated phenomena – like the fall of an apple and planetary orbits or environmental niches and the biogeographical distribution of species – are subsumed under a set of well-confirmed principles. By tracing the history of homologies to descent with modification, driven by natural selection, Darwinian explanations achieve unification. Unification is also enhanced by tracing the history of analogies to convergent evolution. The Copernican model makes a step toward unification of terrestrial and celestial phenomena by assigning a heliocentric structure to the observations and adopting an impetus theory of motion. Unification can be described in terms of embedding. [Friedman 1981; Kitcher 1989] Observable phenomena are embedded into some larger, relatively abstract theoretical structure. We explain terrestrial and celestial motions by embedding them into Newtonian mechanics. We explain various biological phenomena by embedding them into Darwinian evolution theory. These theoretical structures have considerable unifying power.

Structural explanation is a form of unification. Structural models emphasize structural features of explanation. In its most fundamental sense, a structural explanation occurs when the properties or behavior of a complex system can be modeled. That is, we can construct a model that represents the structure of the system. We assign an underlying structure to the observable phenomena. [McMullin 1985; Hughes 1989, 256–8] This is more easily said than done. We often have a host of observations. They need to be assigned to a coherent structure to make sense. The solar system has a structure, so does the DNA molecule. Generally, physical systems are manifestations of structure. A structure consists of a set of components *and* regular relationships between them. The solar system consists of nine planets and a central sun. These elements in themselves do not yet constitute a structure. For once these elements are collected, the question arises how they are to be arranged. The Greeks arranged the planets in a geocentric, the Copernicans in a heliocentric order. To arrive at a structure, a question needs to be answered: How are these elements arranged with respect to each other? The relationships between the planets and the sun are expressed in Kepler's laws. Typically, we look for quantifiable relationships. In the Darwinian model, the elements are the organisms and they are related by trajectories of descent. Once we have identified the elements of the structure and the regular relationships between these elements, we have identified the structure of the system. The job of a model is to represent some aspect of the structure of the physical system. As we saw in Chapter I, this observation gives rise to questions of structural realism and scientific representation. Above, we distinguished between the *topologic* and the *algebraic* structure of models. The emphasis may at first be on the *topologic* structure, the spatial arrangement of the components. Models always raise the question of how the elements are related to each other. Our choice of topologic structure has consequences for the algebraic arrangement of the components.

The algebraic structure puts more emphasis on the mathematical relationships between the elements of the structure. This step from the topologic to the algebraic structure can be observed in both Copernicanism [Figures 2.12, 2.14] and Darwinism [Figure 2.9, 2.11]. The heliocentric model and the branching model of life are topologic models. They put a strong emphasis on the geometric arrangement of elements. [Figures 2.13, 2.6, 2.10].

The topologic structures can be augmented by planetary laws and genetic trajectories, respectively. They are mathematically more elaborate and add algebraic structure to the model. A structural explanation consists in assigning a model structure to a system in the real world. As in the case of unification, the mere assignment of a structure is not sufficient to reach genuine explanatory levels. The model structure must approximately fit the structure of the real system. The geocentric model must fail because the structure it assigns to the real world does not fit the real structure of the planetary system. Design models also fail because the hierarchical, teleological structures they assign to the organic world fail to fit the facts. The requirement that the model structure must represent the structure of the system modeled is based on the realist assumption that scientific models must bear some structural resemblance to a part of the real world. What does it mean that the model fits the world? We can approach this question in terms of our notion of constraints. The model structure must satisfy the constraints in terms of observational data, mathematical theorems, and theoretical principles. Both heliocentric

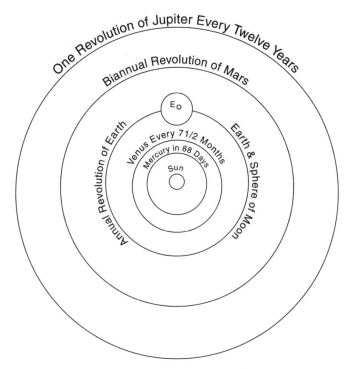

Figure 2.13 A simplified Copernican model of the solar system. It concentrates on the topologic structure, but includes some algebraic elements because of the inclusion of the length of orbital revolutions

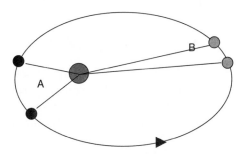

Figure 2.14 Kepler's area law. Areas *A* and *B* have the same size. In its orbit around the sun the planet sweeps equal areas in equal times. Nearer the sun, the planet moves faster than further away from it

and evolutionary models satisfy the constraints more comfortably than their rival models. A model is a better representation of an empirical domain, if it satisfies more constraints than its rival models and *if* the constraints are justified.

Structural models are typical responses to structural questions. Causal models are typical responses to "why-questions." In the ideal case, we can trace a causal mechanism. When this fails, we may still be able to identify a cluster of causal conditions. Causal and structural models are perfectly compatible with each other. We can ask: Why do planetary systems exist? This question sounds like a request for a causal explanation. Although the Newtonian model is unable to offer a traceable causal mechanism, its answer comes in terms of a cluster of conditions, which include Newton's law of inertia and the law of gravitation. Why do species adapt? This too seems to be a request for a causal explanation. Again there is no traceable causal mechanism. An answer can still be derived from the Darwinian model. If a change in the environment occurs, which does not threaten the living conditions of a species, and the condition of isotropic variation is satisfied, natural selection provides a causal answer to the question.

We may not be interested in cause–effect relationships. Our focus may be on understanding the structure of a system. This is a request is for a structural explanation. Copernicus asked: What is the structure of the solar system? Two hundred years later Kant asked: What is the structure of galaxies? And what is the structure of the cosmos? The Copernican–Newtonian model can provide structural explanations. Lamarck asked: What is the structure of evolution? He answered, structurally, with his model of progressive evolution, which includes the inheritance of acquired characteristics. Darwin, too, desired to know what was the structure behind the diversity of life. His theory of descent with modification provided a structural explanation. Once structural models reach a satisfactory level of fit, they provide unification.

It is not surprising that models of explanation always return us to the philosophical question of realism. For science is about the real world. Any stipulation of structure, causal mechanism, and unification poses the question of fit between the stipulated structure and the structure of systems in the natural world. We must briefly return to the question of realism.

6.8   A brief return to realism

I have often personified the word nature; for I found it difficult to avoid this ambiguity; but I mean by nature only the aggregate action and product of many natural laws – and by laws only the ascertained sequences of events. [Darwin quoted in Crombie 1994, Vol. III, 1751; Ellegård 1958, 181 for further references]

We have spoken of the explanatory values of Darwinism. They reside in the proposal of a structure, which unifies many hitherto unrelated phenomena. Darwin's theory also proposed a causal account, in terms of necessary and sufficient conditions. Through random mutation and preservation of favorable modifications, both the existence of lineages and the adaptiveness of species to their respective environments can be explained. But now the question of realism and instrumentalism concerning this mechanism returns.

Darwin himself was cautious: the theory of natural selection could not be tested by direct inference from the evidence. [Darwin 1859, Editor's Introduction, 15; Lloyd 1983] Darwin was not in a position to demonstrate conclusively that evolution had occurred. He tried to show that it was the most adequate explanation, which fitted the facts better than rival theories. On the strength of the available evidence descent with modification was the only admissible inference. Yet when he descended into the details he could not show how random mutation occurred, why some species changed while others did not, what was the precise extent of the operation of natural selection. Darwin added sexual selection as an additional principle to account for apparent nonadaptive change. He even began to talk of the inheritance of acquired characteristics, believing that he had overemphasized the role of natural selection. [Darwin 1871, Part I, Ch. 2, 81]

On the threshold of the twenty-first century, many of Darwin's original difficulties have been cleared up. The theory has also increased in scope to explain such diverse phenomena as sexual and asexual reproduction, the sex ratio problem (why there are approximately equal numbers of males and females), and the evolutionary advantage of altruism. Evolutionary psychology claims an even greater scope for the theory, for it wants to explain mental facts by reference to evolutionary principles. [See Chapter III, Section 5.2]

In Darwin's time the model of natural selection had mere **empirical validity**. From the theory a family of models could be derived, which shared the same fundamental structure. They inhabit the same constraint space. These models allow Darwinian inferences to address specific problems, like the biogeographical distribution of species and their varieties. The models give results that conform to the empirical data: the evolutionary theory produces a good fit between its models and the data. The theory is empirically valid because its basic structures provide a plausible explanation and discredit rival explanations. However, we want more than an agreement of its models with the empirical data. The history of Copernicanism and Darwinism shows that scientists wish to know whether the basic principles – the algebraic structures – are valid. They aspire to **theoretical validity**. There must be an accurate representation of the model structure with the structure of the empirical domain. But at the end of the nineteenth century it would have been difficult to claim that the theory was also theoretically valid. It had not yet been established beyond reasonable doubt that the mechanism and structure assigned to the data also represented the mechanism and structure operative in nature. Recall that Darwin's basic mechanism – natural selection – was in doubt; equally in doubt were the source of random genetic variation and the time frame required for gradualist evolution to take effect. To improve the explanatory value of the theory, and to make a step from empirical validity to theoretical validity, independent tests of both the model assumptions (extrapolationism, gradualism, isotropic variation) and the assumed mechanism as a sufficient causal condition must become available. Luckily for Darwinism, new dating techniques soon confirmed that the Earth was billions of years old. Industrial melanism – the adaptation of moths in nineteenth-century England to the blackening of tree barks in heavily polluted industrial areas – provided early indications that natural selection was at work in nature. Today the observable mutations of the AIDS virus, under laboratory conditions, supply one of the best pieces of evidence that natural selection operates in nature. [Jones 1999] Unluckily for Darwinism, all of its model assumptions have been challenged, both by those who stand in the Darwinian tradition [Gould 2002; Kauffman 2004] and by those who dismiss its inferences as inadequate [Behe 1996; Dembski 1998]. Darwinism, like Copernicanism, was at first compatible with instrumentalism. Evolution was accepted as a fact of nature. But natural selection as an exclusive causal mechanism fell under a cloud of suspicions. Nevertheless, Darwinism slowly began to shed the instrumentalist trappings. The merger of genetics and Darwinism and the development of molecular biology were further promptings which nudged evolutionary theory toward a realist interpretation.

6.9   Darwin and scientific revolutions

When the views entertained in this volume on the origin of species, or when analogous views are generally admitted, we can dimly foresee that there will be a considerable revolution in natural history. [Darwin, Origin of Species (1859), 455]

There is unanimous agreement that Darwin was the author of a significant revolution in science. Darwin satisfies the criteria for a scientific revolutionary.

Firstly, he switched perspectives. Do not explain the observable biological phenomena from the point of view of an intelligent designer. Explain them from the point of view of naturalism. The switch in perspective leads to massive shifts in the conceptual network: the rejection of design and the constancy of species; a heavy emphasis on the impact of the environment on the trajectory of species; the importance of random mutation and the beneficial effect of cumulative selection. To qualify as a revolution more is needed than a switch in perspective and an attendant realignment of conceptual links in the underlying theory. After all, Lamarck, too, switched perspectives with respect to the design argument. The difference between Lamarck and Darwin resides in the testability of the proposed natural mechanism. Darwin proposed natural selection as a testable mechanism.

Secondly, the new theory must also be explanatory. It must solve some outstanding problems. It does so by adopting new methods and techniques. Certainly, Darwin provided an adequate solution of the fundamental problem of his time: the diversity of species. It also provided an elegant account of the hominoid fossil record. Lamarck failed to solve these problems, because his proposed mechanism could not claim much credibility. The new model may at first command no more than empirical adequacy. That is, it may be one among a number of theoretical accounts that can cope with the evidence. But testability requires that the model structure be both empirically and theoretically valid. That means that the theory fits the evidence, has survived independent tests for its principles, and has gained this credit by discrediting some rival model. A consequence of theory change is often the commitment to alternative methods of explanation. The Darwinians made heavy use of historical inferences. They practiced Baconian induction, understood in its proper sense as eliminative inductivism. Darwinism is not a good example of hypothetical deductivism. It is not true that all serious science must proceed by Popperian falsification. Do we require that predictive power is a necessary ingredient of revolutionary change in science? This would be too restrictive, since it would exclude Darwin's revolution. But we should require the accommodation of already known facts. The new theory must accommodate some independently known facts. It must have some new deductive consequences, even if it fails to make precise predictions. Darwinism satisfied this criterion to perfection. The accommodation must succeed on the restrictive condition that any rival models fail to incorporate the known facts.

Thirdly, a new tradition emerges through a chain-of-reasoning process. Darwin certainly started such a chain of reasoning, which would ultimately lead to neo-Darwinism. But *fourthly,* the evolutionary tradition does not yet command unanimous consensus. In the nineteenth century other paradigms in biology were being discussed, which constituted parallel developments (German *Naturphilosophie*, Buffon's system). The emphasis of the German *Naturphilosophen* (Goethe, Oken) on form, rather than function – "form shapes function" – has left echoes in modern structuralist approaches to evolution. The Darwinian tradition is opposed by ID and within this tradition there is no consensus on all the elements of the disciplinary matrix. There are a few rival models but a number of alternatives. At first, all agreed on branching evolution and the non-fixity of species. But the central mechanism – natural selection – was either not fully accepted or its extent was questioned. Around the turn of the century, some Darwinians (Huxley, de Vries) preferred *saltationism* to Darwin's gradualism. Today some Darwinians propose to replace the idea of gradual, imperceptible changes with the concept of *punctuated equilibrium.* [Gould 2002] Nor is there agreement on the unit of natural selection. Darwinians traditionally regard the individual organism as the unit of selection. But further candidates have been proposed: the gene [Dawkins 1976; 1988] and species selection. [Gould 2002; see Mayr 2001] There is no general consensus among the Darwinians with respect to the extent of adaptation. We have already observed that Darwin accepted nonadaptive changes, especially in his views of human nature. But some Darwinians are more radical and envisage alternatives such as structuralism and internal constraints. These approaches resist the attempt of orthodox Darwinism to explain as many features as possible as adaptations. They emphasize instead that adaptationism should acknowledge the existence of constraints in the form of physical limits to organisms' ability to adapt to new environmental pressures. There is a physical limit, for instance, on how tall two- and four-legged creatures can grow. [See Gould (2002) for an overview] Generally, then, there is still not as much convergence on the details of the Darwinian tradition as there is on the Copernican–Newtonian model of the solar system.

A revolution is not necessarily succeeded by a new form of normal science, as Kuhn claimed. We should note this difference with the Copernican revolution. Although Darwinism accommodates a number of alternative models within a common constraint space, Darwin's work can still be regarded as a true revolution in science.

But Darwin's revolution does not really fit Kuhn's picture. [Greene 1980] There is not much evidence in the Darwinian revolution of a *gestalt switch* or of a breakdown in communication or even a period of normal science. Darwin focuses much of his critical attention on the theory of special creations and, to a lesser extent, on

Lamarck's progressive evolutionism. Darwin carefully considered the rival models and found them inadequate, on empirical and theoretical grounds. Darwin's model had its opponents and its proponents. They reasoned with each other. Darwin's friends were engaged in argument with Darwin's detractors. Among themselves, opponents and proponents deliberated the pros and cons of the theory. Just as in the history of Copernicanism, there is differential agreement on fundamental principles and reasoned transitions. [Shapere 1966, 1989; Cohen 1985a; Chen/Barker 2000]

This chapter has reconstructed the chain of reasoning that leads from the pre-revolutionary to the post-revolutionary period. If we regard the elements of a model as units, which are bound into a structure by specific relationships, we can imagine playing games with the model structure. Add and reject elements, move others to different locations, and we change the algebraic and topologic structure of the model. If we are willing to cross the constraints of the logical spaces, we can work backward from a heliocentric to a geocentric model, from an evolutionary to a design model. This conceptual game underscores a conceptual point: revolutions in science are more like *chain-of-reasoning transitions* than knock-down-and-rebuild demolitions. Chain-of-reasoning transitions between pre- and post-revolutionary periods allow us to trace the changes that link the old and the new. The changes are based on arguments that lead to modifications of the old traditions. This procedure results in traceable lines of descent between theories and models. The operations will include additions, deletions, and replacements in the conceptual network. Chain-of-reasoning transitions show lines of continuity and discontinuity through periods of scientific revolutions. They are concerned with a reconstruction of the problem situation at hand. They evaluate the solutions in the light of the accepted problem.

6.9.1 Philosophical consequences What were the **philosophical consequences** of the Darwinian revolution?[18] [Dewey 1909; Haeckel 1866, Ch. IV; Dennett 1995, 201–2] Quite generally, the *loss of rational design*. Darwin replaced teleology with naturalism; design with natural selection. Darwin did not regard species as essences. They were populations of individuals with slight variations. The variations could be beneficial or nefarious. Evolution could only get under way under this basic assumption of isotropic variation. Darwin liberated biology from the dogma of essentialism and replaced it with population thinking. Materialism, though not in a crude sense, was a further consequence of Darwinism. Darwin stressed the importance of empirical evidence in biological explanations. Darwin did not share the belief in determinism to which both Huxley and Haeckel swore their allegiance. Materialism and biological indeterminism lead to a fluid view of the biological world. The Darwinians certainly believed that the organismic world was in a state of flux.

---

[18] Recall that we distinguished above (Section 6) between *deductive, inductive,* and *philosophical* consequences of scientific theories. Deductive consequences can be deduced from the principles of a theory and are in general testable. Inductive consequences follow from the principles with a certain probability. Philosophical consequences do not follow from the theory with mathematical rigor or probability but suggest themselves as being philosophically compatible with the presuppositions of the theory. They are not strictly testable.

The Darwinian revolution also had an impact on the philosophy of mind. As the Darwinians included humans in the realm of nature, they were forced, by the logic of their argument, to explain the manifest superiority of the human brain. This included a naturalist explanation of the existence of mental phenomena.

Thereby Darwinism had a significant impact on psychology, for it represented the phenomena of mind as facts of nature. [Darwin 1859, 458; *Nature* **26**, 1882] Throughout his career, Freud was eager to present psycho-analysis as a scientific discipline. It deserves, he claimed, the same scientific credibility as physics. Sometimes the claim is made that Darwin completed the "final stage of the Copernican revolution that had begun in the 16th and 17th centuries under the leadership of men like Copernicus, Galileo and Newton." [*Encyclopaedia Britannica* Vol. **18**, 1991, 856] Freud disagreed. He believed that his discovery of what I shall call the **loss of transparency** opened a third stage of the Copernican revolution, to which we now turn. He saw in it a further assault on the nobility of humankind.

Reading List

Achinstein, P. [1985]: *The Nature of Explanation*. Oxford: Oxford University Press

Ayala, F. [1987]: "The Biological Roots of Morality," *Biology and Philosophy* **2**, 235–52

Ayala, F. [1995]: "The Distinctness of Biology," in F. Weinert *ed*. [1995], 268–85

Ayala, F. [2004]: "Design without a Designer," in W. A. Dembski/M. Ruse *eds.* [2004], 56–73

Argyll, Duke of [1867]: *The Reign of Law*. London: Strahan

Argyll, Duke of [1886]: "Organic Evolution," *Nature* **34**, 335–6

Bachmann, G. [1995]: "Mechanische und Organische Denkformen im viktorianischen Zeitalter," in *Die Mechanische und die Organische Natur*. Konzepte SFB 230 Heft 45 (März 1995), 9–32

Bacon, F. [1620]: *Novum Organum*. L. Jardine/M. Silverstone *eds*. Cambridge: Cambridge University Press 2000

Barlow, H. [1987]: "The Biological Role of Consciousness," in C. Blakemore/S. Greenfield *eds.* [1987], 361–74

Barrow, J. D./F. J. Tipler [1986]: *The Anthropic Cosmological Principle*. Oxford: Oxford University Press

Behe, M. J. [1996]: *Darwin's Black Box*. New York: The Free Press

Behe, M. J. [2004]: "Irreducible Complexity," in W. A. Dembski/M. Ruse *eds.* [2004], 351–78

Blackmore, S. [2003]: *Consciousness*. London: Hodder & Stoughton

Blakemore, C./S. Greenfield *eds.* [1987]: *Mindwaves*. London: Blackwell

Boyle, R. [1688]: *A Disquisition about the Final Causes of Natural Things*. London: John Taylor

Braubach, W. [1869]: *Religion, Moral und Philosophie der Darwin'schen Artlehre nach ihrer Natur und ihrem Charakter*. Leipzig: Neuwied

Browne, J. [2003]: *Charles Darwin – The Power of Place*. London: Pimlico

Bradley, W. L. [2004]: "Information, Entropy and the Origin of Life," in W. A. Dembski/M. Ruse *eds.* [2004], 331–50

Büchner, L. [1868]: *Sechs Vorlesungen über die Darwin'sche Theorie von der Verwandlung der Arten*. Leipzig: Theodor Thomas

Bunge, M. [1977]: "Emergence and the Mind," *Neuroscience* **2**, 501–9

Bunge, M. [1980]: *The Mind–Body Problem*. Oxford: Pergamon

Chalmers, D. [1996]: *The Conscious Mind*. Oxford: Oxford University Press

Chen, X./P. Barker [2000]: "Continuity through Revolutions," *Philosophy of Science* **67**, S208–S233

Cohen, I. B. [1985a]: *Revolution in Science*. Cambridge, MA: The Belknap Press

Cohen, I. B. [1985b]: "Three Notes on the Reception of Darwin's Ideas on Natural Selection," in D. Kohn *ed.* [1985], *The Darwinian Heritage*. Princeton, NJ: Princeton University Press [1985], 589–607

Clark, A. [1997]: *Being There*. Cambridge, MA: MIT Press

Crombie, A. C [1994]: *Styles of Scientific Thinking in the European Tradition* London: Duckworth, Vol. III

Damasio, A. [1999]: "How the Brain creates the Mind," *Scientific American* (December), 74–9

Darwin, Ch. [1859]: *The Origin of Species*. J. W. Burrow *ed.* London: Pelican Classics [1968]

Darwin, Ch. [1871]: *The Descent of Man*. Introduction by James Moore and Adrian Desmond. London: Penguin [2004]

Darwin, Ch. [1876]: "Sexual Selection in Relation to Monkeys," *Nature* **15**, 18–9

Darwin, Ch. [1878–79]: "Fritz Müller on a Frog having Eggs on its Back," *Nature* **19**, 462–3

Darwin, Ch. [1881]: "Inheritance," *Nature* **24**, 257

Darwin, Ch. [1954/1892]: *The Autobiography of Charles Darwin and Selected Letters*. Francis Darwin *ed.* New York: Dover

Dawkins, R. [1976/²1989]: *The Selfish Gene*. Oxford: Oxford University Press

Dawkins, R. [1988]: *The Blind Watchmaker*. London: Penguin

De Beer, G. [1971]: "Homology: an unsolved problem," in M. Ridley *ed.* [1997], 213–21

De Vries, H. [1907]: "Evolution and Mutation," *The Monist* **17**, 6–22

Dembski, W. A. [1998]: *The Design Inference*. Cambridge: Cambridge University Press

Dembski, W. A./M. Ruse *eds.* [2004]: *Debating Design*. Cambridge: Cambridge University Press

Dennett, D. [1995]: *Darwin's Dangerous Idea*. London: Penguin

Descartes, R. [1664]: *Traité de l'Homme*, in *Œuvre et Lettres*. Paris: Gallimard [1953], 807–73

Dewey, J. [1909]: "The Influence of Darwin on Philosophy," in *The Influence of Darwin on Philosophy*. Bloomington: Indiana University Press [1965], 1–19

D'Holbach, P. Thiry [1770]: *Système de la Nature* [English translation: *The System of Nature*, London, 1817; German translation: *System der Natur*. Frankfurt a./M.: Suhrkamp, 1978]

Dowe, Ph. [2000]: *Physical Causation*. Cambridge: Cambridge University Press

du Bois-Reymond, E. [1882–83]: "Darwin and Copernicus," *Nature* **27**, 557–8

Eiseley, L. [1959]: *Darwin's Century*. London: Gollancz

Edelman, G. [1992]: *Bright Air, Brilliant Fire*. London: Penguin

Ellegård, A. [1958]: *Darwin and the General Reader*. Göteburg: Acta Universitatis Gothoburgensis

Evans, J. [1890]: "Anthropology," *Nature* **42**, 507–10

Fisher, R. A. [1930]: "The Nature of Inheritance," in M. Ridley *ed.* [1997], 22–32

Flower, W. H. [1883]: "The Evolutionary Position," *Nature* **28**, 573–5

Friedman, M. [1981]: "Theoretical Explanation," in R. Healey *ed.* [1981] *Reduction, Time and Reality*. Cambridge: Cambridge University Press, 1–16

Freud, S. [1905]: *Three Essays on Sexuality*, in *The Standard Edition of the Complete Psychological Works of Sigmund Freud*. James Strachley *ed.* London: The Hogarth Press, Vol. VII [1953], 130–243

Freud, S. [1931]: "Female Sexuality," in *The Standard Edition of the Complete Psychological Works of Sigmund Freud*. James Strachley *ed.* London: The Hogarth Press, Vol. XXI [1927–31], 221–43

Geoffroy Saint-Hilaire, M. Isodore [1847]: *Vie, Travaux et Doctrine Scientifique d'Étienne Geoffroy Saint-Hilaire*. Paris: P. Bertrand, 1968

Ghiselin, M. T. [1969]: *The Triumph of Darwinian Method*. Berkeley, CA: University of California Press

Gillispie, C. G. [1958]: "Lamarck and Darwin in the History of Science," *American Scientist* **XLVI**, 388–409

Gillispie, C. G. [1959]: *Genesis and Geology*. New York: Harper & Row

Gould, S. J. [1980]: *Even Since Darwin*. London: Pelican

Gould, S. J. [1987]: *The Panda's Thumb*. London: Penguin

Gould, S. J. [1988]: *Time's Arrow, Time's Cycle*. London: Penguin

Gould, S. J. [1991]: *Wonderful Life*. London: Penguin

Gould, S. J. [2001]: "More Things in Heaven and Earth," in H. Rose/St. Rose *eds.* [2001], *Alas Poor Darwin*. London: Vintage, 85–105

Gould, S. J. [2002]: *The Structure of Evolutionary Theory*. Cambridge, MA: The Belknap Press of Harvard University Press

Greene, J. C. [1980]: "The Kuhnian Paradigm and the Darwinian Revolution in Natural History," reprinted in G. Gutting *ed.* [1980], *Paradigms and Revolutions*, 297–32. Notre Dame, IN: University of Notre Dame Press

Haeckel, E. [1866]: *Generelle Morphologie der Organismen* II. Berlin: Georg Reimer

Haeckel, E. [1876]: *The History of Creation*. London: Henry S. King & Co. [Translation of *Natürliche Schöpfungsgeschichte*, 1868]

Haeckel, E. [1877]: "Present Position of Evolutionary Theory," *Nature* **16**, 492–96

Haeckel, E. [1878]: "Prof. Haeckel on the Doctrine of Evolution," *Nature* **18**, 509–10

Haeckel, E. [1882]: "Professor Haeckel on Darwin, Goethe and Lamarck," *Nature* **26**, 534–41

Haeckel, E. [1929]: *The Riddle of the Universe*. London: Watts & Co. [Translation of *Die Welträtsel*, 1899]

Helvétius, C-A. [1758]: *Sur l'Esprit*. Paris: Chez Durand

Hempel, K. [1965]: *Aspects of Scientific Explanation*. New York: The Free Press

Honderich, T. [1987]: "Mind, Brain and Self-conscious Mind," in C. Blakemore/ S. Greenfield *eds.* [1987], 445–58

Howson, C./P. Urbach [²1993]: *Scientific Reasoning – The Bayesian Approach*. Chicago: Open Court

Hoyningen-Huene, P. [1993]: *Reconstructing Scientific Revolutions*. Chicago: Chicago University Press

Hughes, R. I. G. [1989]: *The Structure and Interpretation of Quantum Mechanics*. Cambridge, MA: Harvard University Press

Humphreys, N. [1997]: "The Inner Eye of Consciousness," in C. Blakemore/S. Greenfield *eds.* [1987], 377–81

Huxley, T. H. [1860]: "The Origin of Species," in *Collected Essays*. Vol. II: Darwiniana [1907], 21–79

Huxley, T. H. [1862]: "Geological Contemporaneity and Persistent Types of Life," in *Collected Essays* Vol. VIII [1894], 272–304

Huxley, T. H. [1863a]: *Man's Place in Nature* (Chicago: The University of Chicago Press, 1959); reprinted in *Collected Essays*. Vol. II. London: Macmillan 1910, 77–156

Huxley, T. H. [1863b]: "On Our Knowledge of the Causes of the Phenomena of Organic Nature," in *Collected Essays*. Vol. II: Darwiniana [1907], 303–475

Huxley, T. H. [1864]: "Criticisms of "The Origin of Species"," in *Collected Essays*. Vol. II: Darwiniana [1907], 80–106

Huxley, T. H. [1874]: "On the Hypothesis that Animals are Automata and its History," *Nature* **X**, 362–66

Huxley, T. H. [1876]: "Lectures on Evolution," in *Collected Essays*. Vol. IV [1898], 46–138

Huxley, T. H. [1880–81]: "Professor Huxley on Evolution," *Nature* **23**, 203–4, 227–31

Huxley, T. H. [1886]: "Science and Morals," in *Collected Essays*. Vol. IX [1911], 117–46

Huxley, T. H. [1888]: *Science and Culture and other Essays*. New York: Macmillan & Co.

Huxley, T. H. [1893]: "Evolution and Ethics," in *Collected Essays*. Vol. IX [1911], 46–86

Huxley, T. H. [1894]: *Collected Essays*. Vol. VIII. London: Macmillan & Co.

Huxley, T. H. [1894–95]: "Past & Present," *Nature* **51**, 1–3

Huxley, T. H. [1898]: *Collected Essays*. Vol. IV: Science and the Hebrew Tradition. London: Macmillan & Co.

Huxley, T. H. [1907]: *Collected Essays*. Vol. II: Darwiniana. London: Macmillan & Co.

Huxley, T. H. [1909]: *Collected Essays*. Vol. V: Science and the Christian Tradition. London: Macmillan & Co.

Huxley, T. H. [1910]: *Collected Essays*. Vol. VII. London: Macmillan & Co.

Huxley, T. H. [1911]: *Collected Essays*. Vol. IX. London: Macmillan & Co.

Jacob, F. [1977]: "Evolution and Tinkering," *Science* **196**, 1161–6

Jacob, F. [1981]: *Le Jeu des Possibles*. Paris: Fayard

Jaeger, G. [1869]: *Die Darwin'sche Theorie und ihre Stellung zu Moral und Religion.* Stuttgart: Julius Hoffmann

Jones, S. [1999]: *Almost like a Whale*. London: Doubleday

Kant, I. [1755]: *Allgemeine Naturgeschichte und Theorie des Himmels*, in I. Kant, *Werkausgabe*. *Hrsg.* von W. Weischedel. Frankfurt a./M.: Suhrkamp. Band I, 226–396 [English translation: *General History of Nature and Theory of the Heavens*]

Kant, I. [1790/1968]: *Kritik der Urteilskraft*. Hamburg: Felix Meiner. [English translation: *The Critique of Judgement*, 1952]

Kauffman, St. [2004]: "Prolegomenon to a General Biology," in W. A. Dembski/M. Ruse eds. [2004], 147–72

Kaufmann, W. [⁴1974]: *Nietzsche*. Princeton, NJ: Princeton University Press

Kepler, J. [1618–21]: *Epitome of Copernican Astronomy*, in J. Kepler [1995], 5–164

Kepler, J. [1619]: *Harmonies of the World*, in J. Kepler [1995], 167–245

Kepler, J. [1995]: *Epitome of Copernican Astronomy* & *Harmonies of the World*, transl. C. G. Wallis. Amherst, NY: Prometheus

Kitcher, P. [1993]: *The Advancement of Science*. Oxford: Oxford University Press

Kitcher, P. [1989]: "Explanatory Unification and the Causal Structure of the World," in P. Kitcher/W. C. Salmon eds. [1989], *Scientific Explanation*. Minnesota Studies in the Philosophy of Science Vol. **XIII**. Minneapolis: University of Minnesota Press, 410–506

Kuhn, T. S. [²1970]: *The Structure of Scientific Revolutions*. Chicago: The University of Chicago Press

LaMettrie, Julien Offray de [1747]: *L'Homme Machine*. Paris: Mille et une Nuits 2000

Lamarck, J. B. [1809]: *Philosophie Zoologique*, transl. with an introduction by Hugh Elliot. New York: Hafner, 1963

Lange, F. A. [²1873/⁷1902]: *Geschichte des Materialismus*. Leipzig: Verlag von J. Baedeker

Leibniz, J. G. [1697]: "On the Ultimate Origination of Things," G. H. R. Parkinson *ed.* [1973], *Leibniz, Philosophical Writings*. London: Rowman and Littlefield, 136–44

Lepenies, W. [1978]: *Das Ende der Naturgeschichte*. Frankfurt a./M.: Suhrkamp

Lewis, D. [1986]: "Causal Explanation," *Philosophical Papers* II. Oxford: Oxford University Press [1986], 214–240

Lipton, P. [²2004]: *Inference to the Best Explanation*. London: Routledge

Lloyd, E. [1983]: "The Nature of Darwin's Support for the Theory of Natural Selection," *Philosophy of Science* **50**, 112–29

Lovejoy, A. O. [1936]: *The Great Chain of Being*. Cambridge, MA: Harvard University Press

Lyons, W. [2001]: *Matters of the Mind*. Edinburgh: Edinburgh University Press

Mackie, J. L. [1980]: *The Cement of the Universe*. Oxford: Clarendon Press

Marsh, O. C. [1896–97]: "Dinosaurs," *Nature* **55**, 463–6

Maupertuis, P. L. Moreau de [1750]: *Essai de Cosmologie*, in *Œuvres*. New York: Georg Holms [1974], 4–78

Mayr, E. [1986]: "The Philosopher and the Biologist," *Paleobiology* **12**, 233–39

Mayr, E. [2000]: "Darwin's Influence on Modern Thought," *Scientific American* (July), 67–71

Mayr, E. [2001]: *What Evolution Is*. New York: Basic Books

McMullin, E. [1985]: "Galilean Idealization," *Studies in History and Philosophy of Science* **16**, 247–73

Menand, L. [2001]: *The Metaphysical Club*. New York: Farrar, Strauss and Giroux

Menuge, A. [2004]: "Who's Afraid of ID? A Survey of the Intelligent Design Movement," in W. A. Dembski/M. Ruse *eds.* [2004], 32–49

Meyer, St. [2004]: "The Cambrian Information Explosion," in W. A. Dembski/M. Ruse *eds.* [2004], 371–88

Mill, J. S. [1843/1898]: *A System of Logic*. London: Longmans, Green & Co.

Miller, H. [1849, 1861]: *Footprints of the Creator*. London: Hamilton, Adams & Co., with Memoir by Louis Agassiz

Miller, K. R. [2004]: "The Flagellum Unspun: The Collapse of 'Irreducible Complexity'," in W. A. Dembski/M. Ruse *eds.* [2004], 81–97

Mivart, St. George [1871a]: *On the Genesis of Species*. London: Macmillan & Co.

Mivart, St. George [1871b]: "Ape Resemblances to Man," *Nature* **III**, 481

Mivart, St. George [1896]: "Are Specific Characters the Result of 'Natural Selection?'" *Nature* **54**, 246–47

Monod, J. [1974]: "On the Molecular Theory of Evolution," in M. Ridley *ed.* [1997], 389–95

Müller, F. [1864]: "Für Darwin," in Fritz Müller, *Werke, Briefe und Leben*. A. Moller *ed.* Jena: Gustao Fischer, 200–63 [English translation: *Facts and Arguments for Darwin*. London: John Murray, 1869]

Nietzsche, F. [1887]: *On the Geneology of Morals*. New York: Vintage, 1967 [Translation of *Über die Geneologie der Moral*, 1887]

Nilsson, Dan-E./S. Pelger [1997]: "A Pessimistic Estimate of the Time Required for an Eye to Evolve," in M. Ridley *ed.* [1997], 293–301

Norton, J. D. [1994]: "Science and Certainty," *Synthese* **99**, 3–22

Norton, J. D. [1995]: "Eliminative Induction as a Method of Discovery: How Einstein Discovered General Relativity," in J. Leplin *ed.* [1995], *The Creation of Ideas in Physics*. Dordrecht: Kluwer, 29–69

Nozick, R. [2001]: *Invariances*. Cambridge, MA: The Belknap Press

O'Hear A. [1999]: *Beyond Evolution*. Oxford: Oxford University Press

Pennock, R. T. [2004]: "DNA by Design?" in W. A. Dembski/M. Ruse *eds.* [2004], 130–45

Planck, M. [1933]: "Ursprung und Auswirkung wissenschaftlicher Ideen," in *Vorträge und Erinnerungen*. Darmstadt: Wissenschaftliche Buchgesellschaft, 1965, 270–84

Popper, K. [1959]: *The Logic of Scientific Discovery*. London: Hutchinson [English translation of *Die Logik der Forschung*, 1934]

Popper, K. [1963]: *Conjectures and Refutations*. London: Routledge & Kegan Paul

Popper, K. [1973]: *Objective Knowledge*. Oxford: Clarendon Press

Psillos, St. [1999]: *Scientific Realism*. London: Routledge

Psillos, St. [2002]: *Causation and Explanation*. Chesham: Acumen

Raff, R. A. [1996]: *The Shape of Life*. Chicago: University of Chicago Press

Rice, W. R./E. E. Hostert [1993]: "Laboratory Experiments on Speciation," in M. Ridley *ed.* [1997], 174–86

Ridley, M. *ed.* [1997]: *Evolution*. Oxford: Oxford University Press

Rolle, F. [²1870]: *Der Mensch, seine Abstammung und Gesittung im Lichte der Darwin'schen Lehre*. Prag: Verlag von Friedrich Tempsky

Romanes, G. [1882–83]: "Natural Selection and Natural Theology," *Nature* 27, 362–4, 528–9

Romanes, G. [1883]: "Natural Selection and Natural Theology," *Nature* 28, 100–1

Rosen, E. *ed.* [1959]: *Three Copernican Treatises*. Mineola, NY: Dover

Rosenberg, A. [²1995]: *Philosophy of Social Science*. Boulder, CO: Westview Press

Ruse, M. [2003]: *Darwin and Design*. Cambridge, MA: Harvard University Press

Salmon, W. C. [1984]: *Scientific Explanation and the Causal Structure of the World*. Princeton, NJ: Princeton University Press

Salmon, W. C. [1998]: *Causality and Explanation*. Oxford: Oxford University Press

Seabright, P. [1987]: "The Order of the Mind," in C. Blakemore/S. Greenfield *eds.* [1987], 209–33

Searle, J. [1984]: *Minds, Brains and Science*. London: BBC Publications

Searle, J. [1987]: "Minds and Brains without Programs," in C. Blakemore/S. Greenfield *eds.* [1987], 209–33

Searle, J. [2004]: *Mind – a Brief Introduction*. Oxford: Oxford University Press

Shapere, D. [1966]: "Meaning and Scientific Change," reprinted in I. Hacking *ed.*, *Scientific Revolutions*. Oxford: Oxford University Press, 1981, 28–59

Shapere, D. [1989]: "Evolution and Continuity in Scientific Change," *Philosophy of Science* 56, 419–37

Smolin, L. [1997]: *The Life of Cosmos*. New York: Oxford University Press

Sober, E. [1999]: "Testability," *Proceedings and Addresses of the American Philosophical Association* 73, 47–76

Sober, E. [2002]: "Intelligent Design and Probability Reasoning," *International Journal for the Philosophy of Religion* 52, 65–80

Sober, E. [2004]: "The Design Argument," in W. Dembski/M. Ruse *eds.* [2004], 99–129

Sperry, R. [1983]: *Science and Moral Priority*. New York: Columbia University Press

Tillyard, E. M. W. [1943/1972]: *The Elizabethan World Picture*. Harmondsworth: Penguin

Tyler E. B. [1881]: *Anthropology*. London: Macmillan & Co.

Vogt, C. [1864]: *Lectures on Man*. London: Longman, Green, Longman, & Roberts [English translation of *Vorlesungen über den Menschen*, 1863]

Wallace, A. R. [1855]: "On the Law which has regulated the Introduction of New Species," in A. R. Wallace [1891], 3–19

Wallace, A. R. [1858]: "On the Tendency of Varieties to Depart Indefinitely from the original Type," in A. R. Wallace [1891], 20–33

Wallace, A. R. [1870]: "The Limits of Natural Selection as Applied to Man," in A. R. Wallace [1891], 186–214

Wallace, A. R. [1891]: *Natural Selection and Tropical Nature*. London: Macmillan and Co.

Wallace, A. R. [1893]: "Reason *versus* Instinct," in *Nature* **48**, 73–4, 267

Wallace, A. R. [1903]: *Man's Place in the Universe*. London: Chapman & Hall

Weber, B. H./D. J. Depew [2004]: "Darwinism, Design and Complex Systems Dynamics," in W. A. Dembski/M. Ruse *eds.* [2004], 173–86

Weber, M. [1948]: "The Social Psychology of the World Religions" (1913), in *From Max Weber*, with an introduction by H. H. Gerth/C. Wright Mills *eds.* Boston, MA: Routledge & Kegan Paul [1948], 267–301

Weismann, A. [1882]: "On the Mechanical Conception of Nature," in *Studies in the Theory of Descent*. London: Sampson Low, Marston, Searle & Rivington. 2 vols. [1882], 634–718

Weinert, F. *ed.* [1995]: *Laws of Nature*. Berlin: Walter de Gruyter

Weinert, F. [2000]: "The Construction of Atom Models: Eliminative Inductivism and its Relation to Falsificationism," *Foundations of Science* **5**, 491–531

Weinert, F. [2004]: *The Scientist as Philosopher*. New York: Springer

Weinert, F. [2007]: "A Conditional View of Causality," in *Causality and Probability in the Sciences*. F. Russo/J. Williamson *eds.* London: College Publications 2007, 415–37

Weinberg, St. [1993]: *Dreams of a Final Theory*. London: Vintage

Wendorff, R. [³1985]: *Zeit und Kultur*. Opladen: Westdeutscher Verlag

Williams, G. C. [1996]: *Plan and Purpose in Nature*. London: Weidenfeld & Nicolson

Williams, M. B. [1973]: "Falsifiable Predictions of Evolutionary Theory," *Philosophy of Science* **40**, 518–37

Woodward, J. [2003]: *Making Things Happen*. Oxford: Oxford University Press

Essay Questions

1 Explain the major differences between **Lamarck** and **Darwin**. In what way does the Darwinian view constitute progress over Lamarck's view?

2 Explain and evaluate the difference between **design arguments** and **evolution arguments**.

3 Discuss the philosophical **presuppositions** inherent in Darwinism. What role do they play?

4 Explain the difference between **progressive** and **branching** evolution.

5 Why did the **gaps** in the **fossil record** constitute a problem for Darwin's theory?

6 In what way does the **antiquity of man** constitute a problem for creationism?

7   How did the view of the emergence of **species** change under the impact of Darwinism?

8   Explain what it means that **Darwin** produced explanatory structures and what they achieved.

9   What's wrong with the widespread view that "we descend from the apes"? What is the correct **Darwinian view**?

10   What is the structure of **Darwinism** and **Darwinian explanations**? Consider the difference between explanation and prediction.

11   Explain the major achievements of the **Darwinian revolution**.

12   What do we understand by a **scientific revolution**? Were *Copernicanism* and *Darwinism* scientific revolutions?

13   Critically discuss the applicability of **Kuhn's paradigm model** of scientific revolutions in the context of the Darwinian model.

14   In which sense could you use *Darwinism* to support, respectively, **instrumentalism** and **realism**?

15   Explain how the *Origin of Species* leads to the *Descent of Man*. What aspects of **evolution** were particularly important?

16   What did the Darwinians mean by a **materialist explanation** of humans' mental and moral capacities?

17   Can Darwinism be regarded as supporting an **emergentist** theory of the mind?

18   Is the distinction between **falsifiability** and **testability** justified in the context of Darwinism?

19   Should Darwin's theory be characterized as **eliminative inductivist** or as **hypothetical deductivist**?

20   In what sense is the Darwinian theory a **structuralist account** of the natural world?

21   In what sense is the Darwinian theory a **causal account** of the natural world?

22   Does the notion of **Darwinian trajectories** make sense?

23   What are **Darwinian inferences**?

24   What arguments does Darwin employ against **Creationism**?

25   Is **Intelligent Design** a serious competitor to Darwinism?

26   Explain the emphasis in **evolutionary psychology** on both particular and universal aspects of the human mind.

27   How does **Darwinism** explain the emergence of the mind from the brain?

28   How do the Darwinians apply **natural selection** to human nature?

29   What did Darwin mean when he wrote that "**Psychology** will be based on a new foundation"?

30   If **models** are ways of representing the natural and social world, how is this representation achieved?

31   The DN model assumes the **symmetry** of **explanation** and **prediction**. Use examples from biology to evaluate the appropriateness of this assumption.

III

Sigmund Freud: The Loss
of Transparency

Psychology will be based on a new foundation, that of the necessary acquirement of each mental power and capacity by gradation. [Darwin, The Origin of Species *(1859), 458]*

1 Copernicus, Darwin, and Freud

Human beings have an instinctive tendency to fend off intellectual novelties. [Freud, Introductory Lectures on Psycho-Analysis *(1916a), 214]*

Freud was eminently aware of his illustrious predecessors, Copernicus and Darwin. The twentieth-century dominant self-image as *psychological man* is greatly due to Freud, who had a significant influence on our language and thought. Freudian terminology has penetrated our vocabulary. It has provided us with coherent ways of "understanding" dreams and Freudian slips. Despite Freud's claims to scientific respectability, his psychoanalytic theory has not been unanimously accepted as a contribution to scientific thought. There are good reasons for this reluctance. Freud belongs to a handful of thinkers in the Western tradition, whose impact on the way we think has been as significant as his contribution to science has been judged insignificant. If the Copernican and Darwinian theories are reasonable representatives of scientific revolutions, Freud's theory is a candidate for a revolution in thought. The doubts about Freud are regularly aired in the features articles of international news magazines.

In the 1880s Ernst Haeckel and Emile du Bois Reymond spoke of Darwin as the Copernicus of the organic world. In 1877 a new journal was launched in Germany. *Kosmos* was devoted to a unified worldview on the basis of Darwin's evolutionary theory. In its first edition it links Copernicus and Darwin: both robbed humans of their privileged position in the universe. In the 15th edition of the *Encyclopaedia Britannica* we find the claim that Darwin completed the Copernican revolution, because Darwin subjects the diversity and complexity of life to the laws of nature.

Freud could not agree with this characterization. Freud regarded his own contribution as the final piece in the completion of the Copernican revolution. On a superficial level, Freud explains, the resistance to psychoanalysis is due to the unfamiliarity of the subject:

> But in thus emphasizing the unconscious in mental life we have conjured up the most evil spirits of criticism against psycho-analysis. Do not be surprised at this, and do not suppose that the resistance to us rests only on the understandable difficulty of the unconscious or the relative inaccessibility of the experiences, which provide evidence of it. Its source, I think, lies deeper.

It is the deeper level that explains the resistance. The deeper level has to do with cultural history:

> In the course of centuries the *naïve* self-love of men has had to submit to two major blows at the hands of science. The first was when they learnt that our earth was not the centre of the universe but only a tiny fragment of a cosmic system of scarcely imagin-able vastness. This is associated in our minds with the name of Copernicus, though something similar had already been asserted by Alexandrian science. The second blow fell when biological research destroyed man's supposedly privileged place in creation and proved his descent from the animal kingdom and his ineradicable animal nature. This revaluation has been accomplished in our own days by Darwin, Wallace and their predecessors, though not without the most violent contemporary opposition.

These two blows, which we characterized as the **loss of centrality** and the **loss of rational design**, will be reinforced by another blow, delivered by Freud:

> But human megalomania will have suffered its third and most wounding blow from the psychological research of the present time which seeks to prove to the ego that it is not even master in its own house, but must content itself with scanty information of what is going on unconsciously in its mind. We psycho-analysts were not the first and not the only ones to utter this call to introspection; but it seems to be our fate to give it its most forcible expression and to support it with empirical material which affects every individual. Hence arises the general revolt against our science, the disre-gard of all considerations of academic civility and the releasing of the opposition from every restraint of impartial logic. [Freud 1916b, 284–5]

"The ego is not master in its own house" – the ego is driven by subconscious motives, of which it is not aware. The motives are not transparent to it, although psychoanalysis claims to be able to unearth them. Hence humans suffer another blow: the **loss of transparency**.

Freud's followers have understandably been very pleased with this apparent line-age of Freudian ideas. Freud will be hailed, not only for his empirical discoveries, but also for his revolution in thought:

> Like Copernicus and Darwin, the men with whom he compared himself, Freud revolutionized our way of looking at ourselves, and like them he may well come to

be regarded rather as a moulder of thought than as a mere discoverer of facts. [Brown 1962, 2]

The problem is that many people disagree. The lack of consensus on Freud's scientific achievements is an indictor that he should not be treated as a scientific revolutionary without a thorough analysis of his theory. In 1946, Wittgenstein succinctly expressed what 30 years later would become the dominant view:

> Through his fanciful pseudo-explanations (precisely because they are brilliant) Freud has performed a major disservice. [Wittgenstein 1980, 55]

After a hundred years of Copernican and Darwinian theory, respectively, a major consensus on their scientific credibility was established. After a hundred years of Freudian theory there is still a major disagreement about his scientific credibility. The suspicion that falls on Freud's contributions to science is strengthened by a comparison with Copernicus and Darwin. An examination of the scientific status of Freudian theory, especially from the point of view of the inferential practices we have observed before, is one of the aims of this chapter. The others concern the epistemological status of the social sciences. As we shall see, Freud's theory is curiously poised between a commitment to empiricism and hermeneutics. Although Freud wishes to practice psychoanalysis as a natural science, he constantly has to borrow techniques from the social sciences. In this chapter Freudianism serves as a launching pad for a philosophical consideration of the social sciences. As in the previous chapters, we use the concrete material from the study of Freudianism to consider philosophical issues, which arise from its problem situation. We will be interested in establishing similarities and dissimilarities between the natural and the social sciences. It will be necessary to go well beyond Freud and consider Weber's contributions to the methodology of the social sciences. We will also return to the question of human nature, which provides a genuine link between Darwin and Freud, between evolution and psychology.

2   Some Views of Humankind

Thus in general a man is only accessible from the intellectual side too, in so far as he is capable of a libidinal cathexis of objects [Objektbesetzung] (…). [Freud, Introductory Lectures on Psycho-Analysis *(1916b), 446]*

We have interpreted Freud's claim to have completed the Copernican revolution as a loss of transparency. As humans do not seem to be master in their own house, Freud's claim amounts to the view that human behavior is grounded in less rationality than had been hitherto been assumed. Freud does not deny that rational motives play a part in human behavior. But he conceded rational reasons for human behavior less scope than the Enlightenment philosophers, who were much concerned with human nature. Throughout his career Freud consistently attributed to drives a greater role in human behavior than to reasons. In his early theory of sexuality, he writes:

> The behaviour of a human being in sexual matters is often a *prototype* for the whole of his other modes of reaction to life. [Freud 1908, 93; italics in original]

He repeats this view in his Vienna lectures on psychoanalysis:

> But what is reasonable is only part of mental life, a number of things take place which are not sensible (…). [Freud 1916a, 211]

And in his relatively late work on the evolution of culture and the future of religious beliefs, he again emphasizes that

> Men are so little accessible to reasonable arguments and are so entirely governed by instinctual wishes (…). [Freud 1927a, 47]

Given the consistency of his views on the role of unconscious motivations in human behavior, there is little doubt that Freud knowingly departed from the Enlightenment tradition in philosophy. While Freud preached the *loss* of transparency, the Enlightenment philosophers emphasized the *gain* of transparency. Let us briefly consider some Enlightenment views on human nature and consider Nietzsche as a precursor to Freud.

2.1   Enlightenment views of human nature

David Hume held that the motives of human actions are mostly desires, which reason cannot change. With the exception of Hume, most Enlightenment philosophers regarded human beings as mainly ruled by reason. They do not deny that desires play a part in human behavior. Kant admits, for instance, that humans are torn between desires, feelings, and reason. But reason is such that it can overcome the dictate of feelings. Famously, Kant defined the Enlightenment as "the emergence from self-inflicted immaturity." Intellectual immaturity is "the inability to use one's own reason, without external guidance." Kant regards the immaturity of individuals as self-inflicted because it is often due to laziness and cowardice on their part. As a motto for the Enlightenment Kant proposes: *Sapere aude!* – or have the courage to use one's own reason. [Kant 1784] Kant admits that sometimes the immaturity of the unenlightened public is due to shackles imposed by society. There may be a lack of political freedom, as in a dictatorship. Or social and cultural inhibitions, imposed by religious tyrannies or ideologies, may stifle the blossoming of maturity. Nevertheless, Kant

Immanuel Kant (1724–1804)

believes that the freedom to use one's own reason, rather than a political revolution, "allows an entire public to enlighten itself." But such enlightenment processes take time. On the personal level, the individuals must embrace the motto "sapere aude." On a societal level, the flourishing of education and culture are preconditions to move a society toward an enlightened age. Kant regards the obstruction of this cultural and political progress by an authoritarian ruler as a violation of human nature. The fulfillment of the Enlightenment project – overcoming ignorance, blocking blind traditions, and increasing knowledge – are for Kant a fundamental human right.

A revealing metaphor of the Enlightenment appeals to a property of light. Throwing light on a situation means that the situation gets illuminated. The general thrust behind the Enlightenment is that the acquisition of knowledge removes the dark presuppositions, which hinder progress. Consider the question of disease. As long as there is no empirical knowledge of what causes a particular disease, the treatment of disease is a haphazard affair. It is entirely dependent on wild assumptions about the causes of disease. But when Pasteur discovered in 1879 that bacteria cause disease, vaccination became an effective control against certain types of disease. The situation that requires illumination also may be political, societal, or religious. The Enlightenment demands that light be thrown on unjustified authority, on unwarranted assumptions. To demand justification for political or religious authority is to demand reasons for its existence. To ask for justification of unwarranted assumptions is to ask for the reasons behind the assumption. The provision of reasons makes assumptions more transparent. As we have seen, the Parisian school questioned the unquestioned assumptions behind the Aristotelian view of motion. Darwin questioned the assumptions behind the *Great Chain of Being*. In Kant's view the rule of reason bestows transparency on human behavior. On a personal level the courage to use one's own reason will make the motivations behind one's behavior less opaque. On the societal level the use of reason will make the assumptions, on which the life of society is based, more transparent. As Kant was aware, the Enlightenment project was postulated as a philosophical ideal. He did not believe that his eighteenth-century Prussian society had reached an enlightened age. But on the strength of the development of science and the Enlightenment philosophy he believed he lived in an age of Enlightenment. Kant thought, rather naively, that the general march toward an enlightened age was unstoppable. But Freud cast a serious doubt on the rationality assumption of the Enlightenment. It seems that light cannot penetrate the darker recesses of an individual's motivational space. According to Freud, we are driven by unconscious dynamic motivations. Reasons seem to play only a minor role in the husbandry of our mental affairs. If Freud is right about the unconscious forces lurking behind our actions, then we suffer a loss of transparency as individual agents. But is the transparency not regained by psychological knowledge? It may be true that each and every individual fails to be master in their own house. Has Freudian psychology not the power to turn on the light? Freud, the Freudians, and many cultural commentators think so. Unfortunately, such claims provoke disagreement. The existence of unconscious forces, the therapeutic value of psychoanalysis, and the Freudian methods of investigation have all come under attack.

2.2   Nietzsche's view of human nature

Before my time there was no psychology. [Nietzsche, Ecce Homo *(1888), "Why I am a Fatality," §6]*

By his own admission, Freud avoided reading Nietzsche. Although he held him in high esteem for his incisive psychological insights, he was keen to develop his own psychological views in an unbiased fashion. [Kaufmann 1974, 182 n. 2] Nietzsche portrays himself as a psychological and historical thinker whose target is Christian

morality, its values and ideals. He describes himself as the "first immoralist," an annihilator *par excellence.* [Nietzsche 1888, "Why I am a Fatality," §2] What Nietzsche aims to destroy are the unquestioned ideals of his society. He regards Christian morality as a decadent morality. And he rejects the ideal of the Good Samaritan. [Nietzsche 1888, §4] Nietzsche, famously, announces the Death of the Christian God. The belief in Christian morality has become doubtful: Good and Evil should not be regarded as absolute values. Such an absolute morality presents a tyranny against nature. Nietzsche offers three objections against Christianity: it is a dogma; it denigrates man's life on Earth; it preaches the subordination of man to God. It is a mistake to think that Nietzsche denies all values. He denigrates only traditional values, which in his eyes have brought about the

Friedrich Nietzsche
(1844–1900)

downfall of Western man. Nietzsche argues for a reevaluation of values, not an abandonment of all values. The job of the philosopher is to *create* values. They are not to be regarded as God-given and unchanging. Morality, in Nietzsche's eyes, is a convenience, not an ideal. All values are contingent. Moral values should enhance the lives of human beings and not hamper their existence. Nietzsche, the philosophical psychologist, sees in the Good Samaritan, the good Christian, a weak and decadent person. He claims that no psychology existed before him, because he regards himself as the first philosopher who questions Christian morality and the human types it has created:

> No one hitherto has felt Christian morality beneath him; to that end there were needed height, a remoteness of vision, and an abysmal psychological depth, not believed to be possible hitherto. Up to the present Christian morality has been the Circe of all thinkers – they stood at her service. What man, before my time, had descended into the underground caverns from out of which the poisonous fumes of this ideal – of this slandering of the world – burst forth? What man had even dared to suppose that there were underground caverns? Was a single one of the philosophers who preceded me a psychologist at all, and not the reverse of a psychologist – that is to say, a "superior swindler," an "Idealist"? [Nietzsche 1888, "Why I am a Fatality," §6]

On the one hand, Nietzsche is closer to Kant's Enlightenment than to Freud's psychoanalysis. He exposes a lack of transparency about the moral foundations of Western society at the end of the nineteenth century. He argues for a reevaluation of the value system, which would be closer to human nature (as Nietzsche sees it). On the other hand, he sees himself as a lone voice, the first psychologist with Freudian insights:

> "I did that," says my memory. "I could not have done that" says my pride, and remains inexorable. Finally – my memory yields. [Nietzsche 1886, IV, §68; cf. Gellner 1993, Ch. I]

In a similar vein, Freud held that culture imposed almost superhuman burdens of sexual abstention on human beings. According to Freud, it made them ill. While Nietzsche was trained as a classicist, Freud was trained as a scientist. He aimed to establish psychoanalysis as a new science.

3   Scientism and the Freudian Model of Personality

It might be said of psycho-analysis that if anyone holds out a little finger to it, it quickly grasps its whole hand. [Freud 1916a, 193]

Freud received his medical training in the late nineteenth century. It was the century of scientism – the belief that most problems in the natural and social sciences could be tackled using the methods of physics. It is no surprise that Freud, from his earliest inroads into the field of psychology, adopted a scientific attitude to the study of mental phenomena. From his "Project for a Scientific Psychology" (1895) to his *Outline of Psycho-analysis* (1938), Freud consistently compares psychoanalysis to science.[1] But Freud was not the first to envisage psychology as a branch of natural science. Both Darwin and Haeckel (1929, Ch. VI) regarded the evolutionary theory as having paved the way for a scientific approach to psychology. In his Vienna lectures Freud describes the doctrine of psychoanalysis as a superstructure, which will one day be anchored in an organic foundation, which is still unknown. [Freud 1916b, 389] In various ways, Freud sought to give his theory scientific credibility. Once such a commitment is taken, certain consequences follow. These consequences must be compatible with the original commitment. If psychoanalysis is a science, according to Freud, then we would expect a certain materialistic view of the mind. We expect the presentation of mechanical models, the postulation of lawful regularities, and a certain amount of determinism about mental events. Freud was consistent in accepting these consequences of his scientific commitments. Yet, as we summarize his model of the mind, we observe that Freud was unable to avoid humanistic endorsements in his

---

[1] In a rare BBC voice recording, which can be heard at the Freud Museum in London, Freud, looking back on his life's work, still insists that he discovered the new science of psychoanalysis.

theory. As we shall later see, Freud demanded explanation of mental phenomena but could not forego intentional notions.

3.1   Freud's model of the mind

Man is a tireless pleasure-seeker. [Freud 1905b, 126]

The twentieth-century century dominant self-image as *psychological man* is greatly due to the influence of Freud's work. Freud was born on May 6, 1856 in Freiberg

Sigmund Freud (1856–1938)

(Moravia). In 1859, the year Darwin published *The Origin of Species*, the family moved to Leipzig and one year later settled in Vienna. Freud remained in Vienna until 1937, the year of the Nazi occupation. Freud disliked Vienna, partly because of its anti-Semitism. But he was also influenced by Viennese society and its attitude to sexuality. In 1873 Freud started a medical career and became lecturer in neuropathology in 1885. In late 1885, Freud went to Paris to work under Jean-Martin Charcot. Charcot's work with so-called "hysterics" introduced Freud to the possibility that psychological disorders might have their source in the mind, rather than the brain. Charcot believed that he had demonstrated a link between hysterical symptoms, such as a paralysis of a limb, and hypnotic sugges-tion. This link implied the power of the mental.

Later Freud struck up a friendship with Joseph Breuer, whose patient was Bertha Pappenheim ("Anna O") who suffered from hysterical symptoms. Breuer used a technique of verbalization (also known as "talking cure" or "chimney sweeping") to treat her symptoms. In 1896 Freud coined the phrase *psychoanalysis*.

3.1.1   A summary of psychoanalytic theory   Freud abandoned hypnosis and deve-loped a new technique of free association. [Freud/Breuer 1895] This technique encourages patients to express any random thoughts that come associatively to their mind. The technique aims at uncovering hitherto unarticulated material from the realm of the psyche. Following a long tradition, Freud baptized this hidden realm the *Unconscious*. Freud noted that his patients experienced difficulties in freely associating – there were sudden silences, stutterings, and refusals to continue the psychoanalytic session. These symptoms suggested to Freud that the material struggling to be expressed was important. The patients also possessed powerful defenses against the expression of the hidden material. Such blockages (or resis-tances) had to be broken down in order to reveal hidden conflicts. Unlike Charcot and Breuer, Freud came to the conclusion that the most insistent source of resisted material was sexual in nature. Even more importantly, he linked neurotic symptoms to the same struggle between sexual feelings and psychic defenses. In an important letter to Fliess he abandoned the so-called Seduction Theory. [Freud 1897] He concluded that the material recalled under psychoanalytic promptings were fantasies, hiding primitive wishes, rather than real experiences. That is, fantasies and instinctual yearnings of the child lay at the root of later conflicts. Over the following years, Freud's system was enhanced and modified in a number of ways:

- Freud experienced the death of his father (1896) as the trauma that permitted him to delve into his own psyche. Freud's self-analysis delivered important insights for his psychoanalytic theory.
- In his *Interpretation of Dreams* (1899) Freud presented dreams as a royal road to knowledge of the Unconscious.

The study of dreams therefore becomes the most convenient means of access to a knowledge of the repressed unconscious, of which the libido withdrawn from the ego forms a part. [Freud 1916b, 456]

All dreams, except those of children, are interpreted as fulfillments of mostly sexual wishes. "Dreaming is evidently mental life during sleep!" [Freud 1916a, 88; 1916a, Ch. II] Freud advances a causal claim: most dreams of adults can be traced to infantile sexual wishes (Freud 1901b, 682–3), even that all inclinations to perversion have their roots in childhood (Freud 1916b, 311). As dreams are the disguised expression of wish fulfillments, Freud distinguished between the *manifest* and *latent* content of dreams. The manifest dream content is the actual, often strange sequence of images, which dreamers see in their dreams. The latent dream content comprises the hidden, unconscious motivation of the dream. Freud suspects that a lawful connection exists between the confused nature of latent dreams and the difficulties experienced in expressing the dream thoughts. [Freud 1901a, 643] Dream work consists in the transformation of latent into manifest dreams – and the job of psychoanalysis is to reverse this transformation.

Dreams are obliged to conceal things and only surrender their secrets to interpretation (…). [Freud/Oppenheim 1911, 181]

Dream work can transform dreams in a number of ways: (1) condensation (*Verdichtung*), e.g., the manifest dream is an abridged, condensed form of the latent dream; (2) displacement (*Verschiebung*) means that the latent meaning of the dream remains hidden; the dream work puts emphasis on unimportant or remote elements; the manifest dreams only vaguely allude at their real meaning; the latent content is disguised; (3) representation (*Dramatisierung*) means that a transformation of thoughts into images occurs. [Freud 1916a, Ch. XI]
- The *Psychopathology of Everyday Life* (1900) examines what has become known as "Freudian slips." These are not just slips of the tongue but also slips of the pen, the misreading of words, and the forgetting of names.

- In his *Three Essays on Sexuality* (1904–5) Freud attributed sexual drives to children, put emphasis on the causal power of fantasies, and established the importance of repressed desires. In these essays, Freud extended the concept of sexuality beyond its conventional usage to include a raft of erotic impulses. Sexuality became, if not the sole, at least the prime mover in a great deal of human affairs. Freud also distinguished three phases in a child's sexual development: oral phase, anal phase, and phallic phase. Freud stipulates a rather direct causal relationship between infantile instinct components and adult character traits:

  > We can lay down a formula for the way in which character in its final shape is formed out of the constituent instincts; the permanent character traits are either unchanged prolongations of the original instincts or sublimations of these instincts or reaction formations against them. [Quoted in Webster 1996, 288]

- In 1908 the Vienna Psychoanalytic Society was established.
- In his later work the tripartite model of mind replaces the dichotomy of the Conscious and the Unconscious: the **Id** is governed by the pleasure principle, the **Ego** by the reality principle, and the **Superego** represents the internalization of social norms and values, instilled in the child's mind through cultural education. In his later work, Freud stipulated that the Id consisted of two parts: a *Death Wish* in addition to the existing *Eros*. [Freud 1938; see Webster 1996, 334–5]
- His Social and Cultural Studies occupy the last stage of Freud's work. Freud examines the origin of cultural phenomena, which he traces to the mechanism of sublimation. The appreciation or creation of cultural products, Freud contended, is rooted in primitive sexual urges that are transformed in culturally acceptable ways. Sublimation is a conflict-free resolution of repression, which leads to intersubjectively available cultural works. In his later works *Totem and Taboo* (1913) and *Group Psychology* (1921), Freud expresses a hefty dose of skepticism toward religious beliefs. Such beliefs in divinity can ultimately be attributed to the need to worship human ancestors. All civilizations, no matter how well planned, can provide only partial relief from sexual urges. For aggression among men is not due to unequal property relations or political injustice. Such problems could be rectified by social reform. Rather, aggression is due to a deep-seated death instinct. A reconciliation of nature and culture is impossible, for the price of any civilization process is the production of guilt on the part of the members of the civilization. Any civilization curbs human sexual instincts; this thwarting of pleasure will leave feelings of resentment toward society.

We have seen that the Copernican world-picture contributed to the mechanization of the world. The Copernicans participated in the movement, which treated nature as an abstract mathematical entity, obeying quantifiable law-like relationships. The belief that all forms of knowledge should approximate the ideal form of quantized knowledge is known as *scientism*.

At the end of the nineteenth century the success of science was so impressive that many people adopted *scientism* as their official philosophy. It is hardly surprising to see Freud pursue the same line of argument. In fact, he claims that psychology can be put on the same footing as physics. How does he justify this claim?

3.1.2 Analogy with physics Freud starts with the assumption that our psyche can be modeled as an extended, functional apparatus, which has three parts: the Super-ego, the Ego, and the Id. Each of these components of the mind has its specific functions, which Freud set out to describe.

Let us analyze the plausibility of the view that psychology is as secure a science as physics. Freud writes:

> There is in the expressions of the psyche nothing trifling, nothing arbitrary and law-less. [Freud 1910, 22]

In other words, Freud believed that it would be possible to discover psychoanalytic laws. [Freud 1910, 19] This is his first postulate:

1. The determination of the psychic life, according to which no mental happen-ings are accidental. [Brown 1962, 3] *Psychic determinism* is Freud's assump-tion that everything we do, think, or feel has meaning and purpose. All events of psychic life are determined: slips of the tongue, gestures, dreams, and neuroses – all have meaning and specific origins in the experience of the individual.

   Nothing in the mind is arbitrary or undetermined. [Freud 1901b, 242]

   But there is, of course, no such thing as arbitrary determination in the mind. [Freud 1901a, 680]

   (Strictly speaking, most behavior has multiple determinants, according to Freud: the drives are a mixture of two primary forces: the Eros and Death Instincts.)

2. The second assumption is that the Unconscious is a dynamic force, not just a wastepaper basket of ideas and memories:

   "Unconscious" is no longer the name of what is latent at the moment; the uncon-scious is a particular realm of the mind with its own wishful impulses, its own mode of expression and its peculiar mental mechanisms which are not in force elsewhere. [Freud 1916a, 212]

   With this assertion Freud moves beyond the traditional understanding of the Unconscious as a passive receptacle for unwanted material. He also overcomes the traditional identification of mind with consciousness, which finds its char-acteristic expression in Descartes's dualism. Freud's dynamic view of the Unconscious means that the Unconscious plays a predominant part in mental life. It bestows meaning on such apparently random events as Freudian slips and strange dreams.

3. The third assumption is that much of human behavior is driven by unconscious motivation. *Unconscious motivation* expresses Freud's conviction that a major

portion of our behavior, thoughts, and feelings is determined by motives about which we are completely unaware. We are not master in our own house. Humans have suffered another blow: the loss of transparency. According to the Libido Theory, all behavior is directed toward the satisfaction of biological needs, either in a directly sexual or a sublimated way. Many motives of human behavior are buried in the Unconscious (Id), and therefore hidden from the individual agent. It is the function of the Ego to channel this energy into modes of expression which are more in accord with the demands of society (Superego). The Id represents the deep inaccessible part of the personality. It is in direct contact with the somatic processes and is the repository for everything inherited and fixed in the real world. The Id has no connection with the real world. We learn about the Id via the analysis of dreams and through various forms of neurotic behavior.[2] There is a reason for its existence: the immediate, unhampered gratification of the instincts. The Id obeys the Pleasure Principle.

The Ego is the manager of the personality. The Ego is the organized, rational, reality-oriented system of personality. It operates according to the Reality Principle: it defers gratification of instinctual urges until a suitable object and method is found. The Ego is entirely pragmatic and without values. While its goal is to satisfy the Id, it will do so only in the context of the demands of reality. It is also charged with maintaining the integrity of the organism.

The Superego develops from the Ego out of a resolution of the Oedipus complex. It represents the ideas and values of society as they are presented to the child through the words and actions of the parents. (Punishments lead to conscience; rewarded behavior leads to the ego-ideal.) In summary, the job of the Superego is to inform the Ego of the value of morality; rather than succumbing to lust or expediency it reminds the person to strive toward perfection.

4. The fourth assumption is that only a developmental or historical approach can reveal the cause of patients' present behavior patterns. Present symptoms appear to be connected to past experiences (whether imagined or real).[3] This fourth assumption is important for Freud's inferential practices, for Freud infers the work of the Unconscious from its apparent manifestations in the present lives of his patients.

> For the infantile is the source of the unconscious, and the unconscious thought processes are none other than those – the one and only ones – produced in early childhood. [Freud 1905b, 170]

Freud is interested in not only the "how" but also the "why" of human behavior. The causes of human behavior seem to reside in past experiences. If this is the case,

---

[2] Freud [1910, 17–18] discusses three ways of widening the consciousness: free association, interpretation of dreams, and Freudian slips.

[3] Consider a funny image: imagine a sex-starved hedonist, a black-frock-coated Puritan minister, and a totally humorless computer scientist chained together and turned loose in the world and you have a good approximation of what Freud was trying to show us about human personality. [Phares 1984, 82–4]

Freud's theory will have to rely heavily on inferences from present observational data to some underlying mental structure. [Brown 1962, Ch. I] In order to understand his patients' ailments Freud makes inferences from present symptoms to past events. Not all forms of human behavior give rise to such inferences. By insisting that dreams are the royal road to the Unconscious, Freud tells us that only certain forms of symptoms – apart from dreams, Freudian slips and neurotic behavior are given preferential status – offer glimpses into the unconscious workings of the mind. He assumes that normal behavior offers few glimpses into the hidden motives of human behavior.

> (To) our study of neuroses (...) we owe the most valuable pointers to an understanding of normal conditions. [Freud 1930, 135]

On the basis of these assumptions, Freud argues that psychoanalysis stands on the same "footing" as physics. Let's see how the psychoanalyst argues by comparison with the physicist. [Box 3.1a, 3.1b]

The first premise in the psychoanalyst's argument leads to the belief that psychology was a new branch of science. [Freud 1938, 19, 52] Is this assumption correct? It is important to note that Freud argues by analogy. In fact, as we shall see, analogical reasoning was an important part of Freud's method.

In the first few steps the analogy between psychology and physics seems to hold. But where there are analogies there are also disanalogies, which have to be taken into account. With Freud's assumption of the unknowability of reality in itself (*an sich*) the argument by analogy begins to show signs of strain. While the physicist can continue the argument, especially appealing to independent tests, the psychoanalyst typically has difficulty satisfying the criterion of independent testability of the fundamental assumptions of the theory. (Recall from Chapter II, Section 6.5 that we interpreted falsifiability as testability.)

Box 3.1a The Psychoanalyst and the Physicist debate the methodological status of psychoanalysis

The Psychoanalyst	The Physicist
• The Unconscious is governed by laws.	• Physical systems are governed by laws.
• We can reconstruct these laws from behavioural peculiarities, for instance people suffering from neuroses. [Freud 1938, 32, 40]	• We can reconstruct these laws from observations and experiments.
• We can construct models of the Unconscious.	• We can construct models of physical systems.
• We can never know the Unconscious directly.	• We cannot know "physical reality" directly either, says Freud.

Box 3.1b   The Psychoanalyst and the Physicist continue their debate about the methodological status of psychoanalysis

The Psychologist	The Physicist
• Even if we cannot know the Unconscious directly, we have techniques for probing it.	• We have different and independent methods of confirming the models and laws.
• These techniques include dream analysis, neurotic behavior, and Freudian slips.	• For instance, the diameter of a human hair can be measured directly under the microscope, and indirectly from the diffraction patterns it creates when held in the passage of a laser light.

But the physicist will take exception now.

• These techniques do not uniquely and independently reveal the Unconscious.	• So we have independent tests for the models,
	• We can make precise numerical predictions,
	• We can give genuine explanations of the behavior of physical systems.
• We can and do have a plurality of divergent, sometimes even contradictory models of the Unconscious.	• We often have conclusive evidence that one model is much more credible than a rival model.
	• As *examples* the physicist can cite Copernicanism and Darwinism.

Has the physicist been unfair to the psychoanalyst? Is it not obvious that physics is unlike psychoanalysis? We should blame not the physicist but Freud. Freud claimed throughout his working life that psychoanalysis can be put on the same footing as any natural science. So while the physicist may not have wanted to be drawn into a debate, it is the Freudian claim that invites the comparison. The physicist baulks at the psychologist's suggestion that it is possible to infer the reality of the Unconscious from the techniques on offer. The physicist stresses the need for *independent* testability. It is at this point that the analogy breaks down. Independent testability is an important feature of scientific reasoning. It is a way of avoiding *ad hoc* explanations. These are explanations in which an explanatory device is introduced to reconcile the explanation with the appearances. For instance, Ptolemy used epicycles, deferents, and the equant to reconcile the a priori assumption of uniform circular motion with the observed nonuniform velocities of the planets. But there is no independent evidence for these geometric devices. Creationism saves itself against refutation by the geological fossil evidence by claiming

---

Box 3.2   Popper's version of *ad hoc* arguments

Here is Popper's example of the need for independent testability:	Here is a psychoanalytic analogy:
The sea is stormy today.	*The patient suffers from neurotic symptoms.*
Why?	*Why?*
Because (God) Jupiter is furious!	*Because past experiences have affected the Unconscious of the patient.*
How do you know that Jupiter is furious?	*How do you know that the Unconscious influences the patient's present existence?*
Don't you see that the sea is stormy?	*Don't you see that the patient suffers from neurotic symptoms?*

---

that God planted the fossil records at the moment of creation 6,000 years ago. Freud seems to struggle with the same problem, for (in Popper's words) the only evidence he has for the explanans (the Unconscious) is the explanandum itself (Freudian slips, weird dreams, and neurotic behavior). [See Box 3.2] According to Popper, what makes a theory – a coherent body of statements – scientific is its falsifiability. Popper requires that the logical form of a theory be such that it can be refuted by experience: it must be possible for an empirical scientific system to be in disagreement with the physical state of the world.

What we want is an independent empirical confirmation that Jupiter is actually causally responsible for the state of the sea. This will be difficult to obtain. Unfortunately, it is equally difficult to obtain independent confirmation of Freud's idea of psychic determination. For the central piece in this explanatory account is the assumption of the Unconscious and its properties. This stipulation of the Unconscious invites several problems. Firstly, it looks ad hoc from the point of view of testability. Secondly, it fails the method of eliminative induction (see below). This is where Freud's analysis with physics sadly breaks down. As one critical writer puts it:

> Freud and his successors have pretended that they listen to the "unconscious" of their patients; in fact they made it talk as others make the spirits talk. [Borch-Jacobsen 2005, 388; translated by the author]

This failure is not such a disaster if we are ready to give up the claim that psychoanalysis is a strict science. We can treat it the way Osiander treated heliocentrism and geocentrism. We can say that orthodox psychoanalysis is *one* possible model of mental life. It provides a coherent story but coherence is not a sufficient condition for successful explanation. This concession amounts to an admittance that there are alternative and rival models. We are not in a position to determine which of the competing models is the most adequate in the light of various constraints. The available empirical evidence is compatible with several competing models of the mind. But our desire, which we felt already in the case of Copernicanism, to probe further and weigh our models in terms of their credibility must remain unsatisfied. It is possible that one day evidence that will help us to redistribute the credibility over the space of psychological models will come to light. But at the moment an instrumentalist-pragmatic attitude is the best advice. If the Freudian model of the mind has some therapeutic value, it can be employed to put people's psyche back in order. It does not matter whether the model is correct, as long as people suffering from neuroses show improvement. We conclude: the Freudian model has empirical adequacy but no theoretical validity. We cannot claim that the model is *true* or *explains* the symptoms. At best it helps us understand what might be going on in the patient. We take it as an analogue model. We say, pragmatically, that it works. But we cannot say *why* it works. If we do not place too much demand on its well-functioning, the Freudian model may be said to share a platform with the Ptolemaic model of the universe. If we do not require too much of the geocentric model, it helps us understand, in a tentative way, the planetary appearances. In a similar way the Freudian model gives us a *picture* of what may be going on in a neurotic mind. But in neither case are we given any genuine explanation. For where rival models claim to explain the evidence and the evidence cannot decide between them, we have no genuine explanation. Let us decide then not to follow Freud in his confident claim that psychoanalysis can acquire the lofty status of a hard-core science like physics. This decision only reflects negatively on his model of the mind if we cling to the nineteenth-century creed of scientism. But we may be interested in the model without thinking that it must explain the structure of the human mind.

3.1.3 Freud as an Enlightenment thinker As Freud mentions in his Copernican claim, there was a *hostile* reaction to his doctrine in Vienna of the early twentieth century. Several features of the doctrine outraged people:

1. the postulation of infantile sexuality, as part of the development of personality;
2. the postulation of sexual and aggressive urges as the principal motivators of human behavior, i.e., psychic determinism;
3. the postulation of the Unconscious and the dynamic role it played in our behavior.

Freud was understood as chaining the rational mind to instinctual drives. The postulation of the dynamic Unconscious rendered the human mind non-transparent. People have spoken of a **Freudian revolution**. It must be located in his idea that our behavior is not the result of (purely) rational motives. The Freudian revolution can therefore be seen as a curtailment of the Enlightenment image of the rationality of human beings.

This image of rationality can be attributed to several factors, two of which are

1. the development of science and its paradigm of rationality;
2. the legacy of the Enlightenment and its view of human nature.

The development of science from the sixteenth to the nineteenth centuries was a triumphant march of Newtonianism from mechanical phenomena, like planetary motion, to phenomena like electricity, magnetism, and heat. The mechanization and mathematization of science drew ever-increasing circles, so that science became the paradigmatic model of reason. The Enlightenment extended this paradigm into the realm of society.

As we have seen with Copernicus, Newton, and Darwin, traditional beliefs had been eroded by their scientific discoveries, i.e., beliefs in the centrality of the Earth and in the superiority of human beings over all other creatures. Scientific activity stresses the rational side of human beings; there was thus a feeling that with an increase in science, human beings would increase their mastery of the world. While science destroyed traditional beliefs, it also gave rise to more and more domination of nature, as for instance in:

1. the Industrial Revolution and its development of steam engines; and
2. Pasteur's germ theory of disease, which led to the first vaccination of a human against rabies.

So the success of science emphasized the importance of rationality in human beings. And the rationality had already found a philosophical interpretation in what is called the *age of reason* or the *age of Enlightenment*.

Although Freud constructed no more than an analogue model of the mind, we do not need to be dismissive about his achievements. According to some of his critics, he built a solid, non-conjectural, support-providing worldview. [Gellner 1993, 126] If it is not subjected to rigorous testing, it can serve cultural functions. Freud's theory has had a liberating effect on Western culture. [Gellner 1993, 83; Webster 1996, 283; Cioffi 2005, 46; Rillaer 2005b, 241] With his frank and unbiased approach to sexuality, Freud helped to free humankind from the chains of sexual repression. Insofar as he throws light on the darker motives of human behavior, he became an Enlightenment figure. In his *Three Essays on Sexuality* [1905] Freud reveals a candor and honesty about sexual practices which would do honor to any enlightened thinker. He describes forms of infantile sexuality in great, shocking details for a Viennese audience. He discusses the practice of "normal" and "deviant" sexuality with admirable openness. In an early essay on sexual morality he already speculates that "our culture is founded on the suppression of our instinctual drives." [Freud 1908, 82] He proposes that sexual energies can be channeled into non-sexual forms of satisfaction. He suspects that at the root of cultural achievements lies a process he calls "sublimation." In *The Future of an Illusion* (1927) he returns to the opposition between nature and culture. He holds that every culture is based on coercion and renunciation of instincts. [Freud 1927a, 7]

Culture protects us against nature. For students of Freud there is also a surprising admission. He had always emphasized the significance of the instinctual drives in human affairs. But now he demands a rational justification of cultural obedience. We can only contain our instinctual nature through our intelligence. The primacy of the intellect is a psychological ideal. In the long run nothing can withstand reason and experience. The conditions of human existence can be improved through the employment of scientific knowledge. [Freud 1927a, §IX]

For Freud psychoanalytic theory was part of scientific knowledge. We have already seen that one of Freud's claims is ill-founded: psychoanalysis is not on the same footing as physics. The analogy between psychoanalysis and physics, which Freud constructs, is misconceived. But this analogical argument was based on a misconception of how science works in the first place. Evolutionary biology, as we have seen, is not on the same footing as physics either. Is there another sense in which Freud's psychoanalysis can be shown to be scientific or nonscientific? We should analyze Freud's own procedures, not his rhetorical claims. We should apply to the corpus of psychoanalysis a logic of evaluation.

3.1.4   The scientific status of the Freudian model
  It is not easy to deal scientifically with feelings. [Freud 1930, 65]

How can the Freudians show that psychoanalysis is scientific? How can their critics show that the Freudian theory does not satisfy criteria of scientificity? Consider, first, Freud's own methods and then the procedure of eliminative induction.

3.1.4.1   Freud's methods   Given Freud's medical training, it should come as no surprise that he was quite aware of methodological issues. It would not be wrong to say that Freud, at least unconsciously, defended a form of eliminative induction as the best method of science. Freud certainly often speaks of psychological *inferences* and appeals to *alternative* explanations. Of the inferences, which the psychoanalyst must make, he sometimes claims that they must lead to inevitable conclusions.

> It is gratifying to be able to report that direct observation [on children] has fully confirmed the conclusions arrived at by psychoanalysis – which is incidentally good evidence of the trustworthiness of that method of research. [Freud 1905a, 193–4 n.; cf. 201, 205 n.]

> The possibility of giving a sense to neurotic symptoms by analytic interpretation is an unshakeable proof of the existence – or, if you prefer, of the necessity for the hypothesis – of unconscious mental processes. [Freud 1916b, 279]

But at other times he no longer makes Cartesian claims about the certainty of psychological knowledge. He even shows willingness to accept the Unconscious as a mere hypothesis. [Freud 1905b, 177–8] He complains, again stressing the analogy with science, that inferences in chemistry are accepted unquestionably but psychoanalytic inferences are contested. Freud accepts that there are no direct proofs of the meaning of psychic events; however, there are *degrees of probability* of their respective analyses. He clearly regards the psychoanalytic interpretation of symptoms

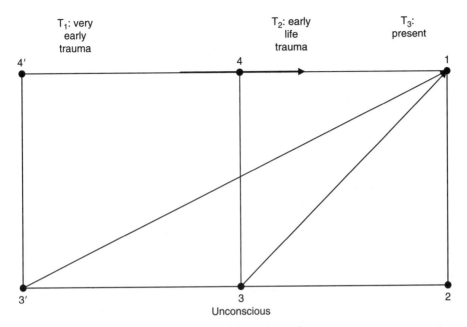

Figure 3.1 Direction of time, from left to right. This diagram depicts the various levels involved in Freudian inferences. [Gellner 1993, 224]: (2) and (3) are not directly observable and hence not independently testable, in principle, because (a) the theory asserts the uniqueness of the psychoanalytic tool of inquiry; (b) they are subject to interpretations which cannot be checked independently against raw data. (4) is not easily testable: what matters was not public events but their meaning for the patient. The links (4) to (3) and (3) to (2) can easily be amputated and abandoned. *Source*: Gellner [1993], 224

as more probable than rival interpretations. The reader may recall from the preceding chapters that Copernicus and Darwin also made the claim that it is possible to draw inferences from the available evidence to the most likely model of explanation. They regarded them as admissible inferences. Freud too, although less explicitly, regards inferences from the evidence to the psychoanalytic model of the mind as the most probable admissible inference. [See Figure 3.1]

> The mental life of human individuals, when subjected to psychoanalytic investigation offers us the explanations with the help of which we are able to solve a number of riddles in the life of human communities or at least to set them in a true light. [Freud 1916a, 168]

On the one hand, the inferences serve to ascribe a mental structure to the appearances, just like in any other natural science (Freud 1938, 196); on the other hand, the inferences do not all possesses equal degrees of probability. Freud clearly thinks that some inferences are more probable, in the light of the evidence, than others. [Freud 1916a, 51, 238; 1916b, 300–2]

Freud at least displayed a partial understanding of the need for a consideration of competing models. In his work on dreams, jokes, sexuality, slips of the tongue, and taboos he considers rival models but finds them inadequate. According to Freud, they do not explain the evidence sufficiently. From this rejection of rival models Freud concludes, hastily, that his own psychoanalytic model must be the correct model. Unfortunately the evidential basis for his claim is very narrow: it consists of the interpretation of dreams, Freudian slips, and analysis of the results of free association, which inspired the construction of the psychoanalytic model in the first place. [Freud 1910, 17–18] In terms of inferential practices the Freudian model is not on the same level as the Copernican and Darwinian models. It fails to derive new testable consequences from the model of the mind. It is built to accommodate previously independently known facts: neurotic behavior, slips of the tongue, and strange dreams. On the model of eliminative induction there is as such nothing wrong with building a model to accommodate well-known facts. But Freud's model of the mind should not claim them as supportive evidence. The cited evidence cannot be supportive for the following reasons: (a) the central idea of the Unconscious is not independently testable; (b) no independently known facts, other than those which form the basis of the model, are accommodated in the Freudian models; Freud never shows that the psychoanalytic model is clearly favored by the available evidence at the expense of competing models; it is not shown that competing models are incompatible with the evidence; (c) no deductive consequences follow from the Freudian model because of its lack of coherence.

Recall the methodological views of Haeckel and Huxley about appropriate methods in evolutionary biology. We may use inductive steps from available evidence to the construction of a model. But this makes the model only a hypothesis. This model must then be tested. The tests may take the form of precise predictions, of deductive or inductive consequences, or the accommodation of independently known facts. Freud was a master of formulating a plausible model of the mind but a third-rate pupil when it came to testing the model. Freud claims for his model an inference to the best explanation. But it is not an inference to the best *contrastive* explanation. [See Sulloway 2005a, b]

Freud thought that he could boost the credibility of psychoanalysis by establishing an analogy with physics. But analogies always come in tandem with disanalogies. Freud's reasoning failed because he did not pay enough attention to the disanalogies with physics.

Freud's model of the mind is an *analogue* model (in terms of our earlier discussion of the role of models). But Freud treats his model of the mind as a realistic mechanical model. In his Vienna lectures he holds that the psychoanalyst must operate with the Unconscious as if it were "something palpable to the senses." [Freud 1916b, 279] He depicts his tripartite model of the mind as an extended structure, a mental apparatus, in which the Id, the Ego, and the Superego occupy well-delineated regions. [Freud 1916b, 283; 1938, Ch. VIII; see Figure 3.2] The Freudian conception of depth, as one writer puts it,

leads into error because it encourages us to transform dispositions, cognitive and emotional mechanisms into substances. [Rillaer 2005a, 224, cf. 219, 228; translated by the author]

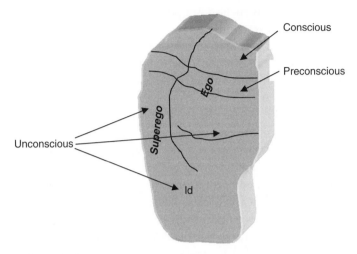

Figure 3.2 A textbook representation of the Freudian model of the mind. *Source*: Adapted from Phares [1988], 80

The role of analogical reasoning is quite widespread in Freud's writings. For instance, Freud claims that the study of neuroses offers the psychoanalyst the most valuable pointers to an understanding of normal conditions. [Freud 1930, 135; 1927b, 165] The role of analogical thinking has not been sufficiently stressed in the Freud literature. Let us look at some examples:

1. There is a comparison of "psychic economy" with a business enterprise. [Freud 1905b, 156]
2. There is an analogy between dream work and joke work because of similarities of techniques. [Freud 1905b, 165, 171; 1927b, 165]
3. There is an analogy between religion and neuroses in children: a child, Freud declares, cannot develop toward culture without passing through a phase of neurosis; similarly humankind, in its development toward civilization, must pass through a religious phase as a neurosis. [Freud 1927a, 43, 53]
4. The system of the Unconscious is compared to a large entrance hall, in which the mental impulses jostle one another like individuals. [Freud 1916b, 295]

There is of course nothing wrong with the use of analogical reasoning if it serves heuristic purposes. But Freud has the tendency to start his reasoning with an analogy that surreptitiously turns into an affirmation. He tends to forget that analogical reasoning is not an *ersatz* for proof.

→ *3.1.4.2 The method of eliminative induction, again* Recall the general procedure of *eliminative induction*. Out of a number of competing explanatory accounts it selects the ones that best agree with the available constraints. This is based on the central idea that not all *positive* instances of a theory are also *supportive* instances of that theory. The positive instances are merely concrete applications of the theory's general principles. Competing theories, however, also claim that they can explain them. Any new bridge construction is a positive instance of Newton's mechanics. The supportive instances test the claims of the theories. For falsificationists this comes in the form of new, precise, testable predictions of a theory. The existence of Neptune was predicted from Newton's theory. When the planet was discovered, it counted as a supportive instance of Newton's celestial mechanics. Imagine that no such precise predictions can be made, because only a number of approximate models are available. It will still be possible for one model to find support in the evidence at the expense of the others. (An analogy is a list of suspects in a murder case. The mounting evidence often converges on one suspect and eliminates all the others.) The evidence may come in the form of crucial experiments. The credibility of a number of rival models is evaluated against the weight of the empirical evidence, derived from crucial experiments. Ideally, only one of the models is compatible with the evidence, while the others are not. In this case the evidence is said to *support* the surviving model and to discredit its rivals. The evidence discredits one model, because it cannot account for it, and it accredits another model, because it can account for the evidence. For instance, if there is well-confirmed evidence of highly elliptical orbits of comets, any planetary models that cannot explain this evidence suffer in their credibility. The basic ideas go back to Francis Bacon and John Stuart Mill.

We discussed a cluster of **features** of this sophisticated form of inductivism in an earlier chapter. The central features are:

- the distinction between **positive** and **supportive evidence**;
- the exploration of the **space** of **possibilities**; this is the conceptual domain of actual or possible rival accounts, which could equally well account for the empirical data; these accounts may be incompatible with each other;
- the problem of the **accommodation** of **known facts** as against the prediction of **novel facts**;
- a conception of what constitutes **rival accounts,** involving different constraint spaces, as opposed to mere **alternative accounts**, within the same constraint space.

Let us measure the standard of Freud's theory by the method of eliminative induction. We can do so by looking at the constraint structures, which get the elimination off the ground. Consider, for instance, how the therapeutic aspect of Freud's theory fares.

Recall that Freud claims at least some therapeutic success of his method. But the method of eliminative induction forbids us to maintain that psychoanalysis is therapeutic merely because some patients improve after treatment. [Grünbaum 1977a, 222; 1985, Pt. I.1, 2, Pt. II.4] Such a strong claim cannot be vindicated by the method of eliminative induction. *Positive* instances of patients recovering from neuroses after having received therapy do not lend support to the therapeutic value of Freudian treatment. It may simply be a spurious correlation, like that between coffee consumption and recovery from a cold. To avoid such spurious correlations we have to exclude competing accounts from the range of possible explanations. Hence to vindicate the strong assertions made by Freudians concerning therapeutic efficacy a number of control procedures should be put in place. [Grünbaum 1977a, §III]

- placebo treatment in control groups (Z);
- non-psychoanalytic forms of treatment (e.g., medication) (X);
- the study of spontaneous remission (Y).

Freudians have not put such control procedures into place. If these possible explanations are not excluded, Freudian therapy cannot claim credibility over its rivals. Freud's tendency to infer from the failure of rival models to the accuracy of his own approach is a non-sequitur as long as there is no supportive evidence for psychoanalysis.

Clinical studies have shown that the reliability of psychoanalytic treatment is not as impressive as the Freudians claim. [See Meyer [2005], Part III, IV] But it would not be scientific by the method of eliminative induction because the positive instances of therapeutic success have not been shown to be supportive instances (of psychoanalysis). Hence the credibility of the Freudian therapeutic theory has not been increased and that of its rivals decreased. Only the combination of positive instances with instances of non-X (i.e., non-psychoanalytic treatment modalities fail) and non-Y (i.e., there is no spontaneous remission) and non-Z (placebo treatments do not have the same effects) could constitute inductively supportive instances of Freudian therapeutic theory. So the logical possibility of such supportive instances goes hand in hand with the logical possibility of refuting instances. [Grünbaum 1977a, 232; Gellner 1993, 199–203, 208–9]

Freud seems to have forgotten a lesson which he spelled out clearly at the beginning of his career:

> Anyone (…) who is engaged scientifically in the construction of hypotheses will only begin to take his theories seriously if they can be fitted into our knowledge from more than one direction and if the arbitrariness of a *construction ad hoc* can be mitigated in relation to them. [Freud 1895, 302]

There is yet another area where Freud's analogical reasoning covers rather than uncovers the deep structure of his system. Reasoning by analogy with the natural sciences helps Freud to hide the fact that his system cannot rely on empirical evidence alone; it must borrow from meaning relations, characteristic of the human sciences. This need to borrow symbolic relations is evident from the fact that neurotic symptoms, dreams, and slips of the tongue are subject to interpretations.[4]

---

[4] Toward the end of his life Freud proposed to replace the term "interpretation" by the term "construction": "If, in accounts of analytic technique, so little is said about 'constructions,' that is because 'interpretations' and their effect are spoken of instead. But, I think that 'construction' is by far the more appropriate description. 'Interpretation' applies to something that one does to some single element of the material, such as an association or a parapraxis. But it is a 'construction' when one lays before the subject of the analysis a piece of his early history that he has forgotten …" [Freud 1937, 261] But this switch in terminology is not accompanied by an enhancement of the evidential basis of psychoanalysis. "If the construction is wrong, there is no change in the patient; but if it is right or given an approximation to the truth, he reacts to it with an unmistakable aggravation of his symptoms and of his general condition." [Freud 1937, 265]

The interpretations are guided by the pre-accepted "validity" of psychoanalytic theory.

3.1.5 Freud stands between the empirical and the hermeneutic models   In an age of scientism Freud was understandably eager to establish the scientific credentials of psychoanalysis. He never ceases to compare psychoanalysis with the respectable sciences. If the comparison is not with physics, it is with mathematics:

> In fact psychoanalysis is a method of research, an impartial instrument, like the infinitesimal calculus. [Freud 1927a, 36]

So Freud positions psychoanalysis firmly on the side of the empirical model of the social sciences. It is a science, like the most respectable natural sciences, which operates with hypothesis, empirical evidence, laws, and confirmation. According to Freud, this positioning of psychoanalysis is possible because his science rests on psychic determinism as one of its pillars. "Psychic arbitrariness does not exist." [Freud 1901a, 680] If no randomness exists in mental phenomena, we must expect that they are governed by laws of nature, like the planets. As we shall see, the conviction that human behavior is governed by social and psychological laws was common to the founding fathers of the empirical model of the social sciences. From his famous *Dream Interpretation* (1901) to his late *Outline* (1938), Freud insists that mental life is based on lawful regularities. It is the job of psychoanalysis to discover the psychic laws. The problem for Freud is that neither dreams nor slips of the tongue nor behavior patterns constitute raw objective data. The data need to be interpreted. Psychoanalysis is just one tool to provide the interpretation. The need to *interpret* psychic material points Freud clearly in the direction of the hermeneutic model of the social sciences. According to this approach, human behavior has symbolic dimensions, which requires the need for understanding. Which format does understanding take in Freud? It takes the form of inferences. Freud infers from the observable symptoms to an underlying cause, which is represented by the model of the Unconscious. [Figure 3.2] Although Freud pretends that the Unconscious exists as part of an extended apparatus, this model is already an interpretation. But

> interpretation doesn't burrow down to an ultimate kernel of unconscious material, which reveals the full and incontestable meaning of the dream. [Cohen, 2005, 41]

Grudgingly, Freud admits this difficulty when he concedes

> that psychoanalysis is made difficult by the fact that it can only reach its data as well as its conclusions after long detours. [Freud 1905a, 201]

Recall that Freud compares the psychoanalytic procedure to a similar procedure in the physical sciences. In both areas we attempt to reach understanding of some phenomenon through model construction. The trouble is that the core assumption of the Freudian model is not independently testable. It is by definition hidden in the recesses of the mind. As evidence for the Unconscious, psychoanalysis cannot offer hard facts, only pre-interpreted data, as they reveal themselves in dreams, Freudian slips, and abnormal behavior. Freud is correct in pointing out that the natural sciences also make inferences to unobservables. For instance, scientists infer the nuclear processes in the sun from the spectral analysis of sunlight on Earth. Kepler inferred elliptical orbits from Brahe's data. Why should the Freudian inferences attract such hostility? On the one hand, Freud offers only a narrow range of techniques and fails to show how the evidence credits his approach while simultaneously discrediting other approaches. On the other hand, Freud cannot establish the scientific credentials of his model of the mind without borrowing techniques that belong to the analysis of meaningful utterances in human affairs. Habermas has drawn the conclusion that Freudian psychoanalysis is essentially an interpretational affair. It is not an empirical science but a hermeneutic enterprise. [Habermas 1968, Ch. III.10; 1983, 214, 230] But, as we shall argue, the social sciences generally, not just Freud, stand between the empirical and the hermeneutic models.

3.1.6 The role of mind in the social world
The problem of other Minds is central to the social sciences. [Hollis 1994, 160]

The evaluation of psychoanalysis as an empirical science has led to a negative judgment. Rather than pointing to the natural sciences, psychoanalysis points us to the social sciences. In the social sciences we encounter a rather different philosophical atmosphere. Here the fundamental question is whether the social sciences constitute a sphere of inquiry of their own – as the hermeneutic model claims – or whether this sphere of inquiry shows some overlap with the natural sciences – as the empirical model holds. We are therefore justified to consider Freud in relation to the social sciences. As we have seen, the unconscious mind played a central role in Freud's view of human affairs. It served to explain individual human behavior and was the basis of culture. Whether we adopt a hermeneutic approach or an empirical approach to the social sciences, it is difficult to deny that symbolic dimensions play a significant part in them. Whether it consists of simple gestures in daily encounters or of whole institutional frameworks, human behavior tends to be more than a mere physical act. It tends to have meaning such that social actors are constantly, yet unwittingly, engaged in the interpretation of the behavior of fellow humans. As Searle points out:

> When we engage in voluntary human actions we typically engage on the basis of reasons and these reasons function causally in explaining our behaviour, but the logical form of the explanation of human behaviour in terms of reasons is radically different from standard forms of causation. [Searle 2004, 212]

With the awareness of the role of Other Minds in the social world it is also time to move beyond Freud. The social world consists of symbolic actors whose interactions are the object of study of disciplines as diverse as anthropology, economics, psychology, and sociology. They form the cluster of disciplines called the social sciences. Just as the natural sciences, they engender distinct philosophical problems.

4   The Social Sciences beyond Freud

The stuff of the mind is the stuff of the world, and so the investigation of the rich structure of the world provides a clearly observable and empirically tractable – if not royal – road into the hidden countries of the mind. [Tooby/Cosmides, "The Psychological Foundations of Culture" (1995), 72–3]

The tension that runs through Freud's work marks the history of the social sciences. Just like Freudian analysis, the social sciences emerged in the climate of scientism at the end of the nineteenth century. This problem situation immediately posed a philosophical challenge to social scientists. Should the social sciences, whose object of study is society, lean toward the natural sciences or the human sciences? This unavoidable philosophical question still divides minds. In order to appreciate the philosophical dimensions of the social sciences, to see the similarities and dissimilarities with the natural sciences, it is important to move beyond psychoanalysis and consider other views of social life. We will discuss the roots and principles of two standard models in the social sciences. They are conceptual models, formulated in sympathy for and opposition to the natural sciences. Following this discussion we will turn our attention to some typical philosophical problems in the social sciences.

4.1   Two standard models of the social sciences – some history[5]

It is fundamental to science that opinions be evidence-driven. [Earman, Bayes or Bust (1996), 201]

The social sciences became emancipated from philosophical thought toward the end of the nineteenth century. The intellectual climate of the nineteenth century was dominated by classical physics, Darwinism, and philosophical reactions to the Enlightenment. Philosophical models of the social sciences – their ontology, their methods and objectives – reflect these currents in thought. Two competing models emerged. (Between the 1930s and 1970s a third model, the critical model or Frankfurt School, emerged, which forged a link between the Marxist tradition and Weber's methodology.) We will call the competing models the *naturalistic* and the *hermeneutic* model, respectively. They descend from different ancestry. The naturalistic model – sometimes also the *empirical* model – saw its roots in classical

---

[5] For introductions to the philosophy of the social sciences, see Delanty/Strydom [2003]; Smith [1998]; Rosenberg [1995]; Hollis [1994]; Little [1991]; Braybrooke [1987]; Papineau [1979]; Thomas [1979]; Wright [1971]; Ryan [1970]; Weber [1968], Ch. I; Habermas [1968]; Habermas [1970]; Nagel [1961], Ch. 14; Winch [1958]; Popper [1957].

physics. The hermeneutic model – sometimes also the *interpretative* model – saw its ancestor in the discipline of history. In this section we shall focus on their respective roots and fundamental principles.

4.1.1   The naturalistic model   The intellectual roots of the naturalistic model lie in France, at the beginning of the nineteenth century. [Hayek 1955, Ch. II] France had gone through its own version of the Enlightenment. French Enlightenment philosophers had put emphasis on empiricism as the source of all knowledge, the division of political powers, and had expressed a strong admiration for Newtonian physics. The French Revolution imposed a radical upheaval on the institutional structure of French society. In this climate two social thinkers stamped their ideas on a philosophical model of the social sciences: Henri de Saint Simon (1760–1825) and Auguste Comte (1798–1857).

We find in Saint-Simon's sketch the main characteristics of a naturalistic program for the social sciences:

- *the unity of method*: the unification of scientific knowledge and methods under one program;
- *physicalism*: the use of the physical sciences and their toolbox as adequate tools for the natural *and* the social sciences;
- *the unity of approach*: an approach to nature and society based on scientific reasoning.

Saint-Simon's approach was new: the problem of social organization was to be treated "exactly in the same manner as one treats other scientific questions." [Hayek 1955, 135] Saint-Simon clearly anticipated themes that were to dominate social science thinking in the nineteenth century. In 1813 Saint-Simon already observed that all sciences must become "positive." Auguste Comte would later speak of a "philosophie positive." According to the positivist program, sociology had to become a natural science. [Hayek 1955, 138–42, 168–88; Jonas 1968 Vol. II, 95–115] For sociology to become positive means that it must deal with its subject matter as a system of natural facts. As such it abstracts from intentions, purposes, and values. One of the central themes of this epoch was the search for natural laws of human behavior (J. S. Mill) or societal development (A. Comte, K. Marx, A. Smith).

August Comte nevertheless stresses that the facts to which scientific explanation refers only exist in a given historical order. It is within this order that they are constituted as facts. A positive philosophy must establish the fundamental laws of the respective orders in which facts exist. The key to sociology is for Comte the philosophy of history. Comte distinguishes two parts of sociology: the social static and the social dynamic. Together they constitute the basic law of the societal order on which the positive explanation is based. [Table 3.1]

There is a connection between social static and social physics. Sociology becomes scientific by demonstrating how the present social order is a result of the progression of civilization. Comte holds that "there are laws governing the development

Table 3.1 Auguste Comte: Social static and social physics

	Social static	Social dynamic or social physics
Stage of society	Present	Necessary material and intellectual evolution of humankind
Basic principle	Organic solidarity	Law of Three Stages
Basic units	Individuals, family, society (as totality of families) or social organism	Stages: • *Theological*: all phenomena are governed by supernatural forces • *Metaphysical*: things have essences; nature is governed by unobservable principles • *Positive*: search for the real laws of observable phenomena
Basic mechanism	Division of labor	Unidirectional succession of three stages
Socio-political structure	Submission under governmental authority	From military to industrial existence

of the human race as definite as those determining the fall of the apple." [Comte 1853, Ch. I; Hayek 1955, 178, 138] Furthermore, the fundamental character of all positive philosophy is to regard all phenomena as subject to "invariable natural laws." The laws, in turn, had to be unified to the smallest possible number. [Hayek 1955, 171, 175] However, Comte does not believe that all sciences can be unified by reduction to physics. Rather, there exists a positive hierarchy among the sciences. The emergence of sociology presupposes the development, in linear order, of mathematics, astronomy, physics, chemistry, and biology. Thus, sociology has more affinity with biology than with physics, as the use of the terms "social organism" and "organic solidarity" in Table 3.1 may suggest. Comte does not share Saint-Simon's view of society as a real machine, based on industry. In his exposition of positive philosophy, Comte makes repeated appeals to the observable phenomena. He rejects the search for "final" or "primary" causes. All the sciences deal with observed facts. For Comte this appeal constituted the crowning of sociology as a positive science. This view became, however, the bone of contention. Can the objects of the social sciences be treated as external facts? Are social facts identical with natural facts? Can social facts be related in the same way as physical facts? Is the assimilation of the social sciences to the physical sciences appropriate? Saint-Simon and Comte take it for granted that these questions can be answered in the affirmative. However, these questions hide assumptions, which were soon to be cast into doubt. The *hermeneutic* model sought to base the social sciences on the study of history, not the philosophy of history. Proponents of the hermeneutic model at the end of the nineteenth century often referred to the social sciences as moral sciences, human studies, or *Geisteswissenschaften*. These alternative denotations reveal that defenders of the hermeneutic model rejected the equation of social facts with natural facts. While natural facts are devoid of meaning, purpose, and value, such categories are intrinsic features of social facts.

4.1.2   The hermeneutic model

We must conclude, I think, that nomological slack between the mental and the physical is essential as long as we conceive of man as a rational animal. [Davidson 1970, 223]

In order to see the origin of the hermeneutic model, we must move from France to Germany, where the Enlightenment had a different emphasis. In the late eighteenth and early nineteenth centuries, Germany was the home of Idealism. Toward the middle of the nineteenth century, the rise of a historical consciousness occurred: Marx reacted against Hegelian objective idealism and Nietzsche criticized the blind devotion to traditions. Historical consciousness encouraged reflections on the role of language and history in human affairs. It was consistent with the tradition of the German Enlightenment that German philosophers and historians placed strong emphasis on questions of method. The term "hermeneutic" originally refers to the art of understanding, in particular the interpretation of biblical texts and historical documents. But German historians also began to reflect on the nature of their discipline. The second half of the nineteenth century saw the birth of *historicism* in Germany. Historicism emphasized the importance of the historical development of events, which it saw as unique. Historical and social events were seen as particular events, owing to their embeddedness in cultural traditions and nationalistic contexts. The human studies, as they were known, were at that time preoccupied with philo-

sophical discussions of the appropriate methods for their disciplines. For the philosophy of the social sciences, two men stand out for their contributions. (1) Johann Droysen was probably the first to distinguish terminologically between explanation and understanding (*Verstehen*). [Droysen 1858; Apel 1979, 15–6; Gadamer 1975] This distinction served to distinguish the natural sciences, especially mathematical physics – whose aim was taken to be *explanation* – and the historical sciences – whose aim was taken to be *understanding*. According to Droysen, understanding is an inference from external expressions to internal events. Freud, of course, follows a similar procedure. That is, humans express mental events in external or public utterances such that every "utterance mirrors internal occurrences."

Johann Gustav Droysen (1808–84)

Such public utterances use the vehicle of language, just like historical documents. The social scientist infers from public expressions their intentional meaning. Both the public utterances and historical documents include dimensions of meaning.

However, the interpreter of public utterances is caught in a *hermeneutic circle*: a single expression is understood ("derived") from a context of totality; and the meaning of this totality is inferred ("induced") from the single expression (or a number of such expressions). "Totality" may refer to a culture, a language, or even a paradigm or worldview. Thus, a symbol or gesture must be understood within the culture in which it is embedded; and a "culture" must be reconstructed from the single expressions of it. This hermeneutic circle is not just a curiosity of Droysen's approach. It is an important feature of mental phenomena for, as Davidson says, "we make sense of particular beliefs only as they cohere with other beliefs, with preferences, with intentions, hopes, fears, expectations and the rest. (…) the content of a propositional attitude derives from its place in the pattern." [Davidson 1970, 221; cf. Gadamer 1975, Part II, Ch. I]

In direct opposition to positivism, Droysen raises a claim to the *methodological* autonomy of the historical sciences. Historical research does not seek explanation, i.e., the deduction of later events from earlier events with the help of "necessary" laws. If historical research could be reduced to explanation, then historical and social life would be without moral dimensions, without freedom and responsibility.

Wilhem Dilthey (1833–1911)

(2) But it was Wilhelm Dilthey who set the tone of the explanation–understanding debate. [Dilthey 1883; 1894; 1906] Dilthey's thinking on the subject of "understanding" is important for two reasons: (1) Dilthey's philosophy seeks to justify an autonomy claim for the social sciences; (2) Dilthey's position shifts from an early subjectivist to a later objectivist view of the notion of understanding. As a result of this transition his model for the social sciences shifts from introspective psychology to objective history.

Let us consider first Dilthey's justification for the autonomy claim of the social sciences. There exists a fundamental difference, he claims, between the natural and the social sciences. The difference can be captured by reflecting on the structural relationships which the natural and the social scientists entertain, respectively, with their subject matter. The details of the matrix provide Dilthey's justification for the autonomy of the social sciences. [Table 3.2]

Table 3.2 Dilthey's matrix

Science	Subject of study	Object of study	Structural relation	Method
Natural science	Natural scientist	Animate and inanimate matter	Subject–object relation	Explanation
Social science	Social scientist	Symbolic human affairs	Subject–subject relation	Understanding

On the one hand, there is the *subject–object* relationship, considered to be characteristic of the natural sciences. The natural scientist forms part of a group of people who share a common paradigm. According to Kuhn, a paradigm consists of symbolic generalizations, exemplary problem solutions, particular views of nature and reality, and a science-specific value system. By using a paradigm, the scientist approaches the natural world with the aid of some symbolic mapping in order to describe and explain natural processes. We can think of the scientist's effort as an attempt to "fit" symbolic thought, which is expressed in theories, models, and mathematical laws, to a natural system. As we know from Kepler, natural systems are subject to certain regularities, which we call laws of nature. Scientists study these regularities and express them in symbolic form (e.g., $F = m\vec{a}; F = -F; A^3 \propto P^2$). These symbolic expressions are the laws of science. Scientists also study the causal order of nature: why stars remain stable, why planets move in orbits, and why species diversify.

What distinguishes the natural from the social sciences is not, according to Dilthey, that they "cover different ranges of facts." [Dilthey 1906, §II (171); 1883, §III (163)] Biology, chemistry, and physiology deal with aspects of human beings, like the human sciences, and yet they are natural sciences. It is not the case that the social and the natural sciences deal with completely disparate objects of study. Rather, it is the *way* these disciplines are related to the objects of study, their facts, that determines the difference.

The study of the *physical* world involves only its external aspects, e.g., the study of its lawful regularities and its causal order. Dilthey calls these aspects "outer reality." [Dilthey 1900, 247]

The human studies, on the other hand, are engaged in a *subject–subject* relationship. There is a social scientist, also forming part of a group of people sharing a common paradigm. But in the social sciences, these paradigms are much less monolithic and command less authority than in the natural sciences. The social scientist also approaches the social world with the aid of some symbolic mapping (ideal types) in order to describe and understand social processes. The approach of the social scientist is heavily influenced by a pre-adopted methodological position. These methodological positions are spelled out in the naturalistic, hermeneutic, or critical models respectively. But the social sciences, in Dilthey's view, are related differently to the social world than the natural sciences are related to the natural world.

The study of the *human world* involves its internal aspects: the study of its symbolic meaning and its purposive order. By purposive order Dilthey means the "inner mental reality" of human beings, the intentional nature of human actions. Freud located much of the purposive order of human behavior in the hidden Unconscious. The aim of the human studies or social sciences resides in the *understanding* of social phenomena. This exercise involves quite different categories: meaning, purpose, value, development, and ideals. [Dilthey 1906, 172, 235–6] Dilthey conceives of understanding as a new and important *technique* in the social sciences. First, there is a reason for the reliance of the human studies on understanding. Only in understanding can the relationship between the inner life and its outer expressions be comprehended. For human actions are influenced – "determined" as Freud would say – by the structure of mental life. It is the mental structure, according to Dilthey, that forms the logical subject of the social sciences. The social scientist abstracts "this structure of experiences from the pattern of man's life." [Dilthey 1906, §1 (170–1)] It is because of the inner reality behind the outer expressions of life that the social sciences are involved in a subject–subject relation. Symbolic meaning meets symbolic meaning.

Both the natural and the social scientists deal with order. Both attempt to order the observational data into a conceptual scheme (a model or theory). [Dilthey 1906, §III (173–5)] The difference between them appears on the object side. [Table 3.3]

Dilthey's insistence on a *mind-constructed* social world marks an important difference between the natural and the social sciences. [Dilthey 1906, 191–5] The appeal to a mind-constructed social world is important for our purposes. In the characterization of the social world as a mind-constructed world, Dilthey reveals an important transformation from a purely subjectivist to a more objectivist view of the notion of understanding.

Droysen had claimed that the study of the human world involves an inference from external expressions to internal events. It refers to "inner aspects" of our mental life, as Dilthey said. If the social sciences must make inferences to the intentional structure of human action, then the notion of understanding takes on a psychologistic connotation. [See Abel 1948] The early Dilthey stressed that the human studies found a firm anchorage only in inner experience, in the facts of consciousness. [Dilthey 1883, 161] Understanding of other social actors was based on lived experience and self-understanding. [Dilthey 1927, 123] However, later Dilthey loosened his fixation on introspective self-knowledge as a model for the human studies. [Dilthey 1906, 176] Rather, humans had to be understood by what is common to them and by virtue of their interrelatedness. [Dilthey 1894, 131] The shared context, in which humans exist, takes on objective forms, owing to the use of symbolic language and the creation of social institutions. The objective forms, which Dilthey calls "objective spirit" [Dilthey 1927, 126], comprise: styles of life, forms of economic interactions, public displays of morality, the existence of law, state, and religion, of art, science, and philosophy. Some of these forms are more objective than others: it may be hard to define the French way of life but forms of economic interactions can be arranged along a continuum from free

Table 3.3 The difference between the natural and the social sciences, according to Dilthey

Natural sciences	Social sciences
• Deal with pre-given world of interrelated physical systems	• Deal with pre-given *mind-constructed* world of interrelated social systems
• Deal with causal order	• Deal with purposive order
• Physical changes occur as a result of blind forces and are described in the laws of science	• Social changes occur as a result of conscious human interventions but they are not based on laws

markets to directed economies. Moral and legal regulations are often written in public, coded form and the state, religions, the arts, science, and philosophy operate in observable institutional frameworks. These forms of collective life are maintained and modified in the present but their roots are grounded in the past and their existence extends into the future. It must be said that Dilthey never freed himself completely from the view that "understanding rests on the relationship of expressions of life to the inner meaning, which comes to expression in them." [Dilthey 1927, 137] Yet he did not relapse into psychologism. Dilthey stresses that the mind-constructed world is a system of interactions, which creates values and realizes purposes. In this respect the social world is not a causal order of nature. Every social fact is a human artifact; it has a historical dimension. [Dilthey 1906, 192; cf. Gadamer 1975] The operation of understanding reveals the "objectifications" of life, that is, the external rather than the internal aspects. Human intentions, the "inner aspects" of human existence, can become externalized. The externalizations are visible in many kinds of structural systems: cultural, economic, and political systems are embedded in observable institutional frameworks.

Dilthey's position in the social sciences anticipated many of the leading themes in twentieth-century philosophy of the social sciences. Many philosophers affirm the autonomy of the social sciences with respect to the natural sciences by virtue of the reason-based nature of human existence. [Hollis 1994, 160; Davidson 1970, 223; Searle 2004, 212] But social scientists and philosophers began to question Dilthey's view that understanding is a *technique* which defines the procedure in the social sciences. Rather they characterized understanding as a *mode* of social existence (Habermas, Gadamer, Giddens). What consequences does this have for the methodological status of the social sciences? Second, Dilthey's inability to shed the psychologistic trappings of his notion of understanding was seen as leading to the threat of relativism in the social sciences. [Apel 1979, 36] Is relativism an insurmountable problem in the social sciences? Third, it was asked to what extent Dilthey's dichotomy between explanation and understanding was an appropriate characterization of the similarities and dissimilarities between the natural and the social sciences. In the light of this discussion we can see that Freud could not avoid his balancing act between an empirical and a hermeneutic model. In view of the historical roots of the social sciences (in France and Germany, respectively) the balancing act is unavoidable. It lies in the logic of the problem situation in which the social sciences are situated. A number of *classical* questions arise from this situation:

1. Can the tension between explanation and understanding be resolved?
2. Do the social sciences have to adopt relativism or can they embrace a position of realism?
3. Can the social sciences be a value-free activity?
4. From which point of view must societal relations be explained (individualism or holism)?

We approach these questions through a consideration of the essential features of the modern versions of the naturalistic and the hermeneutic models, respectively.

Once we have discussed these essential features, further methodological questions appear as consequences of the problem situation of the social sciences:

5. The question of the autonomy of the social sciences in view of the role of mind in the social world.
6. The role of predictions and explanations in the social sciences.
7. The role of models in the social sciences.
8. The question of the existence of laws in the social sciences.
9. The question of underdetermination of social science models with respect to empirical evidence.
10. The existence of causal relationships in the social sciences.

4.2   Essential features of social science models

The hermeneutic view of psychoanalysis has been given philosophical backing by the distinction between reasons *and* causes. *[Stevenson/Haberman,* Ten Theories *(2004), 170]*

The naturalist and hermeneutic models emerged in a climate of scientism. They are reactions to the success of the natural sciences, especially to classical physics and, to a lesser degree, evolutionary biology. The vacillation of Freud's position between explanation and understanding bears witness to this problem situation. The modern social scientist is in the same situation as Freud – a philosophical commitment to a model of the social sciences is unavoidable. [Rosenberg 1995, xiii, 3] It lies in the logic of the social sciences. Their object matter is, broadly speaking, society. So the question of how to approach this object matter immediately arises. As we have seen, the question divided social scientists from the beginning. For the naturalistic model the right approach was to follow the footpath of the natural sciences and adopt, with modifications, the methods of the natural sciences. For the hermeneutic model, the methods of the natural sciences are incompatible with the object matter of the social sciences. The social sciences require a distinctive method, which the proponents of this model identified as *understanding*. All the philosophical problems of the social sciences immediately follow from this basic choice of direction. As we shall see, Max Weber and later writers like Popper attempted a compromise position, which leads to a weak form of naturalism. Their basic methodology is that of ideal types.

4.2.1   Essential features of the naturalistic model   A *first* decisive feature of the naturalistic or empirical model is a commitment to the **unity of method**. This is the belief that the methods of the natural and social sciences are similar. This similarity implies that suitable modifications can transform the methods of the natural sciences into appropriate methods for the social sciences. Thus Durkheim stipulates, as the first and basic rule of sociology, to consider social phenomena as things, i.e., as external objects. [Durkheim 1895, Ch. II. 1] For Durkheim the intentional states of individual actors – their motivations, reasons – need not be considered.

The analysis of society must be objective. It must employ objective concepts, which are possibly quantified. The social scientist must strive to describe the general rather than the individual manifestations of social life. Social, political, and economic structures as well as legal and religious institutions are the objects of social research. A commitment to the unity of method is a common feature of many proponents of the empirical model. Popper stresses that there may be differences in degree between the social and the natural sciences but no differences in kind. [Popper 1957] Both adopt a hypothetico-deductive method, inspired by Hempel's *DN* or *IS* model of explanation.

We cannot assume, as the founding fathers of the empirical model did, that the social sciences have law-like generalizations at their disposal. The social sciences often work with statistical generalizations; for instance, economists can state what percentage of the households in a particular country earns an average salary and what that average salary is. These generalizations are often local in character and riddled with exceptions. Consider, for instance, a medical explanation according to which the probability, p, of recovery, R, for sufferers, S, who are given penicillin treatment, P, is very high: $p(R, S.P) > 0.8$. Envisage two time frames: T_1 is a moment in 1950 and T_2 is a moment in 2000. It is a well-known fact that the probability of recovery for treatment with penicillin has decreased from T_1 to T_2. The decrease is due to the evolution of penicillin-resistant bacteria. While at T_1 there indeed existed a relation of high probability between P, S, and R, this probability may have fallen by as much as 50 percent. Similar situations exist in the social sciences. Nevertheless, they are statements of a general kind, e.g., economic cycles, aggregate demand and supply curves, mobility trends in a population, and the classification of households according to socioeconomic criteria. On the basis of such generalizations, social scientists will make pattern predictions. These are

> predictions of some of the general attributes of the structures that will form themselves but not containing specific statements about the individual elements of which the structures will be made up. [Hayek, quoted in Barrow/Tipler 1986, 140; cf. Woodward 2003, Ch. 6.4]

But the status of these patterns must be established: are they like laws in the natural sciences?

A *second* decisive feature of the empirical model is its insistence on **causal analysis**. Durkheim, the father of modern sociology, had already insisted on the need for causal analysis. The proponents of the empirical model stress that the social sciences, just like the natural sciences, must be able to provide causal explanations of social phenomena. Freud certainly treated dreams, slips of the tongue, and neurotic symptoms as the effects of past causal conditions in an individual's life. Similarly, sociologists seek to explain the existence of the division of labor in many societies and forms of social stratification. Economists construct models to explain consumer behavior and types of market societies. The purpose of these constructions is to answer "why"-questions. The causal analysis in the social sciences can cover very diverse phenomena. For instance:

- In 1999 the Home Office in Britain predicted that the number of burglaries and thefts would increase by almost a third in a short time span of two to three years. The expected rise in crime is a consequence of the rise in the young male population in Britain. In this instance we have a causal analysis and a prediction rolled into one. The increase in the number of burglaries and thefts is blamed on an increase in the number of young adults and a growth in the amount of stealable goods. The latter are the causal conditions, which are said to lead to the effect, if no other conditions interfere.

In an earlier chapter we discussed causal models of explanation. But which model of causation is most appropriate for the social sciences? A moment's reflection shows that we cannot causally explain social events along the line of a causal–mechanical model of physical events. Such a model seeks to trace the physical conditions which lead from antecedent causal conditions to subsequent mechanical effects. A causal–mechanical model may explain why a planet orbits the sun. But it cannot explain why World War II occurred. Nevertheless, the above example demonstrates that social scientists seek causal explanations of social events. It is a separate issue to formulate an appropriate philosophical model of causation, in terms of which these findings can be expressed.

Before we consider such questions as to whether "social" laws exist or how causation happens in the social world, another question arises. Do the changes we observe, as a result of long-term or short-term historical or social developments, originate on the individual or collective level? This *third* feature refers us to the distinction between **individualism** and **holism**. Many authors adopt a position of individualism. The thesis is that there exist (causal) regularities underlying social phenomena, which reflect facts about individual agents. These facts about individual agents express their intentions, reasons, and motives. Some authors, like J. S. Mill and S. Freud, assume that psychological "laws" exist on the level of individuals. But the discussion of Freud has made us suspicious of such an assumption. [See Davidson 1970; Little 1991, 18] Individualism is therefore a bottom-up approach. Any large-scale macro-social effects are said to be reducible to facts about individual members of society. At the individual level, psychological mechanisms may be at work, like the avoidance of cognitive dissonance or an individual's desire to improve their social conditions.

Durkheim did not believe that an individualist account of causation would work. He pursued a holist line. He rejected the idea that all macro-social changes could be reduced to individual factors. The economic activity of a society, Durkheim suggested, could not be reduced to individuals' desire to acquire wealth. [Durkheim 1895, Ch. V.2] Social facts cannot be explained by reference to individual facts or even a collection of individual facts. If large-scale social phenomena cannot be explained "from the bottom up" then individualism cannot account for social phenomena. Durkheim prefers a top-down approach. He points out that social phenomena exert a force on the individuals, so that they must possess a nature of their own. The explanation of social facts must be sought in the nature of society. Society transcends the individual both in space and in time. The whole is more than the sum of its parts. For instance:

- A causal relationship between the drop in crime rate and legalization of abortion has been claimed. [*Scientific American* December 1999] Two American economists proposed the causal hypothesis that the annual drop in crime rates, observed across the US since the early 1990s, is causally related to the legalization of abortion in 1973. The legalization reduced the number of unwanted children; according to the economists, unwanted children are more likely to commit crime.

Freud appealed to similar explanations to understand crowd behavior. As crowds exercise an influence over the behavior of individuals, crowd behavior is qualitatively different from individual behavior. In a crowd, says Freud, individuals display "lack of initiative, weakness of intellectual ability, lack of emotional restraint and delay and the inclination to exceed every limit in the expression of emotions." "Man is (not) a herd animal but rather a horde animal," e.g., individuals are led by a chief. [Freud 1921, 117, 121] It remains to be discussed whether holism must treat society and social institutions as separate entities. [Section 4.4.6]

4.2.2 Essential features of the hermeneutic model

In truth, it is hardly possible to give a description which has general validity. We find the most different reactions in different individuals, and in the same individual the contrary attitudes exist side by side. [Freud 1931, 233]

As we observed before, the hermeneutic (or interpretative) model stands in opposition to the empirical model on many specific issues. This opposition can be derived, in part at least, from the general opposition to the unity of method. Proponents of the hermeneutic model argue that it is wrong to analyze human societies and human relations in terms of general laws and causal relationships. It is wrong because the conceptions in terms of which we explain social events are logically incompatible with concepts in terms of which we explain natural events. [Winch 1958, 95] Already Wilhelm Dilthey, one of the founding fathers of this approach, had argued that human life can only be understood in categories which are useless in knowledge of the physical world. Human behavior has to be understood in terms of meaningful categories, i.e., "purpose," "value," "development," and "ideal." [Dilthey quoted in Hollis 1994, 17]

From this general stance, we can derive several points, which stand in contrast to those made on behalf on the naturalistic model.

1.   Social phenomena must not be regarded as external objects, as Durkheim had urged. Social relations are internal relations. Social actors conceive of their own behavior in terms of meaningful concepts. The social scientist must include these concepts in her/his description of social life. The concept of understanding (*Verstehen*) is essential in this description. If you do not understand the meaning of the social action, you do not understand social life. Social phenomena are to be analyzed in terms of intentions, reasons, social rules, norms, and conventions.

[Winch 1958, 42] The symbolic concepts by which social scientists describe them are part of the social action itself. The concept of some social action, like intentional greeting – and also the network of concepts which surrounds the concept of civil behavior – is internal to the social action of the individuals, whose existence the social scientists describe as "being civil." There must therefore exist an overlap between the vocabularies of the social scientist and the social agent. The social scientist must be a participant observer. Social actors already conceptualize social action, irrespective of the presence or absence of the social scientist; social scientists must capture this conceptualization.

2.   In this connection, proponents of the hermeneutic model emphasize a further distinction between, on the one hand, relations, like cause and effect, which are *external*, and, on the other hand, relations, like reason and action, which are *internal*. Natural events are said to be governed by cause–effect relations, while human behavior is said to be governed by reason–action relations. The thesis is that what prompts human behavior are reasons, not mechanical causes. [Ryan 1970, 117–22] The relation between cause and effect in natural events is quite different from the relation between a reason and an action. This distinction again serves to emphasize the difference between a naturalistic and an interpretative approach. According to naturalism, the connection between reasons (desires, beliefs) and action is a *causal* connection. According to interpretivism, the connection between action and reason is a *logical* connection.

The logical connection argument links reasons and actions. The reason serves to explain why an individual action happened: "She raised her arm because she desired to greet someone." "He spoke rudely because he intended to insult someone." Giving a reason is a re-description of the action. It seems that reason and action are not logically independent of each other, as they must be if they were cause and effect. But even though it may be true that an action cannot be divorced from its intentional content, when we re-describe it, this does not mean that the reasons cannot play a causal role in the occurrence of the action. The reason causes the agent to perform the action. Reasons often have a causal component. [Rosenberg 1995, Ch. 2; Davidson 1963; Mackie 1980, 120–1, 287–96; Levison 1974, Chs. 4, 5]

The proponents of the hermeneutic model are right that there are differences between reasons and causes. We should not infer from these differences, however, that human behavior cannot be explained causally. Recall the examples of causal explanations mentioned in connection with the discussion of the empirical model (the link between an expected rise in crime rates and the increase in the young male population or the drop in crime rate and the legalization of abortion, respectively). The causal explanations hypothesize about probabilistic, rather than deterministic, relations, referring to social groups rather than individual social agents. But we seem to have a cluster of sufficient conditions, probably incomplete, which are candidates for antecedent causal conditions which bring about the observable consequent effects (drops or rises in crime rates), under ceteris paribus assumptions. The examples suggest that it is perfectly possible in the social sciences to state sufficient conditions, which tend to increase the probability of the effect.

In normal intentional actions we have two necessary conditions: (1) there must be a logical relation between the reason and the action. The intention to do X and doing X are logically related and we capture this relation in our re-description of the action. But our ascription of the intention to the agent can be mistaken. (2) The reason must have played a causal role in the occurrence of the action. This causal component of the reason may not be identical with the publicly stated motive of the agent. If the action occurs, there is a causal component in reason-explanations. Furthermore: (3) we can have the intention to do X and not do X. "The way to hell is paved with good intentions." (4) Often individuals and social groups act because they simply follow social norms and rules. [Braybrooke 1987, Ch. 3; Salmon 2003] They exist independently of individual social actors.

Durkheim's strict empirical model is mistaken in neglecting the meaning component in causal explanations. The pure interpretative model is also mistaken in neglecting the causal component in reason-explanations. There is a need for a more considered model of the social sciences, which is able to combine meaning relations and causal ascriptions. We shall encounter such a compromise model in Weber's methodology of ideal types.

There are some further notable differences between the naturalistic and the hermeneutic approaches.

3.   The interpretative approach denies that there are universal or cross-cultural social phenomena. It will admit only the cultural variability of symbolic systems. [Little 1991, Ch. 4] Instead of looking for the existence of social laws, or social regularities, this approach insists on the importance of local rules. It often explains individual behavior by reference to particular intentions. But social behavior is explained by reference to social rules, norms, and conventions, which are particular to individual societies. Social action, for instance, may be a case of rule-following. [Braybrooke 1987; Little 1991] This opposition creates a contrast between the universalism inherent in the naturalistic model and the particularism of the hermeneutic model.

4.   Even though we are now firmly ensconced on the side of understanding, it is still possible to inquire whether the analysis should be pitched at the holistic or the individual level. The question is whether our *understanding* of social phenomena should be grounded in individual actions or in social institutions. Weber makes a distinction between meaningful and social action and regards the social action as the sum of the meaningful actions of the individuals.[6] It is also possible to take the

---

[6] Weber defines social action as "an action in which the meaning intended by the agent or agents involves a relation to another person's behaviour and in which that relation determines the way in which the action proceeds." [Weber 1978, 7, cf. Weber 1968, 4, 22; italics in original] "Not every kind of human contact is social in character: it is only social when one person's behaviour is related in its *meaning* to the behaviour of other people. For example, a collision between two cyclists is a mere occurrence, like a natural event. But when they try to give way to each other, or when they engage in insults, fisticuffs, or peaceful discussion after the collision, this does count as 'social action'." [Weber 1978, 26; italics in original; cf. Weber 1968, 23; Winch 1958, 45–6]. Elster [2007], with his "belief-desire-model of action" defends a contemporary version of individualism.

holist line and to determine all meaningful behavior as social in nature. It can be meaningful only if governed by rules, and rules presuppose a social setting. [Winch 1958, 116] There is thus the same division between *individualism* and *holism* as on the empirical approach.

Looking back over the differences between the naturalistic and the hermeneutic models we see to what extent Freud is poised between these two positions. On the one hand, Freud displays naturalistic tendencies. Psychological facts are external facts, subject to the laws of the dynamic Unconscious. But this approach does not work because Freud makes faulty inferences and cannot rely on a pool of objective raw data. On the other hand, Freud has to resort to techniques of interpretation. Freud's theory of hidden meaning was an attempt to interpret individual behavior – neurotic symptoms, Freudian slips, and dream symbols – in terms of underlying sexual motivations. Thus a neurosis reflects the unhealthy repression of the Id by the Ego. A person's sexual desires must be unearthed in meaning analysis to give an account of their present behavior. Freud extended this approach to explain the existence of many social institutions and cultural forms. Freud interpreted the existence of cultural products – Dilthey's objectifications – as a result of the work of sublimation. Sublimation is characterized as the ability to divert sexual energy from direct sexual satisfaction to indirect forms of satisfaction. These indirect forms manifest themselves as cultural products.

There is clearly a need for a compromise position, which will be able to weave strands from the empirical and the hermeneutic traditions into a coherent model for the social sciences. As we turn to questions of methodology, we shall see that Weber's ideal-type approach to the social sciences is such a constructive compromise.

4.3   Questions of methodology

It is more promising in scientific work to attack whatever is immediately before one and offers an opportunity for research. [Freud 1916a, 27]

From our discussion it has emerged that the use of symbolic mappings is as essential for the social scientist as it is for the natural scientist. The social world is extraordinarily complex, so the social scientist must employ idealizations and abstractions in order to describe and explain social reality. The symbolic mappings used by the social scientist invite a number of methodological questions, which are quite specific to the social sciences. One issue concerns attempts to clarify the notion of *Verstehen*. Another issue to be addressed is what kind of models the social sciences employ. We have already illustrated the use of causal explanations in the social sciences, but what shape does a philosophical model of causation for the social sciences take? This question is rather tightly connected with the issue of social laws. Whether such laws exist or not in turn has an impact on the role of explanations and prediction in the social sciences. Finally, questions of realism, relativism, and underdetermination arise, as well as the issue of functionalism and reductionism.

→ *4.3.1 What is* Verstehen?   We have already pointed out that understanding in the interpretative model means that social action must be analyzed by reference to social rules (norms, conventions) and individual action by reference to intentions (motive, reasons). But if *Verstehen* is to be a method in the social sciences, as Dilthey intended, the question that arises is "What is this operation called *Verstehen*?" There are three ways of characterizing *Verstehen*: (a) in terms of empathy; (b) in terms of ideal types; (c) in terms of social conditions.

Let us briefly consider these three ways of understanding "understanding." The empathy model (a) has its roots in nineteenth-century historicism. As we have stressed, historicism is the view that different ages and cultures have to be understood on their own terms. The method recommended to achieve this aim was *empathy*. Empathy is the presumed ability to immerse oneself retrospectively in a culture and its bearers and understand them from within. [Abel 1948] We understand a given human situation – be it an individual action or a correlation between general events – if we can apply to it a generalization based upon personal experience. We have already seen the weakness of this approach from the early phase of Dilthey's notion of *Verstehen*. Dependence upon personal knowledge is the first limitation of this approach. Most interpretations will remain mere expressions of opinion, subject only to "tests" of plausibility. A second limitation to this use of *Verstehen* is that it is not a method of confirmation. As we have learned from our consideration of Freud, we cannot conclude from the affirmation of a permissible inference that it is also an admissible inference. Freud offers a coherent account of psychic phenomena, a possible connection between present symptoms and past conditions, but he fails to show that it is also a probable connection. From the point of view of the empathy model, any possible connection between motives and actions tends to be a certain connection. In any case, the test of the actual probability of the explanation calls for the application of objective methods of observations, e.g., experiments, comparative studies, statistical calculations. [Abel 1948, §III]

Max Weber formulated a different model of *Verstehen* (b), which does not rely on empathy. What is interesting, however, is that the concept of understanding is not simply abandoned. Weber insists that this concept is central to the social sciences. Weber starts out by saying that social action manifests regularities. But what distinguishes regularities in human action is that they display symbolic meaning, which must be interpreted. But, as Weber points out, ascribing meaning to human action, and claiming to have "understood" this action, does not yet show that it has empirical validity. For two identical types of action, which are observable, may be based on quite different motives, the most evident of which is not necessarily the actual one. Therefore Weber requires that the method of understanding must be

Max Weber (1864–1920)

counterbalanced by the method of causal ascription (*kausale Zurechnung*). It consists in the ascription of rational types of motives to the action. [Weber 1913, 97] To achieve these causal explanations we need rational constructions or ideal types with reference to which empirical reality is compared. [Weber 1914, 303–4] Consider, for instance, an economist who wishes to understand the purchasing behavior of economic agents. The economist will develop a model, based on her/his understanding of human behavior in a particular society, which assumes that typical economic agents have stable preferences, avail themselves of optimal information about the market, and that this market is in equilibrium. [Becker 1986] Note that the model of the economic agent makes a ceteris paribus statement about what typical consumers in such idealized situations would do. The social scientist is in the business of building idealized, hypothetical constructions, to which empirical reality can be compared in terms of its deviations. [Weber 1914, 304–5] The assumptions the economist makes must be confirmed as they may not be correct in all markets. In a nutshell, this is Weber's methodology of ideal types. [See Section 4.3.2]

According to Weber, then, the social sciences need to understand the rules and values of a particular society; on this basis they construct ideal-typical behavior. These ideal types are then checked against empirical reality. If they deviate too much, the models need to be modified. Weber proposes a gamut of idealizations with which to approach empirical cases. They provide "grades of understandability" of social actions. [Mommsen 1974, 221] Consider one of the most important ideal types of social action (apart from emotional and traditional behavior):

- Purposive-rational action (*zweckrationales Handeln*). In this type of action,

    the agent may use his expectations of the behavior of external objects and other human beings as "conditions" or "means" to achieve as the outcome his own rationally pursued and calculated purposes. [Weber 1978, 28, cf. Weber 1968, 24–6]

    According to rational choice theory, an economic agent clearly perceives the objectives and chooses the means which s/he subjectively considers the most adequate to achieve the objective. The maximizing utility behavior of economic agents or individual firms in the concrete marketplace is treated *as if* they were seeking rationally to maximize their expected returns. [Friedman 1953, 21] The understanding of human action in terms of purposive rationality has, according to Weber, the highest degree of "evidence." [Weber 1913, 107; Mommsen 1974, 221]

The function of ideal types is to understand social action in terms of these models. The messy behavior of actual social agents must be assessed in the light of the ideal types:

    For example, in explaining a panic on the stock exchange, it is first convenient to decide how the individuals concerned would have acted if they had not been influenced by irrational emotional impulses; then these irrational elements can be brought in to the explanation as disturbances. [Weber 1978, 9; 1968, 6]

The ideal type serves as a hypothetical model, which can explain empirical cases, when abstraction is made from deviations, due to irrational factors (emotional impulses, errors). [Weber 1913, §2] A social science, which aims to describe and explain general aspects of social life, cannot sum over what actual individuals do; it must average over what typical individuals do and think. Social science is interested in the typical consumer, motorist, voter, who are modeled in ideal types.

By combining explanation and understanding, Weber hopes to achieve *explanatory understanding*. [Weber 1913, 107–9; Weber 1968, 8; cf. 1978, 11] The aim of Weber's work is to formulate a compromise position between the empirical and the hermeneutic models. Weber rejects the Diltheyian contrast between explanation and understanding, which was supposed to be the distinctive difference between the natural and the social sciences. Consider a social action which occurs with some frequency, like a large number of people entering and leaving a particular building with some predictable regularity. Stating this frequency in terms of numbers does not help to understand that action. Constructing an ideal type of purposive rational behavior explains nothing about the frequency of the action in society. The social scientist must combine both aspects. The statistical frequency of some action is to be regarded as "explained" when its meaning is understood and interpreted. The particular building may be a university; this sets the symbolic frame, in which the movement of people makes sense. The interpretation of meaningful behavior, even when it seems evident, is at first only a hypothesis, which is subject to tests. [Weber 1913, 108] We achieve explanatory understanding of social action in two ways: either in terms of the rational motives, which the model assigns to individuals and which are confirmed in general; or in terms of social norms and values, which hold in a society and guide social action on average. Weber writes:

> In all these cases, the term "understanding" refers to the interpretative grasp of the meaning or pattern of meanings, which are either a) really intended in a particular case (as is normal in historical inquiry), b) intended by the average agent to some degree of approximation (as in sociological studies of large groups) or c) constructed scientifically for the "pure" or "ideal" type of a frequently occurring phenomenon (this can be called "ideal-typical" meaning). [Weber 1978, 12; 1968, 9]

(c) In recent contributions to the methodology of the social sciences, the emphasis on *Verstehen* has shifted from the earlier Diltheyian conception as a method which was supposed to be peculiar to the social sciences, to a new understanding of *Verstehen* as generic to all social interactions. [Giddens 1993; Habermas 1970 (1988); 1981 (1984); Gadamer 1975] There is still a contrast between the natural and the social sciences. It is not a contrast of methods but one between a *single* and a *double* hermeneutic.

If social action has symbolic meaning then the social sciences must possess features which are absent from the natural sciences. Consider the contrast between *nature* and *society* and by extension between the natural and the social sciences. [Giddens 1993, 20–1, 9, 86–7; Habermas 1970, 87 (1988, 13–4)] If by *nature* we understand the set of natural processes which evolve irrespective of human volition and planning, then nature is not a human product; it is not created by human action.

Darwin wrote that by nature he meant only the "aggregate action and product of many natural laws." [Crombie 1994, Vol. III, 1752] If by *society* we understand the ensemble of social relations which humans entertain, then society is created and re-created by the participants in every human encounter. This production of social life by lay actors involves frames of meaning. This symbolic meaning is derived from cultural traditions in society; it is also continued and changed by the interaction of social actors. [Gadamer 1975, Ch. II]

Hence the data with which the natural and the social sciences deal, respectively, do not have the same structure. The social sciences deal with a pre-interpreted world; they stand in a *subject–subject* relation to their "field of study." Their field of study is the pre-interpreted world of the social actors. The social actors uphold and develop the symbolic social world; the social actors' symbolic world enters into the construction and production of that world. In this sense the construction of social theory involves a *double hermeneutic*. [Giddens 1993, 154; Habermas 1981, 159, 162 (1984, 107, 110)] The social scientist must interpret a social world which already exhibits symbolic meaning. In this sense the social scientist must become a participant observer. The problem of *Verstehen* arises already beneath the threshold of a theoretical reconstruction of the data. It arises on the level of producing and obtaining the data. The situation of the double hermeneutic, with which the social sciences are faced, has a bearing on the attitude of the social scientists toward their subject matter. *First*, the social scientist must map her/his symbolic constructs (ideal types) onto the frames of meaning involved in the production of social life by lay actors. *Second*, technical concepts and theories invented by social scientists can in turn be fed back into the social world. They become constituting elements of that very subject matter they were coined to characterize; by that token they alter the context of their application. [Giddens 1993, 86] That is, technical terms (social class, globalization), invented by the social scientist to grasp social reality, can be appropriated in the social world, becoming part of the symbolic dealings of social agents. The concepts employed by social scientists are dependent upon a prior understanding of those used by lay people in sustaining a meaningful social world. [Giddens 1993, 59, 155–62] Hence, the sociological observer cannot construct a technical metalanguage that is unconnected with the categories of natural language. [Habermas, 1970, 202, 263 (1988, 105, 152)]

This double hermeneutic has, according to Giddens, no parallel in the natural sciences. [Giddens 1993, 154] They stand in a *subject–object* relation to their field of study. The natural world is not a pre-interpreted world awash with symbolic meanings. The natural world consists of natural systems, biological and physical processes, devoid of symbolic dimensions. The natural scientist maps symbolic constructs (equations, models, theories) onto this world. For this reason it is a *single hermeneutic*. The equations, models, and theories are symbolic constructs but they are mapped onto a world without symbolic dimensions. [Figures 3.3a, b]

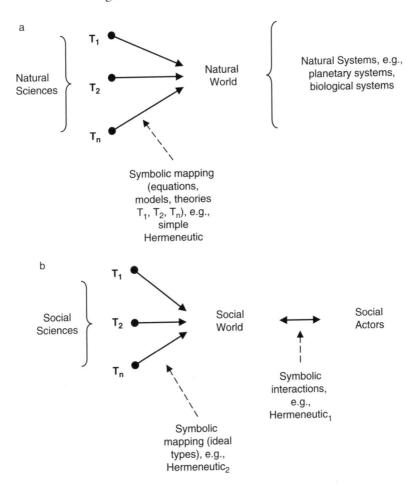

Figure 3.3 (a) The simple hermeneutic; (b) The double hermeneutic

If interpretative understanding is a ubiquitous phenomenon in social life and if this reflects on the work of the social scientist, how can the social sciences remain objective? The need for objective inquiries in the social sciences must be squared with the need for interpretative understanding. It must also be squared, as Weber would add, with explanatory understanding. In the following section we look, first, at Weber's methodology of ideal types, and then address the question of the compatibility of *Verstehen* with objectivity.

→ *4.3.2 Weber's methodology of ideal types*
An "ideal type" in our sense (…) has no connection at all with value-judgements, and it has nothing to do with any type of perfection other than a purely logical one. There are ideal types of brothels as well as of religions (…).
[Weber 1904, 98–9]

According to Weber, the aim of the social sciences is fourfold: (A) The study of social reality in its cultural context, and the explanation of social behavior with the help of hypothetical regularities. Weber seems to deny that there are strict laws in the social sciences, so any social regularities have only hypothetical character. (B) It is characteristic of social action that its regularities are meaningful. We can understand social action when we understand its symbolic meaning. We may, however, misinterpret a social action, i.e., ascribe the wrong meaning or intention to it. The ascription of meaning by a social scientist, or even a lay social actor, must be subject to some empirical test. According to Weber, it is possible to understand some social action by ascribing a rationality motive to it. Weber distinguishes various rationality motives but held that purposive rationality enjoyed the highest measure of evidence. [Weber 1913, 97; Mommsen 1974, 221] If the ascription assigns purposive rationality to the action, according to which the social agent seeks the best means to achieve some end, it should be possible to test this ascription of meaning. This is Weber's combination of the empirical and hermeneutic strands of the social sciences. The empirical strand serves to explain social action with the help of confirmed, hypothetical regularities. The hermeneutic strand serves to provide the meaning of these social regularities. [Weber 1904, 216; Habermas 1970, 83–91 (1988, 10–6)] Later Habermas generalized this idea to the claim that all social actions are associated with validity claims (see below). (C) A further aim of the social sciences is to provide a historical explanation of the emergence of social systems. Weber's famous rationalization thesis is an attempt to explain the rise of modern capitalism out of the adoption of a puritan lifestyle. [Weber 1904–5; Tawney 1922] (D) Finally, Weber holds that the social sciences are able to make limited predictions of the future effects of social actions. Sociologists predict a rise in juvenile crime rates; economists predict the effect of a drop in interest rates on the economy. They are not long-term predictions, and they come attached with a ceteris paribus clause.

Weber clearly sees *Verstehen* as a condition of social life. But even if symbolic dimensions run through social life like a spider's web, *Verstehen* must also be an objective operation. It is not a method specific to the social sciences. Already the work of the natural scientist requires understanding of natural phenomena in terms of model-building. [Weinert 2004, Ch. 3] The social scientist is engaged in a double hermeneutic dimension in that s/he maps meaning onto preexisting symbolic structures. The axis of the single/double hermeneutic gives us a way of interpreting Weber's insistence on *explanatory* understanding. We should conceive of *Verstehen* as model-construction. We find its widespread use in both the natural and the social sciences. In the social sciences it takes the form of ideal types.

Weber regards ideal types as theoretical constructions; both in a logical and a practical sense they are *model types*. [Weber 1904, 97; see Chapter I, Section 6.5] Ideal types are not meant to be descriptions of social reality. They do not serve as schemes, under which a real situation or action can be subsumed as one *instance*. [Weber 1904, 92; Weinert 1996] Rather they serve as models, which allow the social scientist to abstract from irrelevant or particular details in order to concentrate on the essential features of a cultural or social phenomenon. They are idealizations in the sense of focusing on the structure of a social phenomenon under investigation. The social scientist, according to Weber, is interested in what is typical of a certain social agent, situation, or system. The ideal type brings out the essential core of a social phenomenon like authority, capitalism, or feudalism. Ideal types allow a systematic characterization of what is common to many individual cases in the social world. They highlight what is structurally important in social phenomena. Ideal types are characterized by logical consistency and conceptual precision. They are *pure* types. For instance,

> "Charismatic authority" shall refer to a rule over men, whether predominantly external or predominantly internal, to which the governed submit because of their belief in the extraordinary quality of the specific *person*. [Weber 1915, 295–6; italics in original]

Weber suggests that the use of ideal types is mandatory for working out the cultural meanings of social interactions. [Weber 1904, 90–3] The operation called *Verstehen* is therefore best understood as an operation of model-building. The ideal type then becomes a certain type of *model*.

Weber tends to regard ideal types as mere heuristic devices. They are sharply defined but their purity comes with a price. It is immaterial whether the historical or social reality actually corresponds to the respective types. Ideal types are yardsticks, against which the historical or social reality can be assessed as a deviation. According to Weber, ideal types have four functions:

- They highlight the essential components of historical and social phenomena ("feudalism," "imperialism," "authority," "rationality," "secularization").
- They serve as limiting concepts. Modern Western societies are "free markets" or have "democratic political structures" in the limit of ideal conditions. We can define the model of a democratic society, with its fundamental principles of freedom and equality. Concrete societies can be characterized as approximations to the limiting concept. Different societies approximate the limiting concept in different degrees but no society is identical with the ideal of a democratic society.
- They serve to reconstruct the "developmental sequences" of events in history. Weber devoted much of his work to the reconstruction of the growth of modern capitalism out of the Protestant ethic. Marx tried to reconstruct the development of society from feudalism through capitalism to socialism. Comte saw the development of society from a theological to a metaphysical to a positive stage.
- They help to define the notions of *adequate causation* and *objective possibility* in the social sciences. Weber believed with the proponents of the empirical model that it was possible to construct causal accounts in the social sciences. Social scientists seek causal explanations of social events. As we shall see, the belief in causal accounts in the social sciences depends on the appropriateness of philosophical models of causality. Was Hitler's dictatorship the causal condition that plunged the world into World War II? As we shall argue, Weber's notion of adequate causation encapsulates a philosophical model of causation in terms of necessary and sufficient conditions, which allow the historian to speak of the

causal conditions that propelled Germany toward war. In this sense, ideal types spell out objectively possible situations.

Weber was well aware of the changing nature of ideal types. Just as in the natural sciences, models in the social sciences must be adaptable to new evidence. Therefore ideal types cannot be mere impositions on social data on the part of the social scientist. Just like models in the natural sciences, hypothetical models in the social sciences must be adaptable to the changing social world. Ideal types, like "class society," may have been adequate in nineteenth-century industrial Britain, but the social world has changed to such an extent that such terms no longer serve to model British society adequately. [Dahrendorf 1959] By contrast, a new phenomenon like globalization requires the construction of a new ideal type, which is able to model that phenomenon. Societal conditions change and this must be reflected in the social scientist's choice of model. Whereas sociologists in the past were able to characterize European societies in terms of a hierarchy of social classes, their modern counterparts prefer the description of their respective societies in terms of social stratification. This change in conceptualization reflects a change in social reality. It is in principle no different from the modifications of the Copernican or Darwinian models as more evidence came to hand.

Similar to the natural sciences, the social sciences use a number of different models. The supply and demand curve of the economist is a *functional* model, since it shows how one parameter (price) is functionally dependent on the interplay of two other parameters (supply and demand). Where the supply and demand curves meet, the intersection determines the price. [Fig. 3.4] Freud used both *analogue* reasoning and mechanical *scale* models of the mind. He argued, as we have seen, on the basis of an analogy with physics to the scientific validity of psychoanalysis. But Freud also employs the tripartite structure of the mind as a scale model. It is supposed to be a mechanical model, with dynamic connections between its three parts. As Freud often speaks of libido energy, which the psychic apparatus must channel and control, an appropriate metaphor for this apparatus is that of the steam engine. [See Rillaer 2005c, 430; Freud 1916b, Ch. XXII]

What kind of models are ideal types? Ideal types are models in the hypothetical sense. They are *as-if* models. They model the socioeconomic reality *as if* it consisted *only* of the parameters included in the model. *Homo economicus, homo sociologicus, homo psychologicus* – they all "reduce" human beings to a few parameters, deemed relevant for the descriptive and explanatory purposes of the economist, the sociologist, or the psychologist. They all abstract from irrelevant features of human beings, they idealize the relevant features to the level of significance, and systematize the relevant parameters to form a coherent account of aspects of human behavior. Understood as hypothetical models, the ideal types may express an algebraic structure but mostly are concerned with topologic structure, which approximate the structure of empirical reality.

Although we stressed the similarity of models in the social and the natural sciences, the problem of closure in the social sciences imposes some limitation on this similarity. *First,* the distinction between essential and accidental features, between relevant and irrelevant conditions, is much more difficult to draw in the social sciences. The laboratory physicist is often in a situation to define both the experimentally relevant and irrelevant conditions. For instance, at the beginning of the twentieth century Rutherford and his co-workers discovered the atomic nucleus. They did so by measuring the scattering of ionized helium atoms off gold atoms, embedded in a gold foil. They could assume that the presence of electrons within the gold atoms had no impact on the measured trajectory of the scattered helium atoms. This assumption was justified by energy considerations from classical physics. Weber stresses that the function of an ideal type is its "comparison with empirical reality in order to establish its divergences or similarities." [Weber 1914, 43] But Weber's ideal types are not quantitative scientific concepts. There will often be no more than a qualitative consensus among, say, historians, as to which factors are likely to be relevant to concrete historical situations. As we shall see, this problem is particularly serious in causal considerations in the social sciences. Historical evidence will, however, often help the historians to delineate clusters of significant factors, with some overlap between them. We should therefore expect that there exist some empirical constraints on the admissibility of ideal-type models. *Second*, it is difficult to decide whether certain deviations of the model from empirical reality are genuine or apparent deviations. When the wind blows a leaf through the autumn air, it is clear to physicist and layperson alike that the leaf's trajectory is only an apparent deviation from Newton's law of gravitation. The leaf's trajectory is compatible with the law of gravity. Deterministic physical laws do not allow genuine exceptions. While the social sciences deal with social regularities, it is far from clear whether they constitute genuine laws. As we shall argue below, such social regularities are mere trends. But trends are compatible with both apparent and genuine exceptions. When an exception to a trend occurs, it is difficult to say whether it constitutes a violation of the regularity pattern. *Finally*, the degrees of idealization and approximation that the model can achieve are a function of the extent to which relevant factors can be separated from irrelevant factors. In a physics laboratory, the parameters are largely under the control of the experimenter, including the presence of "noise." But the social scientist does not enjoy this level of control over the parameters. This is the problem of closure. Even in laboratory-type experiments involving human subjects it is not possible to define and control all the parameters that enter the experimental situation. Human subjects in such situations have a history and form part of a tradition whose effects on the present situation cannot easily be estimated. The models portray what Weber called "typical" cases. In the absence of quantitative information it is difficult to measure degrees of approximation and idealization. What, for instance, is a typical consumer? And how much does the real consumer deviate from the typical consumer?

Despite these imponderables of model construction in the social sciences it remains nevertheless the case, as Weber stresses, that there are objective ways of constructing ideal types and other models in the social sciences. Functional models depend purely on statistical data. Hypothetical models have been constructed by way of psychological experimentation, by dint of statistical evidence, and by comparison of historical and contemporary processes. [Sherif 1936; Milgram 1974;

Weber 1904–5] Weber regards model-construction as an objective procedure. Explanatory understanding in interpretative sociology is therefore comparable with methods in the natural sciences. Contrary to the findings of the empirical and hermeneutic models, we must conclude that the social sciences are characterized by a **weak naturalism**. [Little 1991, Ch. 11; Popper 1957, 121–3] It is a weak naturalism because (unlike the naturalistic model) it does not assume the existence of social laws and the applicability of the *DN* and *IS* models of explanation. There is a certain continuity between the social and the natural sciences, despite obvious dissimilarities. It lies in the use of certain ideal-typical constructions, which involve hypothetical regularities. They offer the social sciences the chance to provide explanatory understanding of social phenomena.

→ *4.3.3* Verstehen *and objectivity*   Weber constructed a logical relation between understanding and explanation. *Verstehen* served as a method for the construction of ideal types of social action, in particular purposive-rational action. *Verstehen* provides a hypothetical construct, which must empirically be confirmed. The access to social facts must be achieved through interpretative sociology (*verstehende Soziologie*). [Habermas 1970, 164 (1988, 73–4)] Once the ideal type is empirically confirmed, it suggests a social regularity, which serves as an explanation of social actions. [Habermas 1970, 84 (1988, 11)] Thus *Verstehen* is no substitute for the explanation of social action. [Habermas 1970, 146 (1988, 59)]

If social life is pervaded by symbolic meaning, how can the social sciences hope to make objective statements about the social world? Interpretative sociology answers that the social scientist must make rationality assumptions. These are expressed in hypothetical constructions, like Weber's categories of social actions. They are supposed to capture the motives and intentions of a "typical" social agent. In an extension of Weber's approach, Habermas relates different types of social action to different aspects of the world. He distinguishes an *objective*, a *social*, and a *subjective* world. Different types of social action presuppose different types of relations to the respective worlds. And the way the social actor relates to different aspects of the world determines both different aspects of rationality and the rationality of the interpretation of these actions by a social investigator. [Habermas 1981 (1984), Ch. I.4; cf. Weinert 1999]

For instance, *teleological/strategic* action involves a relation between an actor and an objective world of facts. Two types of rational relations to this world are possible: (a) the actor can make assertions about this world, which are true or false, and these can be rationally judged according to their *truth*; (b) the actor can also intervene in the objective world to achieve specific aims, and his/her intervention can fail or succeed. It can be judged according to its *effectiveness*. The rationality involved in these types of action is reminiscent of Weber's objective-purposive and purposive-rational action, respectively. But one of Habermas's criticisms of Weber is that the latter has narrowed types of rationality to purposive or means–end rationality. According to Habermas, Weber's typology of social action suppresses some of the rationality aspects which were part of the Enlightenment project. Thus Habermas further distinguishes norm-regulated, dramaturgic, and communicative actions and relates them to normative rightness, subjective truthfulness, and rational consensus as their standard of validity, respectively.

How can this role of the social scientist as a virtual participant answer to the requirement of objectivity in their understanding of social interaction? [Habermas 1981, 168 (1984, 114)] Habermas postulates a connection between the fundamental types of social action and the methodology of understanding of social actions. He proposes that the problem of rationality poses itself with the interpretative access to the different kinds of worlds. [Habermas 1981, 152, 157 (1984, 102, 106)] The rationality aspects of the fundamental types of social action – the references of teleologic/strategic, norm-regulated, dramaturgic, and communicative action to the objective, social, and subjective worlds, respectively – are principally open to an objective judgment by both the social actor and the social observer. In order to understand a social action or its equivalent linguistic form the interpreter must understand the condition of its validity. He must understand under which conditions the validity claim which is connected with the action – propositional truth, normative rightness, subjective truthfulness, and intersubjective consensus – is acceptable. That is, the interpreter must understand the reasons with which social actors would defend the validity of an action if challenged to do so. [Habermas 1981, 169 (1984, 115)] He must interpret social action rationally with respect to the validity claims which are associated with it. For instance, the effectiveness of strategic action can be measured against the aim which was being pursued. To highlight the objectivity of understanding even further, Habermas claims that the types of rationality (effectiveness, truth and rightness, etc.) have universal dimensions. [Habermas 1981, 188, 193, 197–8 (1984, 130, 134, 137–8)] He makes a distinction between the *universality* of the context-independent validity claims and the concrete, context-dependent linguistic means by which the validity claims are made. [Habermas 1988, 179, 182 (1992, 139, 142)] Social actors make or imply claims to validity in concrete situations, and in concrete speech acts; equally they are accepted or rejected in concrete situations. But in whichever context and by whatever means they are expressed, these validity claims themselves transcend these concrete situations. This is most evident in truth claims: by whatever linguistic means a fact "p" is expressed, it is either true or false "that 'p'" and this fact transcends particular speech situations. In this manner Habermas opposes all forms of relativism, as enshrined in postmodern thinking. He reemphasizes the distinction between context of discovery and context of justification. The validity claims belong to the context of justification. As we shall see, Habermas uses the "universality of reason in the diversity of its voices" as an argument against relativism.

We started this chapter with a discussion of the emergence of two models in the social sciences and their core elements. This discussion naturally followed from a consideration of the balancing act between explanation and understanding in Freud's method. A consideration of the notion of *Verstehen* led from its early embodiment as empathy to its latest incarnation as an inevitable condition of social life. Once the ubiquity of the symbolic dimension of social life is apparent, the question of objectivity of social science knowledge imposes itself. Habermas's answer in terms of validity claims is very much indebted to the Enlightenment view of human nature. It commits social scientists to an investigation of the rationality aspects of human actions, when they are often concerned with the consequences of irrational human behavior.

An interpretive sociology should be able to deal with problems of what Dilthey called the objectifications of social life: the functioning and malfunctioning of institutional organizations, power relationships, and the struggle of various social groups against the "authorities." [Giddens 1993, Chs. IV, V] Perhaps Habermas sees the rationality types and the validity claims, inherent in types of social action, as ideal types (Weber) or a zero method (Popper), against which deviations can be gauged. Such "measurements" will often be of a qualitative rather than a quantitative kind, because of the pervasiveness of symbolic constructs in human life. The presence of other minds or human agency really is central to the social sciences, for it not only affects the interpretative approach of the social sciences to social life but also the construction of ideal types and their objectivity. It also extends its tentacles, as we shall now discuss, to a number of other methodological questions, which arise from the specific problem situation of the social sciences. In the social sciences the concern about whether statements are true of a model world rather than the modeled world are really more pertinent than in the natural sciences.

→ 4.4 Causation in the social sciences

As we observed earlier, the empirical or naturalistic model emphasizes that the causal analysis of social events is not only desirable but essential for the social sciences. We have discussed some examples of causal accounts, on the assumption of both individualism and holism. The question is: How are we to think of causation in the social sciences from a philosophical point of view? If we entertain a mechanistic view of causation, according to which only physical pulls and pushes and their traceable effects constitute proper causation, our assessment of causal accounts in the social sciences will turn out to be rather gloomy. We should not harbor philosophical ideas of causation which produce the wrong results when applied to social science research. It seems that social scientists have no qualms about connecting, say, two events in history or society, such that the prior antecedent event qualifies as a cause of the later consequent event, which qualifies as the effect. Weber made a significant contribution to this topic with his notion of adequate causation as a model of causation in the social sciences.

→ *4.4.1 Weber on causation* As Weber has shown, the theoretical reconstructions or ideal types in the social sciences can cover such diverse phenomena as types of authority, sequences of historical events, and causal chains of events. With his belief that the social sciences can arrange certain events in terms of explanatory sequences, Weber moves away from a pure interpretative model of the social sciences. But explanation is often associated with the availability of regularities. The question then arises as to the nature of such regularities in the social sciences, the existence of social laws, and the availability of causal accounts. Once we have clarified these questions, it will be easier to discuss the nature of explanation and prediction in the social sciences. This aspect raises further questions regarding realism and relativism, as well as reductionism and functionalism.

Weber proposes a notion of *adequate causation*, which he considers to be an appropriate view of causation in the social sciences. For Weber this notion is based on his ideal-type methodology. As we have seen, ideal types can be understood as hypothetical (or *as-if*) models in the social sciences. [Weber 1905; Weinert 1996] The use of ideal types emphasizes the need on the part of the social scientist to use abstractions and idealizations in the causal reconstruction of a social event. While these operations make models refer to the structure of a model world, the model also offers the advantage of relating several parameters in a manageable way. A causal model seeks to relate some occurrence in the social world, which is regarded as an effect, to prior causal conditions. Looking at Weber's own work, his attempt to explain the emergence of capitalism in the West as a result of the adoption of puritan lifestyles is a striking example of causal analysis in the social world. But social scientists also seek to explain, say, the origin of the slave trade, the outbreak of World War II, juvenile delinquency, and differential educational performances. The aim in each case is to isolate as far as possible the actual determinant factors which are likely to have caused some event in history or the social world. Social systems, however, are open-ended. The social scientist is faced with a cluster of potentially determining factors, which could be possible causal conditions. Out of the complex of potential determining factors, the social scientist must distill a complex of causal relations, which "should culminate in a synthesis of the 'real' causal complex." [Weber 1905, 173] Weber speaks of "adequate causation" when the social science model meets several conditions: (a) the social scientist has isolated a number of conditions which are regarded as statistically relevant to the effect in question; (b) the reconstruction of the social or historical event, on the part of the social scientist, probably isolates the "likely cause of an actual historical event or events of that type." The ideal-type model of the causal sequence of social events therefore depicts an objective possibility. That is, it is objectively possible, even likely, that the isolated conditions are causally responsible for the occurrence of the event. The model of the social scientist, which has some claim to probability, provides the most adequate causal conditions which are likely to have brought about the social event in question. How can a social scientist be relatively certain that a proposed causal model of, say, the outbreak of World War II captures the most adequate conditions which are most likely to have brought about this event? Weber insists that "it is possible to determine, even in the field of history, with a certain degree of certainty which conditions are more likely to bring about an effect than others." [Weber 1905, 183] The way to achieve this aim is to submit the ideal-type model of some causal sequence to factual knowledge of a historical or social event. Thus Weber tests the model against reality. In this manner Weber hopes to throw light on the "historical significance" of the actual determinant factors in the emergence of some historical event. It is well known that historians and sociologists disagree about the relevant factors which can be held responsible for some event in history or society. Certain factors will be so improbable that they can be omitted from the causal account. It is implausible, for instance, that the eruption of a volcano will have had an effect on the outbreak of World War II. On the other hand, new empirical data regarding the outbreak of World War II may well credit some factors as the relevant ones and discredit others as irrelevant.

In Chapter II, Section 6.7.3, we discussed causation in terms of a cluster of necessary and sufficient conditions. It is to be expected from the nature of the social sciences and the concern for actual obtaining factors that causal analysis in the social sciences will mostly be concerned with sufficient conditions. These are the actual conditions which occurred prior to an event and from which the causal account must be reconstructed. But Weber felt that his ideal types could also spell out necessary conditions for social phenomena like feudalism and authority. For instance, in the absence of a belief in the "extraordinary qualities" of a leader, Weber would not have spoken of "charismatic authority."

Although social actions can be explained causally in terms of necessary and sufficient conditions, there are disanalogies between causal explanations in the natural and the social sciences.

1. In the social sciences we deal with reasons rather than mechanical causes. Reasons act as causes but we can state neither strictly deterministic nor even probabilistic laws which connect reasons and actions. [Davidson 1974; but see Fay 1994 for an alternative assessment] Any regularities in the social sciences take the form of trends rather than law-like regularities. (As we argue below, trends are inductively generated empirical regularities, which permit both exceptions and reversals.)
2. We cannot state a closed set of jointly necessary and sufficient conditions for acting on a reason. Social scientists will agree on a number of "sufficient" conditions which reasonably explain the occurrence of a social event. There will be disagreement about a "full" list of such sufficient conditions. But it would be impossible to demonstrate that all agreed conditions were jointly necessary and sufficient for the event. The problem of closure prevents a quantitative evaluation of the parameters that enter a causal situation.
3. Let us assume that social scientists have agreed on a cluster of conditions which they consider to be the most likely candidates for a causal explanation of some social event. Such likelihoods are still different from probabilistic occurrences in the natural world. An ensemble of radium atoms has a 50 percent probability of decay in 1,600 years. It is not possible to predict which particular atoms will decay in which particular time span. But it is possible to say, in accordance with the so-called decay law, that half of the collection of atoms will decay in 1,600 years. But in the social world there exist no such laws which would allow us to say that, given certain antecedent conditions, the subsequent event, E, will occur with a given probability p. All we can say is that a certain cluster of conditions makes the social event quantitatively more likely.

We can summarize our discussion by saying that social scientists attempt to isolate a cluster of necessary and sufficient conditions which they regard as adequate to explain the occurrence of a certain event in history or society as the effect which followed from the antecedent conditions. It is a model that depicts an objective possibility. The choice of antecedent conditions will be influenced by the interest of the social scientist. But Weber was adamant that this choice did not render causal accounts in the social sciences subjective. Weber grants that it is not possible to establish numerical relationships between the selected relevant causal conditions, X, and the probability of the effect, Y. Nor is it possible to assess quantitatively the influence of background conditions, U, in the causal field on the effect, Y. Nevertheless,

> we can (...) estimate the *degree* to which a certain effect is "favored" by certain "conditions" – although we cannot do it in a way which will be perfectly unambiguous or even in accordance with the procedures of the calculus of probability. [Weber 1905, 183]

In our estimation of the degree of likelihood of certain causal conditions to be objectively responsible for the effect, we are guided by our empirical knowledge of the facts of a particular historical or social situation. We gain further assistance from

> knowledge of certain known empirical rules, particularly those relating to the ways in which human beings are prone to react under given situations. [Weber 1905, 174]

Weber is obviously referring to the ability of the social scientist to *understand* human behavior through recourse to the operation of *Verstehen*. It enables the social scientist to understand intentions, motivating individual behavior, and norms and values, motivating collective behavior. Humans, then, possess knowledge of general empirical rules which allow them to navigate their way through social life. We have, as Dilthey already stressed, inductive expectations about the relative constancy of the social world around us. What structure do we expect these empirical generalizations to have? Are they similar to lawful regularities in the natural sciences? Or are they more like rules and social conventions by which social actors abide? By speaking of general empirical rules [Weber 1905, 174, 187], Weber seems to lean toward the view that the social scientist deals with empirical generalities rather than abstract "uniformities." [Weber 1905, 168] The question we need to address is whether there are social laws.

→ 4.4.2 *On the existence of social laws* There is a long tradition in social thought which is committed to the view that the social world is as much subject to lawful regularities as the planets in the solar system. Freud, as we have seen, considered that in the "expression of the psyche" there existed "nothing trifling, nothing arbitrary and lawless." The psyche was subject to deterministic psychological laws, which Freud was eager to discover. Freud only stood in a long line of thinkers, starting with the Enlightenment, who considered that the social and individual world was subject to various laws of nature. Marx and Comte believed in a law-like sequence of historical events; so did Adam Smith. It is therefore not an idle question to contemplate whether the empirical regularities that exist in the social world may be regarded as laws of nature. Many other philosophical questions, which emerge in the social sciences, turn on the question of whether there are lawful regularities in the social world. It impacts on questions of causality, on method, on predictions and explanation.

We have already considered philosophical views on laws of nature in a previous chapter. In particular we distinguished between *laws of nature* and *laws of science*. The former are the lawful regularities in the natural world, irrespective of our knowledge about them. The latter are the symbolic representations of the laws of nature as we encounter them in the textbooks of science. We defended the thesis that laws of nature constitute structural constraints on the behavior of natural systems. This view was called the *structural* view of laws. According to this view, laws of nature encode structural features of natural systems. [Weinert 1993; 1995] The *laws of science* express mathematical relations between the relata of the systems. We also argued that various types of models serve to represent the (algebraic and topologic) structure of natural systems. The job of models was to abstract, idealize, and systematize the parameters of the systems. Laws of science can then be regarded as ceteris paribus statements. They hold under certain idealizing conditions or for simplified models. They may lose their validity for very complex systems with conditions that are not represented in the model. Could this characterization be applied to social regularities? The general idea is that social laws must be regarded as ceteris paribus statements. [Kincaid 1994; McIntyre 1994; see Weinert 1997] In an economics textbook we may find such claims:

> Economic science, like the natural sciences and the other social sciences, attempts to find a body of laws of nature. [Parkin/King 1992, 23]

The view is that the social sciences are capable of formulating general statements which are genuine candidates for lawful regularities of the social world. These statements are more than just empirical generalizations or statistical trends. They express law-like generalizations in the social world, albeit of an idealized form.

Social laws may be ceteris paribus statements; however, this would not guarantee that they are genuine laws. Two considerations count against the view that general statements in the social sciences should be accepted as genuine lawful statements. (1) Social "laws" may simply be empirical generalizations, descriptive of a large number of cases. They may simply be inductively generated from initial conditions.[7] [Popper 1957, §§27, 29; Weinert 1997; cf. Woodward 2003, Ch. 4] If this case can be established, we will be looking at social trends rather than at genuine laws.

---

[7] Law statements are universal statements, which state mathematical relationships between variables. Initial conditions provide specific values to these variables. They permit a universal law statement to be applied to a particular situation. Kepler's third law – $A^3 \propto P^2$ – states a mathematical relationship between the orbital period, P, of a planet and its average distance from the sun, A. Initial conditions supply the specific values of the variables, which appear in the universal law statement. If, say, the planet has an orbital period of 8 years, its average distance from the sun is 4 times the Earth–sun distance.

It is characteristic of trends that they are inductive generalizations over a number of observed cases. If they depend on initial conditions, they should not be regarded as genuine laws. There are several indicators that social generalizations bear the hallmarks of trends. One sign is that the interference with initial conditions will change the very nature of the regularities. If this is the case then this constitutes a stark contrast with physical laws. As initial conditions vary from country to country, from culture to culture, social regularities change with them. Genuine laws, like Kepler's laws, do not depend on initial conditions; the latter simply instantiate them. Popper [1957, 128] has taken the dependence of social regularities on initial conditions to defend the thesis that such regularities are *trends* rather than genuine laws. From this dependence a further difference between genuine laws and social trends follows: trends can be changed, even reversed, while genuine laws can be neither changed nor reversed. Knowledge of social conditions can be employed in the service of changing a social trend. If, for instance, a dramatic rise in the crime rate is expected as a result of an increased number of young people, social engineering can be employed to break this link. Educational and employment opportunities for young people may have an effect on the behavior of potential young offenders. What is even more surprising is that knowledge of social trends can be used to reverse this very trend. If social scientists establish a correlation between excessive alcohol consumption and increased road fatalities, political decision makers can then use this very correlation to interfere with this trend and reverse it. But no human intervention can physically alter a law of nature. While laws of nature are structural constraints on the behavior of physical systems, social trends act as *practical* constraints, in most situations. But social regularities are as much affected by the symbolic and social dimensions of human life as other social relations. In particular, social regularities are subject to the double hermeneutic. Consequently, any such regularities fail to qualify as genuine laws.

(2) It is generally accepted that generalities in the social world are "riddled with exceptions." As trends, social regularities are compatible with exceptions in a way that genuine laws are not. Genuine laws permit only apparent exceptions, which can be accounted for in terms of some other regularity. (A feather does not fall to the ground with the same speed as a stone because it is subject to air resistance). Exceptions to genuine laws must remain compatible with the laws. But trends permit genuine exceptions which cannot be accounted for in terms of other generalities. Freud maintained the existence of psychodynamic laws between a person's behavior and their mental states. He accepted, however, that "it is hardly possible to give a description which has general validity." [Freud 1931, 233]

Thus there are genuine differences between the social and the natural sciences. According to John Searle:

> when we engage in voluntary human actions we typically engage on the basis of reasons and these reasons function causally in explaining our behaviour, but the logical form of the explanation of human behavior in terms of reasons is radically different from standard forms of causation. [Searle 2004, 212]

Nevertheless, there are also sufficient similarities between the natural and the social sciences to defend a position of "weak naturalism," as explained in the next section. Despite the lack of genuine social laws the social sciences have the ability to explain and predict social events.

4.4.3 Explanation and prediction in the social sciences As we noted in Chapter II, Section 6.6, it is possible to explain events without being able to predict them. Equally, it is possible to predict events without being able to explain them. The nonexistence of genuine laws in the social sciences will have an impact on their ability to explain and predict social events. If there were no empirical regularities at all, explanation and prediction would be well-nigh impossible. But there are social trends and patterns with fairly good reliability. On the basis of such social generalities, the social sciences can both explain and predict.

As ideal types are models in the social sciences, various types of explanation are open to social scientists. The use of functional models puts the emphasis on *functional* explanations. The most straightforward type of functional explanation is to be found in economics. The fixing of prices through the intersection of the supply and demand curves constitutes the clearest example. [Figure 3.4] There are also functional models of the social stratification of society, according to which social stratification fulfills a positive function for the maintenance of society. The social sciences also make use of *hypothetical* (or *as-if*) model explanations. For instance, Weber proposed to analyze authority in terms of the ideal-typical divisions into charismatic, traditional, and rational authority. Rational authority is based on rules and regulations and institutional structures. In these models authority is treated *as if* it consisted only of elements, which the model ascribes to this type. The social sciences are also capable of using *structural* explanations, which are based on structural models. What is the structure of a free-market society, of capitalism? In such questions it is more the topologic than the algebraic structure which is of interest. As a structure consists of relata and relations, a structural explanation of capitalism will explain how economic agents relate to each other in capitalistic societies. If we accept Weber's notion of adequate causation, the social sciences can lay claim to *causal* explanations of social events. This is made possible through the existence of reliable trends and rationality patterns in the social world. Despite the logical

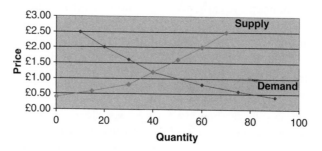

Figure 3.4 Supply and demand curve

distinction between explanation and prediction, there is of course some link between them. The availability of explanations facilitates predictions. If the social world and its actors display patterns of regularities, then the social sciences can construct what the sociologist Robert Merton called "theories of the middle range." [Merton 1968, Pt. I, Ch. II]

Such theories give rise to inferences, e.g., what may be called "middle-range predictions," derived from the patterns of regularity in social life. Economists, for instance, predict economic cycles, market trends, with a certain amount of short-term reliability. In this sense the predictive ability of the social sciences can be compared with evolutionary biology. The predictions are not quantitatively precise, as, say, in astronomy, but they give rise to predictions of patterns of varying degrees of reliability.

But the social sciences have to deal with a specific problem, which Popper called the problem of *unplanned planning.* [Popper 1957, Ch. 21] Popper used the argument from unplanned planning as an objection against the possibility of planning the future direction of a whole society. In the course of pushing society into a certain political direction, say toward a communist society, problems that were not anticipated in the original plan will occur. These unforeseen problems oblige the political authorities to modify their original plan to take into consideration the unforeseen disturbances. Of course the new, modified plan will also run into problems of unplanned planning, which, as Popper argued, makes the idea of the long-term planning of the development of a whole society unfeasible. By implication, long-term predictions are impossible for the same reason: the ever-present repercussion of human agency. Consider specifically problems like *self-fulfilling* and *self-negating* prophecies and decision-making in the absence of stable equilibria. A self-fulfilling prophecy occurs when a prediction causes people to behave in a manner that makes the prediction come true. (A rumor that bread prices will rise causes people to buy more bread as a result of which the bread price rises.) A self-negating prophecy occurs when a prediction causes people to behave in a manner that makes the prediction come out false. (Most people hope that scientific warnings about the effects of global warming will cause people to react in such a way that the catastrophe can be averted.) [Merton 1968, Ch. XIII] Decision making in the absence of stable equilibria can also have an effect on the prediction of behavior. For instance, a lack of clear optimal choices about the best strategy to achieve an aim will render the prediction of human behavior difficult. Rational choices can also be obscured, as Freud pointed out, by the presence of irrational motives in human behavior. [Elster 1989, Ch. IV] These problems occur because of the uncertainty involved in human agency; they occur because predictions in the social sciences are based on trends, not genuine laws; they occur because of the complicated interaction between observers and observed in the social sciences.

4.4.4 Underdetermination We have discussed the problem of underdetermination in a previous chapter. We also proposed certain solutions to this problem in terms of the availability of constraints. Weber, as we have seen, was very aware of the need to fit his ideal-typical constructions to the available data. But the problem of underdetermination is multiplied in the social sciences owing to a number of factors. First, there is the problem of the double hermeneutic. The social scientist maps models onto a preexisting world, which is pregnant with symbolic meaning. Furthermore, social science regularities are dependent on initial conditions and therefore mere trends. These trends are compatible with unaccountable exceptions. And as our consideration of Weber's notion of adequate causation has shown, the social scientist must impose the most likely causal explanation on the available data without being able to deal with closed systems. These problems mean that the social sciences suffer from a greater degree of underdetermination than, say, astronomy or evolutionary biology. There are fewer constraints to eliminate unfit models. There are no genuine laws in the social world and few fundamental theoretical principles on which social scientists will agree. The constraints on the models arise mostly from empirical evidence. Nevertheless, the progress of the social sciences will have to be assessed differentially. There are many social sciences, ranging from anthropology to sociology and economics. In psychology many divergent models compete with each other without any sign of convergence. It appears that on the 150th anniversary of Freud's birthday (May 6, 2006), Freudian ideas are making a comeback. [See *Scientific American* 2006] But they still are far from commanding unanimity of expert opinion. There is thus little cognitive progress regarding the scientific status of psychoanalysis in terms of a differential weighting of the evidence. Arguably, however, the study of social phenomena can experience progress owing to increasing volumes of evidence, against which the ideal-typical models of social reality can be measured. The accumulation of evidence will therefore go some way toward alleviating an extreme form of underdetermination of social science models. As in other areas of science, increasing evidence has a tendency to distribute credibility differentially over the competing models. The subject matter of the social sciences, as we have seen, may well prevent them from ever achieving the situation in physics, where generally a convergence of the evidence toward one probable model takes place. In physics the empirical evidence is complemented by powerful theoretical principles, like the second law of thermodynamics or the principles of relativity. This convergence can be seen at work in the story of Copernicanism and Darwinism. But this is not to say that the social sciences have to forego all objectivity of their models. The situation is more complicated through the nature of the social world. But ideal-typical constructions are as much subject to testability as models in the natural sciences.

The problem of underdetermination has its impact on the issue of realism vs. relativism, owing to the factors that give rise to it. Dilthey never fully escaped the underlying relativism of his position, despite his insistence on the objectifications of social systems. Although the social sciences are more susceptible to the problem of relativism, it is possible to defend a position of realism in the social sciences.

4.4.5 Realism and relativism The problem of relativism in the social sciences essentially arises through the fact that social actors share symbolic meaning. It arises from the double hermeneutic. This makes the social scientist a participant observer, an observer engaged in a double hermeneutic. Although the later Dilthey emphasized the objectifications of social life, he still saw them as culture-specific entities. Every society, he argued, had its own "focal point." Consequently, meaning in social life varies from society to society, from culture to culture, from epoch to epoch. The question arises whether these symbolic meanings are translatable, transferable, and comparable to other social contexts. All forms of relativism relate the beliefs held by people to some nonuniversal framework or background against which these beliefs are held to be valid.

Judgments of validity of these beliefs must be made against the local background beliefs or framework. This framework may be some historical epoch (Antiquity, the Middle Ages) or some form of culture (primitive society, industrial society). There are many forms of relativism (conceptual, ethical, perceptual). Conceptual relativism is relevant in the present context: conceptual schemes organize the natural or social world according to their own criteria; the correctness of these criteria cannot be judged from outside the conceptual scheme. In contemporary eyes, Darwin held rather racist views about the hierarchy of ethnic groups. Freud propagated Victorian views about women. In the view of the relativist, it would be anachronistic to brand these views as "false," since they were inspired by the presuppositions people accepted in their own time. Thus, slavery and the slave trade were accepted as "natural" by most people from Antiquity to the Enlightenment. Equally, what counts as "real" in one culture may not count as "real" in another culture, but there can be no cross-cultural judgments about what is "real." If for Aristotle the heavenly spheres were "real" then we should accept his judgment. This may be a poor example in defense of relativism, since among the Greeks there was no agreement on the "reality" of the spheres. And this belief was heavily criticized by the Arab philosophers. Nevertheless, Peter Winch [1964] offers the following relativist defense of "reality" and "rationality": what counts as "real" and "rational" is context- and culture-dependent.

The relativist position on rationality is that other cultures' beliefs are rational in their own terms because these views are based on context-specific principles. Contrast this view with a realist view of the rationality of beliefs: They are

- subject to the rules of logic (systems of statements must be noncontradictory, consistent, coherent, valid);
- subject to empirical testability and confirmation;
- subject to a distinction to be made between a mind-independent external (natural or social) world and our fallible knowledge of this world; and
- social structures possess a relative independence from the belief states of individuals, as Durkheim stressed, because they have the status of "objectifications."

The realist and the relativist clash in their assessment of true and false beliefs, of rational and irrational beliefs. For the realist there is an important distinction between true and rational beliefs, on the one hand, and false and irrational beliefs on the other, irrespective of local contexts. Both require different explanations. Rational beliefs can be regarded as valid, because they can be justified by reasons.

Irrational beliefs may be credible to those who entertain them, but their beliefs require psychological or social explanations. Freud certainly believed that slips of the tongue, dreams, and neurotic behavior could be explained by an appeal to unconscious motivation. For Freud the beliefs of his patients were credible on the background of their psychological dispositions. But a morbid fear of open spaces or prowling wolves is not a justifying reason for these fears.

The realist therefore makes a sharp distinction between the context of *discovery* and the context of *justification*. The context of discovery comprises the rational or irrational ways in which ideas may emerge in our minds. All kinds of irrational motives may be at work, which produce someone's ideas. These motives cannot be reiterated. Archimedes stepped out of his bath and shouted "Eureka." This was his context of discovery. But his bath cannot be the justification for his discovery that the specific mass of a substance equals its displaced volume. The justification requires a context of rational deliberation, which is in principle open to everyone. This is the context of justification. As we have seen, this partly explains the difficulties of the Freudian model of the mind. For the relativist there is no sharp distinction between the context of discovery and the context of justification. The distinction between credible and valid beliefs must be made against local background assumptions. What is regarded and accepted as reasons is determined by what is accepted in local contexts. For instance, the insistence on *intersubjective testability*, which is so characteristic of modern science, then becomes a specific value in modern science. Also, what counts as evidence is itself based on local assumptions. Consider, for instance, the trust in sense experience across the ages. For the ancient Greeks, in their majority, the evidence of the eye was a trustworthy source in favor of a motionless Earth. For the natural theologians, prior to Darwin, the senses revealed the contrived complexity of the natural world. The context of discovery – the use of naked-eye observations – overlaps with the context of justification – the naked-eye observations apparently bear testimony to the stationary Earth and divine purpose.

The realist may not convince a determined relativist. Yet there are a number of arguments in favor of realism. One general remark may be apposite. The evolution of knowledge is a dynamic process. Even within a particular culture or society there often are dissenting voices. The Greeks did not monolithically believe in the geocentric worldview. Nor was there a general anti-evolutionist consensus among nineteenth-century Europeans. There are always competing models of explanation, which lead to divergent views of the universe.

One point against relativism is the **bridgehead argument**. All cultures have a common core of true beliefs and rationally justified patterns of inference. In a recent study, Harvard psychologists compared the geometric abilities of Americans with those of primitive tribesmen. They found that the geometric judgments of the two groups did not differ significantly. Elementary inferences about natural phenomena – leaden clouds herald rain – are also largely made independently of cultural contexts. There exist independent facts about the world, whatever people's judgment is about them. These independent facts will certainly result in a convergence of judgments from people across the world. Social facts seem to lack the autonomy from human volition that natural facts seem to enjoy. However, the objectification of social institutions means that they possess a relative independence from the daily commerce of social agents. Most social institutions outlive their current incumbents. Most social institutions possess a relative independence from the individuals that sustain them at any one time. [Sayer 2000, 33–5]

The **argument from evolution** appeals to the common ancestry of all humans. A similar response to environmental pressures provides the underlying substance which ensures commensurability between cultures. An analogy resides in the descent of modern languages from ancient languages. We find in Darwin, Huxley, and Tyler a defense of the idea that the human brain has evolved in response to similar environmental conditions. And larger brain capacities lead to higher mental functions, like the development of symbolic languages. As we shall see later, evolutionary psychology argues that the common descent of humankind has led to universal, modular structures in the brain. Humans have evolved problem-solving capacities in response to a common environment.

Finally there is the argument from **validity claims**. Realists point out that there exists a fundamental distinction between the *social acceptability* and the *rational validity* of ideas. [Habermas 1988 (1992), Ch. III.7] Consider some of the fundamental ideas we have discussed. Geocentrism was held to be true for two thousand years; the *Great Chain of Being* can boast of a similar pedigree. Yet it is a fundamental fallacy (the appeal to a majority) to think that some actions or ideas must be true because a majority of people support them. Their social acceptability does not demonstrate their rational validity. Otherwise it would be hard to understand how discoveries contrary to expectation could ever occur. The mind-independence of the world is precisely shown by the fallibility of our knowledge. [Sayer 2000, Ch. I; Bhaskar 1978, Ch. I] The realist draws a fundamental distinction between facts and concepts, between the external world and our fallible knowledge of it. Without this fundamental distinction it would be hard to explain the conceptual improvements which we have seen in the stories of Copernicanism and Darwinism. The realist concedes that our conceptual schemes guide our views of the world. Conceptual schemes often carry fundamental presuppositions, which may remain unquestioned for extended periods of time. Nevertheless, our conceptual schemes do not determine our ways of seeing the world. The pre-Copernican and pre-Darwinian schemes eventually faced large-scale discrepancies between the facts and the concepts. Habermas coins the phrase *situational reason* to label this guidance without determination. [Habermas 1988, 179 (1992, 139)] Reason is context-dependent in its reliance on language and concept formation. The growth of scientific knowledge shows that new evidence often requires new concepts. [Weinert 2004, Ch. I] But in establishing its validity claims reason must be regarded as context-independent. *The unity of reason resides in the diversity of its voices.* Some claims are cognitively more adequate than others in the face of the evidence. This is strikingly illustrated in the story of Copernicanism and Darwinism. Even interpretative social science, as Weber never tired of repeating, is capable of objective statements. Ideal types are required to "fit" the empirical data. As Habermas points out, the cognitive adequacy of conceptual schemes is also mirrored in the practical way of life. [Habermas 1981 (1984), Vol. I, Ch. I.2] A conceptual scheme has to prove its mettle by dealing with the discrepancies and contradictions it may encounter on the practical level.

If the cognitive adequacy of conceptual schemes has a palpable effect on the way we live our lives, this should be an argument for realism in the social sciences. The knowledge produced by social science theories can be used to interfere in social life; for instance, it may help to reverse unpalatable social trends. It therefore matters whether they are interpreted in a realist or instrumentalist sense. As Freud's psychoanalytic theory fails the strictures of a realist view of the mental, we argued that it should be read instrumentally. But generally, if the social sciences aim at explanatory understanding, their theories must be read in a realist sense. They are statements about the real social world; their cognitive adequacy matters greatly for policy decisions.

But there has always been a query regarding the autonomy of the social sciences. For the hermeneutic model the autonomy of the social sciences imposes itself through the double hermeneutic. But matters are less straightforward for the empirical model. The Darwinians anticipated the reduction of psychology to evolutionary biology. Positivists, like A. Comte, did not believe that sociology could be reduced to physics, since he assumed a positive hierarchy among the sciences.

While in earlier sections we looked at individualism and holism as explanatory levels of explanation, the next section considers whether the social sciences can be subjected to the strategies of reductionism.

→ *4.4.6 Reductionism and functionalism* Freud was an extreme reductionist, in a dual sense. Freud thought that all human behavior could be mechanically reduced to underlying psychodynamic forces. Freud also believed that psychology could ultimately be reduced to biology if not physics. We have already discussed Freud's "physics envy": his desire to place psychoanalysis on the same footing as physics. He even regarded it as a method comparable to infinitesimal calculus. [Freud 1927a, 136] We also find in Freud echoes of Darwin's prediction that "psychology will be based on a new foundation." Thus he declares in his Vienna lectures that

> the theoretical structure of psychoanalysis (…) is a superstructure, which will one day have to be set on its organic foundation, but we are still ignorant of this. [Freud 1916b, 389]

We have seen that Freud's crude reductionism must fail. To carry out his psychoanalytic program, Freud must borrow from the interpretative approach. Given the role of mind in the social world, the failure of closure in the social sciences and the existence of mere trends, any reduction of the social sciences to physics looks like a forlorn affair. The erection of psychology on an evolutionary foundation is, however, an option which is actively pursued in the program of evolutionary psychology. [See Section 5] But if this enterprise is to succeed it must account for the emergence of the mind in the material world. The Darwinians clearly saw this challenge, as we have observed. Before we return to this question, there is a more amenable form of reductionism whose theme runs through the social sciences. It is the question of *individualism* and *holism*.

As a psychoanalyst Freud naturally leans toward individualism. [Rillaer 2005c, 438] Generally, he explains group phenomena in terms of what individuals do; and what individuals do he reduces to their unconscious motivations. For instance, he explains cultural achievements as the result of deflection of sexual drives away from immediate satisfaction. This deflection is due to the process of sublimation. Weber, too, is an individualist who reduces "social phenomena" to "individual actions"

Charles Darwin, sporting an impressive beard as a man of his time

[Weber 1913, 110] For the individualist, to use terminology we employed before, "social phenomena supervene upon individual action and belief." There is a "multiple realizability" of social phenomena due to their dependence on individuals. [Elster 1989, Ch. II, XV; Little 1991, Ch. 9; Hollis 1994, Ch. V] But Freud's analysis of group psychology also shows that it is not easy to treat mass behavior as a simple summation over individual behavior. Many social phenomena are qualitatively different from individual behavior. Durkheim concedes that individuals are necessary but not sufficient ingredients for the constitution of social systems. Social entities are made up of individuals *and* the social relations they entertain. The dynamics of group behavior shows that individual behavior is strongly influenced by the behavior of the group to which the individual belongs. In fact, what individuals do and think is also influenced by their surrounding culture. Darwin's imperialist opinions on human races and Freud's antiquated views on female gender simply reflected the prevailing cultural attitudes of their lifetime. The way people appear and present themselves reflects their culture. In his book on crowd psychology, Freud deviates from his individualism. He explains mass behavior through the effect of the mass on the individual's psychic economy. [Freud 1921, 117] He observes that mass behavior is qualitatively different from the behavior of individuals, even if summed over many individuals. The way individuals interact with each other produces a difference in their behavior, which is observable as mass or group behavior. Freud, the author of *Group Psychology* [1921], is not in fundamental disagreement with Durkheim on this point. We can interpret Durkheim's view as saying that collective behavior *emerges* from individual behavior. It follows its own rules, and cannot be reduced to lower-level individual regularities. Thus Durkheim adopts a position that stands between a pure individualism and a dubious form of Platonic holism.

As objectifications, social institutions, traditions, and value systems can exist independently of any particular individual agents. Note that social institutions can exist even in the absence of any incumbent, at least momentarily. Social institutions like democracy can survive in the world of ideas, without any activation on the part of agents at a particular point in time. Durkheim's view is that the existence of social or political institutions requires instantiation on the part of individual social agents. The existence of social agents is necessary but not sufficient for the existence of social systems.

An unwanted kind of holism is often said to be the price to pay for adopting functionalism. [See Nagel 1961, Ch. 14; Habermas 1970, Ch. II.5; Hollis 1994, Ch. 5; Rosenberg 1995, Ch. 5] Functionalism emphasizes that social elements – individuals, groups, and institutions – play contributing roles to the functioning and maintenance of the whole society. Social agents or organizations that fail to contribute to the functioning of the society as a whole are regarded as dysfunctional. Consider, for instance, a functionalist explanation of social stratification or the division of labor. It interprets the social division of groups of individuals according to income, power, and other parameters as a wholesome contribution to the functioning of the whole system. In this sense functionalism seems to require the existence of a supra-individual entity called "society," which distributes the necessary functions for its maintenance. Durkheim [1897] explains suicide rates by reference to levels of social cohesion. The existence of society seems to prescribe the functions that individuals will take up. We seem to face the same problem, which we encountered in the debate between the Deist and the Evolutionist. [Chapter II, Section 2.1.2] The Deist stipulated that the function comes first through divine will, so that the organ will be tailored to it. But the Evolutionist must rely on the mechanism of natural selection in order to explain how organs come to adopt certain functions over time.

Is functionalism committed to a metaphysical kind of holism, which presupposes the existence of social entities in a realm of their own? On the Durkheimian view social entities do not exist as mere aggregates of individuals. Social agents are necessary for social entities. Rather than thinking of society as "organisms" whose individual members are its "organs," biology suggests thinking of them as "self-regulating systems."

> Social organizations can be planned and directed, but afterward they reproduce themselves in the manner of self-regulating systems. (…) social organizations provide the model for the theoretical framework of systems research. [Habermas 1970, 170 (1988), 77]

Such social entities, if we add Dilthey's crucial insight, are held together by symbolic strings. As social entities are largely symbolic in nature, the holist does not need to appeal to some evolutionary account of the growth of societies in order to explain the creation of societal functions. The explanation is inherent in the hermeneutic dimension. Social institutions undoubtedly exercise various functions. The function of legislative power (e.g., the parliament) is to create legislation; the function of the executive power (e.g., the police) is to maintain the laws against their violation; and the function of the judiciary power (e.g., the law courts) is to sanction the perpetrators of the laws. When the philosophers of the Enlightenment began to think about the division of power into its three arms (Locke, Montesquieu), these separate functions existed merely in their minds, or on paper. The human mind can create imaginary scenarios and anticipate the future. The Deist needed to postulate a Superior Being whose creative acts invested the natural world with biological functions. The social philosopher can rely on the creativity of the human mind. The human mind creates the function in anticipation of its realization. The social scientist can observe the installation of functions in society. This seems to be a regular feature of human life. First, ideas create a space of possibilities; then the possibilities, whose consequences can be anticipated, are put into action. If the symbolic dimensions of human existence are taken into consideration, functionalism does not need to postulate a preexistent functional whole. The functions exist as plans and are then ascribed to social roles. Social roles are effective only if concrete social agents occupy them. In carrying out their social roles, social agents fulfill their functions. Once a function is planned – as in legislating, policing, judging or, say, teaching – and social roles are created to fulfill the functions – the parliamentarian, the police officer, the judge, the teacher – concrete individuals are needed to carry out these social roles. But note that a social role is not dependent on any particular individual. That functions are anticipated or planned also implies, according to Popper's problem of unplanned planning, that the planning may go awry. The individual or organizational roles do not rise to the expectations of the anticipated functions. If the function is to be maintained, the individual and organizational roles must be reformed. All this happens on the level of individual members of society, say the Members of Parliament. But Durkheim's point is that once a function and a social role have been "defined," it then transcends the volition of its individual incumbent.

Do societies exist over and above their individual components? For the individualist only individuals and their actions exist. But as Weber's introduction of ideal types illustrates, pure individualism has difficulties in explaining group behavior. It struggles even more to explain complex modern societies. Social structures cannot easily be explained on the basis of individual actions. For the holist social structures exist over and above individual actions. But in which sense do these super-individual structures exist? Marx speaks of a superstructure over a material base; Durkheim appeals to a collective soul. If our analysis is correct, neither Marx's superstructures nor Durkheim's *esprit collectif* are to be understood as Platonic entities. The material base and individual social agents are necessary conditions for the existence of social wholes. Note that a Platonic existence of social wholes in a supra-individual sphere is incompatible with our finding that social regularities exist only as trends. Their dependence on initial conditions and their reversibility require their instantiation by social agents. In this sense individual social agents are necessary for social life.

In view of the distastefulness of Platonic holism, a philosophical position recommends itself: mix *ontological* individualism with *methodological* holism. On the ontological basis only individuals are said to exist. But the existence of individuals is insufficient to explain social regularities. We therefore adopt *methodological* holism. Social scientists obtain social generalizations to explain social facts without mustering appropriate "individual" generalizations. The social sciences have failed to live up to Mill's and Freud's dream of discovering the psychological laws, which presumably govern individual behavior. Weber's ideal-type methodology has shown how the social scientist can nevertheless make general statements about social behavior. But these ideal-typical social generalizations are regarded as expressing "typical" or "average" types of behavior. There is no claim that social generalizations reflect social entities, which would exist over and above the individuals in society. But the views of Marx and Durkheim permit the adoption of an *emergent* holism. Social systems, by virtue of their coercive nature, can claim some independence from their individual members, which make up the relata of the social system. The social system only emerges when individual members of society begin to relate and interrelate with each other.[8] Individual planets do not form a solar system until they, as the relata of the system, enter into lawful relations (symbolically expressed in Kepler's laws). Similarly, individual social agents do not form social systems until they begin to entertain social relations with each other. These relations may get solidified into customs, institutions, trends, and traditions – Dilthey's objectifications. The social systems remain dependent on their individual members but only as necessary constituents. Social entities take many forms: social substances (France), social types (capitalism), events (Wall Street crash of 1929), processes (decline of the Roman Empire) and social states (servitude). [Ruben 1985, 8–9] This may be viewed as nested systems. [See Box 3.3]

Box 3.3   Nested systems
In order to show the irreducibility of social to individual properties, H. D. Ruben spells out four necessary, jointly sufficient components of a nested system. [Ruben 1985, 105–7] He starts with the claim that

> a necessary and sufficient condition for a relational property P being a social relational property is that, if x and y stand in the relation P, then it follows that a nested system of interlocking beliefs and expectations exists.

There follow four necessary and jointly sufficient components of a nested system:

1.   If there are any social relations between x and y, x and y will have interlocking beliefs and expectations about actions.
2.   There exist second-order beliefs and expectations.

These two conditions are sufficient, to return to Winch's example, to explain the difference between gravity and "being at war." The gravitational pull of the Earth on the moon does not involve shared beliefs about gravity nor second-order expectations that the Earth will continue to keep the moon in its gravitational field. However, if country x is at war with country y, there will

---

[8] M. Bunge [1996] proposes a position between Individualism and Holism, called "systemism," according to which individuals and society, agency, and structure interact.

> **Box 3.3** (*cont'd*)
>
> be anticipations of hostile actions and anticipations that possibly the anticipated action could be thwarted, e.g., through the destruction of enemy positions.
>
> 3. Some descending reason-relations exist among the set of beliefs and expectations. For example, the reason a social agent acts in a particular way is at least partly influenced by the way others expect that agent to act. This situation is typical of expectations or norms associated with social roles. A social agent, who exercises a social role, will face social expectations about how to behave in that role. S/he will most conform to these expectations. If the agent fails to perform according to the expected form of behavior associated with the role, say as father or officer, the social group or society will impose sanctions on the agent, which can range from mild admonitions to harsh imprisonment.
> 4. X believes that sometimes Y does what s/he does because of Y's belief about what X believes that Y will do; and Y believes that X sometimes does what s/he does for the same reasons. This refers to second-order expectations of social agents. The agent anticipates that others entertain certain beliefs about how the agent will and should behave and adapts her/his behavior accordingly. These second-order beliefs are essential for the maintenance of trust amongst members of a society.

5 Evolution and the Social Sciences

Evolutionary biology is fundamentally relevant to the study of human behavior and thought because our species is the product of naturalistic terrestrial processes – evolutionary processes – and not of divine creation or extraterrestrial intervention. [Tooby/Cosmides, "The Psychological Foundations of Culture" (1995), 50]

The social sciences remained relatively unaffected by the publication of *The Origin of Species* (1859). The hermeneutic model reacted against the severe naturalism of this model, but still took physics as the paradigmatic science. Proponents of the hermeneutic approach underlined the importance of symbolic dimensions in social life. Their argument for the autonomy of the social sciences meant that no inspiration was to be gained from either the physical or the biological sciences. In France the empirical model had put its faith in the physical sciences. Durkheim accepts that quite general features of human nature play their part in social life. But the mind is only a minor actor in human affairs.

> Collective representations, emotions, and tendencies have not as their causes certain states of consciousness in individuals but the conditions under which the body social as a whole exists.

This is Durkheim's endorsement of holism. He accepts that such social actions can only materialize "if the individual natures are not opposed to them." Individuals are thus a necessary but not a sufficient condition for the formation of social entities.

> But these [individual natures] are merely indeterminate matter which the social factor fashions and transforms. Their contribution is made up exclusively of very general states, vague and thus malleable predispositions which, of themselves, could not assume the definite and complex forms, which characterize social phenomena if other agents did not intervene. [Durkheim 1895, 131]

To a certain extent this neglect of human nature is understandable, even from an empiricist like Durkheim. Although empiricism recommends a science-inspired approach to the social sciences, it does not question the independence of the social sciences. For almost the entire length of the twentieth century, the social sciences, irrespective of the approach they adopted, conducted their affairs as autonomous disciplines. Evolutionary perspectives played no significant role in their research. It is true that Habermas, among others, considered whether the growth of modern societies could be explained in evolutionary terms. [Habermas 1976, Ch. III] But there was no suggestion, until recently, to make evolutionary principles bear on the subject matter of the traditional social sciences. This changed abruptly in 1975, when E. O. Wilson published his book *Sociobiology*. In his *Descent of Man* (1871) Darwin had used evolutionary principles to explain moral and social features of human groups. Wilson's book caused a furor because it updated Darwin's ideas.

5.1   *Sociobiology* – the fourth revolution?

Copernicus resurrected the idea held by a lone Greek – Aristarchos of Samos – that the Earth was not the hub of the universe. Darwin argued persuasively that Earthlings shared a common origin with all other living matter. Freud believed that he had destroyed the Enlightenment image of rational man. Did the American entomologist Edward O. Wilson show that all social behavior in animals and humans has biological roots? If sociobiology had achieved its goal, it would have achieved a fourth revolution. Freud's audacious claim to have competed the Scientific Revolution would have been premature. The accolade for this achievement would have to go to Wilson. Although sociobiologists write confidently as if the biological basis of social behavior had already been established, reality looks less reassuring. Often bold speculations replace hard empirical evidence. It seems fairly uncontroversial that sociobiology is a branch of modern evolutionary theory. [Wilson 1978, 16–7; Kitcher 1985, 118; Ruse 1979, Ch. 2] But it is far from clear whether this branch can carry the weighty assertion it is made to bear. In particular, the precise nature of the link said to exist between social behavior and biological structure has not yet been established. Wilson's research program may have faltered but his ideas live on in evolutionary psychology.

According to E. O. Wilson, sociobiology is "the systematic study of the biological basis of all (forms of) social behavior (in all kinds of organisms, including man)." [Wilson 1980, 4; 1978, 16] These are sweeping general statements, which need to be cashed in by empirical details. Wilson cannot mean that absolutely *all* social behavior is genetically determined or at least has biological roots. Much of our behavior follows norms and conventions, and is inspired by social values. [Salmon 2003] Such conventions, norms, and values tend to vary from society to society. The English have their dessert before their cheese; the French serve their cheese before their dessert. No one would want to claim that this difference in cultural habits is genetically determined. Freud may be correct that some of our behavior is caused by hidden motives. But much human behavior is based on the deliberate weighing of alternative actions. If the motive for the action is also its cause then it may be rational behavior. [Davidson 1982] For Wilson's program to succeed it must stay clear of crude determinism, which is committed to attributing an underlying gene to every type of behavior. Sociobiology must also avoid falling into the platitude that our biological constitution imposes constraints on our social behavior. [Sterelny 1988, 539] As humans cannot climb very well, they prefer to eat their meals under rather than on trees. A viable sociobiology must say more than that biology constrains. It is fairly obvious that every organism's constitution places constraints on its field of action. What is the most fruitful line to adopt?

By his very definition of sociobiology Wilson has committed himself to a strong program, according to which the stipulation of a genetic link is always better than a non-biological explanation. Consequently,

> the influence of genetic factors in the adoption of certain *broad* roles cannot be discounted. [Wilson 1980, 279]

Although there is much vacillation in Wilson's writings about the fixity of human behavior, the most sympathetic characterization of his program lies in the framework idea:

> Ethnographic detail is genetically underprescribed, resulting in great amounts of diversity among societies. Underprescription does not mean that culture has been freed from the genes. What has evolved is the capacity for culture, indeed the overwhelming tendency to develop one culture or another. [Wilson 1980, 284]

This emphasis on a "framework idea" of sociobiology is reminiscent of Darwin's views in the *Descent of Man*.

> Each person is moulded by an interaction of his environment, especially his cultural environment, with the genes that affect social behaviour. [Wilson 1978, 18]

Wilson even grants that despite the (presumably) innate aggressiveness of humans, the evolution of warfare could be reversed and brought under the control of rational conflict management. [Wilson 1978, 119–20] Thus, Wilson himself seems to steer toward a compromise between nature and nurture. [See Pinker 2004; Ridley 2003] Such a compromise position casts the genes into the role of a general framework, which is compatible with different kinds of cultural manifestations. For instance, all humans are capable of learning a language but which language they learn depends on cultural circumstances. Wilson does not pursue the consequences of this compromise position. It would have an impact on his understanding of the biological roots of social behavior. It would also lead to a modification of his radical theses about the integration of the social sciences into sociobiology. Sociobiologists of Wilsonian persuasion tend to misunderstand the subject matter of the social sciences. For instance, they fail to make appropriate distinctions between different types of behavior.

Consider the distinction between *compulsive* and *social* behavior. Only social behavior has a moral and symbolic dimension, because it contains an element of choice of action. Humans can assess the likely outcomes of their actions. Where their choices affect the well-being of others their actions can be regarded as morally right or wrong. Social behavior is also adaptable to the natural and cultural environment, hence it can be learned. Compulsive behavior, however, crucially lacks this dimension of choice and adaptability. It is incorrect to propose (as does Ruse [1979], §7.2) that schizophrenia and manic depression are examples of social behavior which provide positive evidence for sociobiology. It is undoubtedly true that these examples of compulsive behavior have strong genetic components but they are not types of social behavior. They neither change with the environment nor are they the object of learning processes. They are not part of nested systems.

Having rehearsed the arguments of the hermeneutic approach, we can reconstruct the response of the social scientist to sociobiology. The anthropologist Marshall Sahlins has urged against sociobiology that the symbolic event marks a radical discontinuity between culture and nature. [Sahlins 1976, 12] Humans are defined in terms of symbolic attributes, so that there is a basic indeterminacy between human nature and its cultural expression. [Sahlins 1976, 61] If culture is "biology plus symbolic faculty," as Sahlins suggests, then biology is only a necessary, never a sufficient condition for social behavior in humans. [Sahlins 1976, xi, 65–7]

In many of his passages Wilson seems to anticipate the compromise position, which many of his critics propose. [See Ruse 1979, §§7.12, 8.2; Sterelny 1988, 552–4; Pinker 2004; Ridley 2003] The idea is compelling. Genes do not determine social behavior or even types of social behavior. Rather they provide capacities within which different manifestations of behavior can be accommodated. The linguist Noam Chomsky argues that humans possess innate linguistic abilities, which manifest themselves in hundreds of different concrete ways of speaking a human language. Thus, the genes structure human capacities, while culture determines the particulars.

> Human behavior can thus be seen as biologically adaptive, which is what the sociobiologists want, but crucially causally influenced by learning, which is what the culturalists want. [Ruse 1979, 160]

Such statements are echoed in Wilson's work. Thus he states that

> the form and intensity of altruistic acts are to a large extent culturally determined. Human social evolution is obviously more cultural than genetic. The point is that the underlying emotion, powerfully manifested in virtually all human societies, is what is considered to evolve through genes. [Wilson 1978, 153]

The genes provide constraints on human social behavior since they "hold culture on a long leash." [Wilson 1978, 167] The universal has to be separated from the particular just as a universal grammar, shared by all human speakers, differentiates into the particular grammars of our mother tongues. The fatal weakness of Wilson's whole program lay in its vacillation between incompatible positions: does our biological nature impose strong or weak constraints on social behavior? Let us say that the framework idea is retained: the genes are responsible for capacities. But genes do not determine rational behavior. The social scientists can grant this point without conceding that this admission reduces the importance of the social sciences. On the contrary: the capacities need molding and the social sciences will need to spell out all the sufficient conditions required to turn capacities into cultural forms. Contrary to Wilson's suggestion [1980, 4] the social sciences are not "the last branches of biology waiting to be included in the Modern Synthesis." Nevertheless, it is a worthwhile pursuit to enquire whether social behavior has biological roots. At the end of the 1980s the "sociobiological" revolution was not yet in sight. Sociobiology soon lost its appeal; in the popular mind it was too redolent of genetic determinism. But the programmatic ideas did not wither. Soon the program returned under a new name: evolutionary psychology. It took up the cudgels of sociobiology. In this process of adaptation, it made more far-reaching philosophical commitments.

5.2   Evolutionary psychology

(...) everything, from the most delicate nuance of Richard Strauss's last performance of Beethoven's Fifth Symphony to the presence of calcium salts in his bones at birth, is totally and exactly to the same extent genetically and environmentally codetermined. [Tooby/Cosmides 1995, 83–4]

Sociologists like Durkheim and Weber had no room for biology in their sociological theories. By contrast, Freud seems to have listened to Darwin's prediction. He presented psychoanalysis as a theory that would eventually be based on biological foundations. Evolutionary psychology (EP) is heavily indebted to Darwinian principles. It seeks to establish a tight link between evolution and the social sciences. In fact, both sociobiology and evolutionary psychology are committed to a unification of the social sciences with evolutionary biology. The dream of a fourth revolution is still alive. If the social sciences became a specialized branch of biology, the traditional autonomy of the social sciences would come to an end. Does EP succeed where Wilson's program failed through vacillation? Philosophically, EP adopts a much more efficient procedure, which we have seen at work before. EP claims that the traditional social sciences have failed. What's wrong with the Standard Social Science Model (SSSM)? As we have seen, the central postulate of the social sciences is that social facts mold plastic human nature. To SSSM nurture is far more important than nature. SSSM is not necessarily committed to the Lockean view of the human mind as a blank slate, on which culture leaves its imprints. Under Wilson's framework idea SSSM can admit general-purpose mechanisms in the mind. These general genetic capacities do not, however, distract from the role of culture. But according to EP, the traditional social sciences encounter an insuperable problem. SSSM cannot explain how social agents solve an array of complicated tasks. Noam Chomsky employed a similar argument in his critique of behavioristic accounts of language learning. The mastery of a human language by a young child, Chomsky argued, could not be explained in terms of feedback- and response-mechanisms. The obstacle to such learning mechanisms is the syntactic and semantic complexity of human languages. Proponents of EP *infer* from the alleged failure of SSSM to the need for a new model of the social sciences. They call it the Integrated Causal Model. What are the human abilities and capacities that allow them to solve complex tasks in the material and social world? Proponents of EP state, by way of example, the following abilities: [Cosmides/Tooby 1998, 9, 11; Tooby/Cosmides 1995, 121, 123]

- ability to learn symbolic languages;
- ability to reason and draw inferences;
- ability to comprehend the material world;
- ability to anticipate the future;
- ability to organize social life;
- ability to exercise mate and food preferences;
- ability in infants to recognize faces.

EP's epistemological claim is that only the Integrated Causal Model can account for the complex tasks that humans routinely perform.

Before we consider the substantive claims of EP, let us briefly consider a *modified* version of SSSM. We should note that vast areas of social life can be studied irrespective of any biological foundations of human nature (economic life, social structure, cultural institutions). Such a modified version, as it already emerged in response to sociobiology, accepts that human nature may have a larger part to play in the social sciences than Durkheim was willing to concede. Animals are rigidly controlled by their biology (instincts). But human behavior is largely determined by culture, albeit within limits of biological constraints. In this context culture is loosely defined as the existence of largely independent systems of norms, values, symbols. Some of these cultural products will be context-specific but, as the arguments against relativism have shown, humanity also shares a common core of abilities and practices. The genetic structure of humans provides constraints, which present themselves as capacities, dispositions, which require cultural inputs to manifest themselves. There is reason to believe that such a modified view of SSSM could do its job. [Ridley 2003]

It is this common core of human abilities on which EP's substantive claims focus. EP considers that the traditional social sciences have failed because of their neglect

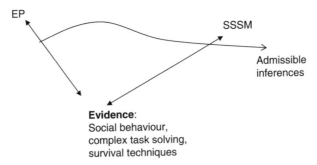

Figure 3.5   Inferences from evidence to competing models

of the biological dimensions of human existence. The social sciences should therefore be based on evolutionary theory. But where sociobiology embraced rather vague philosophical commitments of reductionism, EP is philosophically sounder, for two reasons. First, proponents of EP adopt the correct inferential attitude. From the point of view of eliminative induction the inference to the Integrated Causal Model is the right strategy, *if* SSSM indeed fails. [Figure 3.5] But the failure of one model, like SSSM, does not automatically bestow credibility on a rival model. A rival model, like EP, needs to present independent evidence for its claims, e.g., evidence, which is incompatible with the rival model. In an earlier chapter we distinguished between *admissible* and *permissible* inferences. We regarded inferences to intelligent design as permissible but not admissible, since there can be no empirical proof of an intelligent designer. As EP is grounded in evolutionary theory, it takes itself to be making admissible inferences from *observable* behavior to *unobservable* mental structures. If the evidence can be shown to discredit SSSM but credits a rival model, EP, then this is an admissible inference. However, we recall from the methodological discussions of Huxley and Haeckel that the favored model must find supportive evidence, for instance in the form of deductive or inductive consequences.

There is a second reason why EP has superseded sociobiology. It endorses a philosophy of mind. Its philosophy of mind is a presupposition, open to criticism. Its view of the mind can best be expressed by an analogy: the Swiss-army-knife view of the mind. [Dennett 1995] According to this view, the human brain is equipped with numerous special-purpose mechanisms (neuron circuits), also called "modules." EP also subscribes to the strong materialism of the Darwinians. These mechanisms, they claim, have evolved to solve adaptive problems encountered by our ancestors in our evolutionary past. Consider for instance the reasoning instinct. It has evolved as a response to particular problems in the environment of our remote ancestors.

> The mind is a set of information-processing machines that were designed by natural selection to solve adaptive problems faced by our hunter-gatherer ancestors. [Cosmides/Tooby 1998, 1; see Rose/Rose 2001 for a sustained attack on EP from many different perspectives]

Different neural circuits – or modules – have been honed to solve different adaptive problems. The assumption of a network of neural circuitry in the brain is a far cry from the acceptance of general-purpose mechanisms in the modified version of SSSM. Owing to the slowness of evolution, our modern skulls house a Stone Age mind. Proponents of EP therefore consider that our mental architecture has been installed by evolutionary forces. The gradualness of the evolutionary mechanism makes the 2,000 years of Western civilization a mere blink of an eyelid in evolutionary terms. In an obvious reference to Freud, proponents of EP also consider that most of what goes on in our minds is hidden from our conscious gaze. But the Unconscious is not an evil demon; it is a network of neural circuits.

Freud would have been sympathetic to the universalism inherent in this position. Every species has a universal, species-typical evolved architecture. It is innate in the sense of having been forged by evolution. EP is a self-styled Kantian position in an evolutionary cloak, since it postulates "innate" universal, a priori structures. [See Ratcliffe 2005, 53–4] But EP avoids a crude form of determinism. The universal mental architecture requires a certain particularism for its implementation. Different social circumstances lead to different cultural manifestations of the underlying evolved psychology. The procedure of EP follows a bottom-up approach in which the evolved architecture is the foundation on which behavioral characteristics are erected.

The methodological procedure adopted by proponents of EP is reminiscent of inferential practices in the history of science. EP, it is claimed, is an inference to the best explanation (IBE). As we have seen, Freud made similar inferences: from manifest complex behavior to an unconscious mental architecture. The Freudian case also made clear what is wrong with IBE. It must not be an inference to the best explanation *tout court*, but to the best contrastive explanation.

EP is seriously affected by the problem of unobserved entities, like its Freudian cousin. It stands in serious danger of violating Ockham's razor: "entities are not to be multiplied beyond necessity." In modern terms Occam's razor is the recommendation that researchers should choose the simplest hypothesis that will fit the facts. The defenders of EP multiply mental modules to keep track of human abilities. It is probably true that the mind cannot be understood without biology. But there are alternative ways of accepting this discovery. EP is involved in a genocentric fallacy. [Dupré 2005, 81–7] Recall that one of the criticisms against intelligent design, inspired by Nietzsche, was that we cannot infer from today's function of a biological organ what its original function would have been. In a similar way we cannot easily infer from an *adapted* trait to genetic evolution. In making this inference we exclude the possibility of *cultural* evolution, i.e., transmission of behavior and traits by processes of learning. By adopting a functional view in the philosophy of mind as a presupposition, EP assumes a 1:1 correspondence between neural circuits and particular human problem-solving capacities. One "can view the brain as a collection of dedicated mini-computers whose operations are *functionally integrated* to produce behavior." [Cosmides/Tooby 1998, 8–9; italics in original] However, this presupposes rather than solves the mind–brain dualism. Humans are largely defined in terms of symbolic attributes. We cannot simply postulate a mapping of a piece of neural machinery to a particular type of mental structure, which results in meaningful social behavior. EP is an alternative research program, which still has to establish its credentials.

6   Freud and Revolutions in Thought

The old man (…) had a sharp vision; no illusion lulled him asleep except for an often exaggerated faith in his own ideas. [Einstein on Freud, quoted in Pais 1982, 515]

There are striking contrasts in the public reception of mistaken ideas in the history of science. It is rare for a discredited scientific idea to attract the attention of journalists. Important figures in the history of science, like Ptolemy and Lamarck, are ignored in favor of the official heroes of science. The recent review of Freud's corpus of ideas – on the occasion of the 150th anniversary of his birth (May 6, 1856) – shows that Freud has acquired his own journalistic history. The doubtful scientific status of Freud's ideas has never deterred the magazine and newspaper writers from celebrating Freud's achievements. Over many years international news magazines – *Le Point, Der Spiegel, Die Zeit, Time Magazine* – Sunday newspapers and the *Scientific American* have informed their readers about the fortunes of Freud's theory. For the last three decades the general tenor had been to announce the demise of Freudian theory. The publication of *Le Livre Noir de la Psychanalyse* [Meyer 2005] prompted *Le Point* to ask: "Faut-il en finir avec Freud?" But the articles on the occasion of the 150th anniversary of Freud's birth gave a much more upbeat assessment. Freud's ideas, it seems, are fashionable again. It is agreed that some of his ideas, like his views on female sexuality, are discredited. But some of his core ideas are

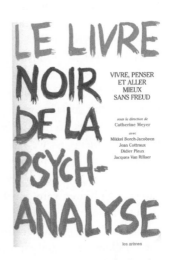

The publication of this book in 2005 marked a significant shift in France away from the unquestioned acceptance of Freud's psychoanalysis

said to be testable. [See Solms 2004; Crews 1997; Grünbaum 1977a, b; 1979] For instance, Freud is credited with the discovery of the unconscious mechanisms of much of our behavior. Some researchers, like Mark Solms [2004], claim that Freud's views on dreams as wish fulfillment may have been right after all. What is salvaged from Freud is a general framework, a paradigm. The evolutionary psychologists, for instance, speak of the neural mechanisms underlying our conscious behavior. But the details of their models are very different from Freud's original ideas. Others remain as hostile to Freud as ever. For them the resurrection of Freudian ideas is not worth the effort; they are discredited and we might as well make a fresh start.

Whatever attitude one adopts toward Freud the *scientist*, it is generally agreed that he was a great figure in the history of ideas. Thomas Mann declared that the "pschoanalytic insight is world-changing." [Mann 1936, 150] Freud's impact as a molder of thought has left its traces in the world of arts (James Joyce, Virginia Woolf). If Freud has redrawn the psychological map of human nature, is he, after all, a pillar of Enlightenment thinking, rather than a pseudo-scientific trickster? [Cioffi 1970] This whole chapter was subtitled "The Loss of Transparency." This subtitle served to indicate that Freud shrank the realm of rational thought, as the theater of human action, to a size which many Enlightenment philosophers would have found unacceptable. But Habermas and cultural commentators often associate Freud's achievement with the project of the Enlightenment. [Habermas 1968; 1970] The importance of Freud's model resides for Habermas in the way it combines understanding and explanation. On the one hand, there is the level of interpretation, which is driven by the symbolic manifestation of the unconscious motives. On the other hand, these motives are not consciously available to the patient; they act, according to Freud, as causes which require an explanation. [Habermas 1968 (1972), Part III, Ch. 10, 11; 1970, 297–8 (1988, 180–1); 1983, 214] *Verstehen* in itself is not an adequate method; in fact it is not a special method of the social sciences at all. But explanation cannot do full justice to the subject matter of the social sciences either, because of the ubiquity of the double hermeneutic in the social world. [Habermas 1970, 281, 289 (1988, 167, 174)] But Habermas takes Freud's insistence of the scientific nature of his personality model far too seriously. As we have seen, the level of interpretation of the data already causes considerable problems, because the interpretation of dreams and other symptoms of troubled psyches is free-floating. The unconscious motives are not subject to uninterpreted evidential relations.

But Freud may still be credited for having shown us how to hold a liberated, guilt-free discourse about human sexuality without blushing. He liberated us from the shackles of sexual suppression. His psychoanalysis opened our eyes to the unconscious, often sexual motivation of human behavior. Shining light on the dark corners of our existence was after all the main focus of the Enlightenment. In which sense then can we speak of a loss of transparency? The answer to this question refers us back the assessment of the scientific status of Freud's theory. Modern Freudians treat Freud's theory at least as a serious framework theory, a paradigm, which offers the additional bonus of self-reflection. [Habermas 1968 (1972), Part III. §10; Solms 2004] The anti-Freudians simply reject Freud's efforts as pseudo-scientific humbug. Who is right? In this chapter we have avoided a predetermined answer to this question. We have examined Freud's inferential practices and found them wanting. Freud argued that his case studies offered supportive evidence for his psychoanalytic model. But his evidence is both slender and ambiguous. Freudian slips, dream analysis, and free association are neither objective nor intersubjective. Dream data are not hard empirical facts. A dream stands in need of interpretation. If the interpretation is provided by psychoanalysis, the dream data can then not serve as confirming or falsifying instances. Freud would have contributed to a gain rather than a loss of transparency if his views had commanded scientific credibility. But as it stands, he has provided no supportive evidence for his theory. Such evidence may one day be procured in a laboratory. But as long as this day has not yet arrived, Freud remains a revolutionary in the history of thought rather than in science.

6.1 Revolutions in thought vs. revolutions in science

In earlier chapters we considered some criteria for a scientific revolution. Freudianism satisfies some of these criteria. *Firstly*, Freud switched perspectives. He argued that we should look at human beings from the point of view of their unconscious motivations, rather than their rationality. He makes a conceptual innovation over his predecessors, for he argues that the Unconscious is dynamic. *Secondly*, in a loose sense it could be claimed that Freud introduced some new techniques into the study of mental phenomena. But talking cures, free associations, and interpretations are qualitative procedures, which are fraught with difficulties. These techniques do not seem to be independent of the theory which uses them to support its claims. In connection with these criteria we argued that a new theory must also be explanatory. It must solve some outstanding problems, make some explanatory gain. It is the problem-solving ability of the Freudian theory that has attracted much criticism. The paucity of the evidence for Freud's claims affects the distribution of credit over the model space. The qualitative evidence fails to credit the Freudian model, while discrediting rival models. Freud's model cannot be called explanatory. The *third* aspect of a scientific revolution related to the emergence of a new tradition through chain-of-reasoning transitions. However, this new tradition lacks the problem-solving capacities of Copernicanism and Darwinism. *Fourthly*, there was the convergence of expert opinion on the new tradition. Freud's case differs from that of Copernicus and Darwin. One hundred and fifty years after Freud's birth the world is still divided over his claims. That is, the central aspect of convergence is still missing from the history of the Freudian theory. There is neither convergence of the evidence in favor of Freud's model nor convergence on at least some alternative models in the scientific community. If Freud has failed as an author of a scientific revolution, he has succeeded as the originator of a revolution in thought. Freud's terminology and ideas have certainly penetrated the cultural sphere. Freud has created *"homo psychologicus."*

Reading List

Abel, Th. [1948]: "The Operation called *Verstehen*," *American Journal of Sociology* **54**, 211–18

Apel, K. O. [1979]: *Die Erklären: Verstehen-Kontroverse in transzendetnal-pragmatischer Sicht*. Frankfurt a./M.: Suhrkamp

Barrow, J. D./F. J. Tipler [1986]: *The Anthropic Cosmological Principle*. Oxford: Oxford University Press

Becker, G. [1986]: "The Economic Approach to Human Behaviour," in J. Elster *ed.* [1986], *Rational Choice*. London: Blackwell, 108–22

Bhaskar, R. [1978]: *A Realist Theory of Science*. Sussex: The Harvester Press

Borch-Jacobsen, M. [2005]: "La querelle de la suggestion," in C. Meyer [2005], 381–8

Braybrooke, D. [1987]: *Philosophy of Social Science*. Englewood Cliffs, NJ: Prentice-Hall

Brown, J. A. C. [1962]: *Freud and the Post-Freudians*. Harmondsworth: Penguin

Bunge, M. [1996]: *Finding Philosophy in Social Science*. New Haven, CT: Yale University Press

Cioffi, C. F. [1970]: "Freud and the Idea of a Pseudo-Science," in *Explanations in the Behavioural Sciences*. F. Cioffi/R. Borger *eds.* Cambridge: Cambridge University Press, 471–515

Cioffi, C. F. [2005]: "Freud, était-il un menteur?", in C. Meyer [2005], 42–6

Cohen, J. [2005]: *How to Read Freud*. London: Granta Books

Comte, A. [1853]: *Cours de Philosophie Positive*. Paris: Bachelier

Cosmides, L./J. Tooby [1998]: "Evolutionary Psychology. A Primer." Available at <http://cogweb.ucla.edu/ep/EP-primer.html>

Crews, F. et al. [1997]: *The Memory Wars – Freud's Legacy in Dispute*. London: Granta Books

Crombie, A. C. [1994]: *Styles of Scientific Reasoning in the European Tradition*. 3 vols. London: Duckworth

Dahrendorf, R. [1959]: *Class and Class Conflict in an Industrial Society*. London: Routledge

Davidson, D. [1963]: "Actions, Reasons and Causes," reprinted in D. Davidson [1980], 3–19

Davidson, D. [1967]: "Causal Relations," reprinted in D. Davidson [1980], 149–62

Davidson, D. [1970]: "Mental Events," reprinted in D. Davidson [1980], 207–25

Davidson, D. [1974]: "Psychology as Philosophy," reprinted in D. Davidson [1980], 229–39

Davidson, D. [1980]: *Essays on Actions and Events*. Oxford: Clarendon Press

Davidson, D. [1982]: "Paradoxes of Irrationality," in R. Wollheim/J. Hopkins *eds.* [1982], *Philosophical Essays on Freud*. Cambridge: Cambridge University Press, 289–305

Delanty, G./P. Strydom *eds.* [2003]: *Philosophy of Social Science*. Philadelphia, PA: McGraw-Hill

Dennett, D. [1995]: *Darwin's Dangerous Idea*. London: Penguin

Dilthey, W. [1883]: *An Introduction to Human Sciences*, in W. Dilthey [1976], 159–67 [Translated from *Gesammelte Schriften*, Vol. I, xv–xix, 14–21]

Dilthey, W. [1894]: "Ideas Concerning a Descriptive and Analytic Psychology," in *Descriptive Psychology and Historical Understanding*, transl. M. Zaner/K. L. Heiges. The Hague: Martinus Nijhoff [1977], 21–120 [Translated from *Gesammelte Schriften*, Vol. V. 139–240]

Dilthey, W. [1900]: "The Development of Hermeneutics," in W. Dilthey [1976], 247–63 [Translated from *Gesammelte Schriften*, Vol. V. 317–37]

Dilthey, W. [1906]: "The Construction of the Historical World in the Human Sciences," in W. Dilthey [1976], 168–245 [Translated from *Gesammelte Schriften*, Vol. II, 79–88, 130–66, 189–200; Vol. VII, 228–45]

Dilthey, W. [1927]: "The Understanding of Other Persons and their Expressions of Life," in *Descriptive Psychology and Historical Understanding*, transl. M. Zaner/K. L. Heiges. The Hague: Martinus Nijhoff [1977], 123–44 [Translated from *Gesammelte Schriften*, Vol. VII, 205–27]

Dilthey, W. [1976]: *Selected Writings*. H. P. Rickman *ed.* Cambridge: Cambridge University Press

Droysen, J. G. [1858]: *Grundriß der Historik*. Reprinted in *Historik*. R. Hübner *ed.* München: Oldenburg

Dupré, J. [2005]: *Darwin's Legacy*. Oxford: Oxford University Press

Durkheim, D. [1895/1982]: *The Rules of Sociological Method*. London: Macmillan

Durkheim, D. [1897]: *Suicide*. Paris: Felix Alcan [English translation: *Suicide*. London: Routledge & Kegan Paul, 1952]

Earman, J. [1996]: *Bayes or Bust? A Critical Examination of Bayesian Confirmation Theory*. Cambridge, MA: MIT Press

Elster, J. [1989]: *Nuts and Bolts for the Social Sciences*. Cambridge: Cambridge University Press

Elster, J. [2007]: *Explaining Social Behavior*. Cambridge: Cambridge University Press

Fay, B. [1994]: "General Laws and Explaining Human Behaviour," in M. Martin/L. C. McIntyre *eds.* [1994], 91–110

Freud, S. [1953–74]: *The Standard Edition of the Complete Psychological Works of Sigmund Freud.*. James Strachley *ed.* London: The Hogarth Press

Freud, S. [1895]: "Project for a Scientific Psychology," Vol. I. [1966], 295–387

Freud, S. [1897]: "Letter to Fliess" (September 21, 1897), Vol. I [1966], 259–60

Freud, S. [1900]: *The Interpretation of Dreams*, Vols. IV–V [1953], 1–338, 339–621

Freud, S. [1901a]: "On Dreams," Vol. V [1953], 631–686

Freud, S. [1901b]: *The Psychopathology of Everyday Life*, Vol. VI [1960], 1–279

Freud, S. [1905a]: *Three Essays on Sexuality*, Vol. VII [1953], 130–243

Freud, S. [1905b]: *Jokes and their Relation to the Unconscious*, Vol. VIII [1960], 9–236

Freud, S. [1908]: "'Civilized' Sexual Morality and Modern Nervousness," *Collected Papers* II, transl. J. Riviere. London: Hogarth Press [1957], 76–99

Freud, S. [1910]: "The Origin and Development of Psychoanalysis," transl. in *A General Selection from the Works of Sigmund Freud*. J. Rickman *ed.* New York: Liveright Publishing Corporation [1957], 3–36

Freud, S. [1913]: *Totem and Taboo*, Vol. XIII [1953], 1–161

Freud, S. [1916a]: *Introductory Lectures on Psycho-Analysis*, Part I, II, Vol. XV [1961], 9–239

Freud, S. [1916b]: *Introductory Lectures on Psycho-Analysis*, Part IIII, Vol. XVI [1963], 243–463

Freud, S. [1921]: *Group Psychology and the Analysis of the Ego*, Vol. XVIII [1955], 69–143

Freud, S. [1927a]: *The Future of an Illusion*, Vol. XXI [1961], 5–56

Freud, S. [1927b]: "Humour," Vol. XXI [1961], 161–6

Freud, S. [1930]: *Civilization and its Discontents*, Vol. XXI [1961], 64–145

Freud, S. [1931]: "Female Sexuality," Vol. XXI [1927–31], 221–43

Freud, S. [1937]: "Constructions in Analysis," Vol. XXIII [1937–39], 257–69

Freud, S. [1938]: *An Outline of Psychoanalysis*, Vol. XXIII [1964], 144–207

Freud, S./J. Breuer [1895]: *Studies on Hysteria*, Vol. II [1955], 1–305

Freud, S./E. Oppenheim [1911]: "Dreams in Folklore," Vol. XII [1911–13], 180–203

Friedman, M. [1953]: "The Methodology of Positive Economics," in *Essays in Positive Economics*. Chicago: University of Chicago Press, 3–43

Gadamer, H. G. [⁴1975]: *Wahrheit und Methode*. Tübingen, Germany: Mohr [English translation: *Truth and Method*. London: Sheed and Ward, 1979]

Gellner, E. [²1993]: *The Psychoanalytic Movement*. London: Fontana Press

Giddens, A. [1993]: *New Rules of Sociological Method*. Cambridge: Polity Press

Grünbaum, A. [1977a]: "How Scientific is Psychoanalysis?" in R. Stern/L. S. Horowitz/J. Lynes *eds.* [1977], *Science and Psychotherapy*. New York: Haven Publishing Corp., 219–54

Grünbaum, A. [1977b]: "Is Psychoanalysis a Pseudo-science?" *Zeitschrift für philosophische Forschung* **31**, 333–53

Grünbaum, A. [1979]: "Is Freudian Psychoanalytic Theory Pseudo-Scientific by Karl Popper's Criterion of Demarcation?" *American Philosophical Quarterly* **16**, 131–41

Grünbaum, A. [1985]: *The Foundations of Psychoanalysis – A Philosophical Critique.* Berkeley: University of California Press

Habermas, J. [1968]: *Erkenntnis und Interesse.* Frankfurt a./M.: Suhrkamp [English translation: *Knowledge and Human Interests.* London: Heinemann, 1972]

Habermas, J. [1970]: *Zur Logik der Sozialwissenschaften.* Frankfurt a./M.: Suhrkamp. [English translation: *On the Logic of the Social Sciences.* Cambridge, MA: MIT Press, 1988]

Habermas, J. [1976]: *Zur Rekonstruktion des historischen Materialismus.* Frankfurt a./M.: Suhrkamp

Habermas, J. [1980]: "Rekonstruktive *versus* verstehende Sozialwissenschaften," in J. Habermas, *Moralbewußtsein und kommunikatives Handeln.* Frankfurt a./M.: Suhrkamp 1983, 29–51

Habermas, J. [1981]: *Theorie des kommunikativen Handelns.* Frankfurt a./M.: Suhrkamp [English translation: *The Theory of Communicative Action. Vol. 1: Reason and the Rationalization of Society.* Boston: Beacon Press, 1984)

Habermas, J. [1983]: *Die Neue Unübersichtlichkeit.* Frankfurt a./M.: Suhrkamp.

Habermas, J. [1988]: *Nachmetaphysisches Denken.* Frankfurt a./M.: Suhrkamp [English translation: *Postmetaphysical Thinking.* Oxford Polity Press, 1992)

Haeckel, E. [1929]: *The Riddle of the Universe.* London: Watts & Co. [Translation of *Die Welträtsel,* 1899]

Hayek, F. A. [1955]: *The Counter-Revolution of Science.* New York: The Free Press

Hollis, M. [1994]: *The Philosophy of Social Science.* Cambridge: Cambridge University Press

Jonas, F. [1968–9]: *Geschichte der Soziologie.* 4 vols. Hamburg: Rowohlt

Kant, I. [1784]: "Was ist Aufklärung?" *Berlinische Monatsschrift* (Dezember 1784), 481–94 [English translation: "What is the Enlightenment?" in Immanuel Kant, *Political Writings.* Cambridge: Cambridge University Press, 1991]

Kaufmann, W. [⁴1974]: *Nietzsche: Philosopher, Psychologist, Antichrist.* Princeton, NJ: Princeton University Press

Kincaid, H. [1994]: "Defending Laws in the Social Sciences," in M. Martin/L. C. McIntyre [1994], 111–30

Kitcher, Ph. [1985]: *Vaulting Ambition: Sociobiology and the Quest for Human Nature.* Cambridge, MA: MIT Press

Levison, A. B. [1974]: *Knowledge and Society.* New York: Pegasus

Little, D. [1991]: *Varieties of Social Explanation.* Boulder, CO: Westview Press

Mackie, J. L. [1980]: *The Cement of the Universe.* Oxford: Clarendon Press

Mann, Th. [1936]: "Freud und die Zukunft," in S. Freud, *Abriß der Psychoanalyse/Das Unbehagen in der Kultur.* Frankfurt a./M.: Fischer 1970, 131–51

Martin, M./L. C. McIntyre eds. [1994]: *Readings in the Philosophy of Social Science.* Cambridge, MA: MIT Press

McIntyre, L. C. [1994]: "Complexity and Social Scientific Law," in M. Martin/L. C. McIntyre [1994], 131–43

Merton, R. K. [1968]: *Social Theory and Social Structure.* New York: The Free Press

Meyer, C. *et al.* [2005]: *Le Livre Noir de la Psychanalyse.* Paris: Les Arènes

Milgram, S. [1974]: *Obedience to Authority.* New York: Harper Torchbooks

Mommsen, W. [1974]: *Max Weber: Gesellschaft, Politik und Geschichte.* Frankfurt a./M.: Suhrkamp

Motley, M. T. [1985]: "Slips of the Tongue," *Scientific American* **253** (September), 114–8

Nagel, E. [1961]: *The Structure of Science.* London: Routledge & Kegan Paul

Nietzsche, F. [1888]: "Ecce Homo," in *The Complete Works of F. Nietzsche.* O. Levy *ed.* Vol. 17. New York: Russell & Russell [1964]

Nietzsche, F. [1886]: "Beyond Good and Evil," in *The Complete Works of F. Nietzsche.* O. Levy *ed.* Vol. 12. New York: Russell & Russell [1964]

Papineau, D. [1979]: *For Science in the Social Sciences.* New York: St. Martin's Press

Parkin, M./D. King [1992]: *Economics.* Reading, MA: Addison-Wesley

Pais, A. [1982]: *Subtle is the Lord.* Oxford: Oxford University Press

Phares, E. J. [1984/²1988]: *Introduction to Personality.* New York: HarperCollins

Pinker, S. [2004]: "Why Nature and Nurture Won't Go Away," *Dædalus* (Fall), 1–13

Popper, K. [1957]: *The Poverty of Historicism.* London: Routledge & Kegan Paul

Popper, K. [³1971]: "Die Logik der Sozialwissenschaften," in Th. W. Adorno et al., *Der Positivismusstreit in der deutschen Soziologie.* Neuwied/Berlin: Luchterhand 1971, 103–23 [English translation: *The Positivist Dispute in German Sociology.* London: Heinemann, 1976, 87–104]

Ratcliffe, M. [2005]: "An Epistemological Problem for Evolutionary Psychology," *International Studies in the Philosophy of Science* **19**, 47–63

Ridley, M. [2003]: *Nature via Nurture.* London: Fourth Estate

Rillaer, J. van [2005a]: "La mythologie de la thérapie en profondeur," in C. Meyer [2005], 216–33

Rillaer, J. van [2005b]: "Psychanalyse populaire et psychanalyse pour initiés," in C. Meyer [2005], 235–41

Rillaer, J. van [2005c]: "Les mécanismes de défense de la psychanalyse," in C. Meyer [2005], 414–41

Rose, H./S. Rose *eds.* [2001]: *Alas, Poor Darwin: Arguments Against Evolutionary Psychology.* London: Vintage

Rosenberg, A. [²1995]: *Philosophy of Social Science.* Boulder, CO: Westview Press

Rosenberg, A. [2004]: "Lessons from Biology for Philosophy of the Human Sciences," *Philosophy of the Social Sciences* **35**, 3–19

Ruse, M. [1979]: *Sociobiology: Sense or Nonsense?* Dordrecht: Reidel

Ruben, D. H. [1985]: *The Metaphysics of the Social World.* London: Routledge & Kegan Paul

Ryan, A. [1970]: *The Philosophy of the Social Sciences.* London: Macmillan

Sahlins, M. [1976]: *The Use and Abuse of Biology. An Anthropological Critique of Sociobiology.* Ann Arbor, MI: University of Michigan Press

Salmon, M. H. [2003]: "Causal Explanations of Behaviour," *Philosophy of Science* **70**, 720–38

Sayer, A. [2000]: *Realism and Social Science.* London: Sage

Searle, J. [2004]: *Mind.* Oxford: Oxford University Press

Sherif, M. [1936]: *The Psychology of Social Norms.* New York: Harper Torchbook, 1966

Shils, E. A./H. A. Finch *eds.* [1949]: *The Methodology of the Social Sciences: Max Weber.* Glencoe, IL: The Free Press

Smith, M. J. [1998]: *Social Science in Question.* London: Sage

Solms, M. [2004]: "Freud Returns," *Scientific American* (May), 56–63

Sterelny, K. [1988]: "Critical Notice. Kitcher, Philip, *Vaulting Ambition* 1985," *Australasian Journal of Philosophy* **66**, 538–55

Stevenson, L./D. L. Haberman [⁴2004]: *Ten Theories of Human Nature*. New York: Oxford University Press

Sulloway, F. J. [1979]: *Freud, Biologist of the Mind*. Cambridge, MA: Harvard University Press 1992

Sulloway, F. J. [1991]: "Reassessing Freud's Case Histories: The Social Construction of Psychoanalysis," *Isis* **82**, 245–75

Sulloway, F. [2005a]: "Freud recycleur: cryptobiologie et pseudoscience," in C. Meyer [2005], 49–66

Sulloway, F. [2005b]: "Qui a peur de l'homme aux loups?" in C. Meyer [2005], 81–6

Sulloway, F. [2005c]: "L'Homme aux rats comme vitrine de la psychanalyse," in C. Meyer [2005], 95–100

Tawney, R. H. [1922/1990]: *Religion and the Rise of Science*. London: Penguin

Thomas, D. [1979]: *Naturalism and Social Science*. Cambridge: Cambridge University Press

Tooby, J./L. Cosmides [1995]: "The Psychological Foundations of Culture," in *The Adapted Mind*. J. H. Barkow/L. Cosmides/J. Tooby *eds*. Oxford: Oxford University Press 1995, 19–136

Weber, M. [1904]: "Die 'Objektivität' sozialwissentschaftlicher Erkenntnis," in M. Weber [1973], 186–262 [English translation: "Objectivity in Social Science," reprinted in E. A. Shils/H. A. Finch, 1949, 50–112]

Weber, M. [1904–5]: *Die protestantische Ethik und der Geist des Kapitalismus* [English translation: *The Protestant Ethics and the Spirit of Capitalism*. London: Unwin University Books, 1930]

Weber, M. [1905]: "Objective Possibility and Adequate Causation in Historical Explanation," reprinted in E. A. Shils/H. A. Finch, 1949, 164–88

Weber, M. [1913]: "Über einige Kategorien der verstehenden Soziologie," in M. Weber [1973], 97–150

Weber, M. [1914]: "Der Sinn der 'Wertfreiheit' in den Sozialwissenschaften," in M. Weber [1973], 263–310 [English translation: "The Meaning of Ethical Neutrality," reprinted in E. A. Shils/H. A. Finch, 1949, 1–49]

Weber, M. [1915]: "The Social Psychology of World Religions," reprinted in *From Max Weber: Essays in Sociology*. H. H. Gerth/C. Wright *eds*. London: Routledge & Kegan Paul, 1948, 267–301

Weber, M. [1968]: *Economy and Society*. New York: Bedminster Press [English translation of *Wirtschaft und Gesellschaft*. Tübingen, Germany: J. C. B. Mohr, ⁵1972]

Weber, M. [⁵1973]: *Soziologie, Universalgeschichtliche Analysen, Politik*. J. Winckelmann *ed*. Stuttgart: Körner

Weber, M. [1978]: *Selections in Translation*. W. G. Runciman *ed*. Cambridge: Cambridge University Press

Webster, R. [1996]: *Why Freud was Wrong*. London: Fontana Press

Weinert, F. [1993]: "Laws of Nature: A Structural Approach," *Philosophia Naturalis* **30/2**, 147–71.

Weinert, F. [1995]: "Laws of Nature – Laws of Science," in F. Weinert *ed*. [1995], *Laws of Nature*. Berlin: Walter de Gruyter, 3–64.

Weinert, F. [1996]: "Weber's Ideal Types as Models in the Social Sciences," in A. O'Hear *ed*. [1996], *Verstehen and Humane Understanding*. Royal Institute of Philosophy Supplement: 41. Cambridge: Cambridge University Press, 73–93

Weinert, F. [1997]: "On the Status of Social Laws," *Dialectica* **51**, 225–42

Weinert, F. [1999]: "Habermas, Science and Modernity," in A. O'Hear *ed.* [1999], *German Philosophy Since Kant.* Royal Institute of Philosophy Supplement: 44. Cambridge: Cambridge University Press, 329–55

Weinert, F. [2004]: *The Scientist as Philosopher: Philosophical Consequences of Great Scientific Discoveries.* New York: Springer

Wilson, E. O. [1978]: *On Human Nature.* Cambridge, MA: Harvard University Press

Wilson, E. O. [1980]: *Sociobiology. The Abridged Edition.* Cambridge, MA: The Belknap Press

Winch, P. [1958]: *The Idea of a Social Science and its Relation to Philosophy.* London: Routledge & Kegan Paul

Winch, P. [1964]: "Understanding a Primitive Society," *American Philosophical Quarterly* I, 307–24, reprinted in *Rationality.* B. R. Wilson *ed.* London: Blackwell 1979, 78–111

Wittgenstein, L. [1980]: *Vermischte Bemerkungen/Culture and Value.* Oxford: Blackwell

Woodward, J. [2003]: *Making Things Happen.* Oxford: Oxford University Press

Wright, G. H. von [1971]: *Explanation and Understanding.* Ithaca, NY: Cornell University Press

Essay Questions

1   Explain Freud's assumption of **psychic determinism**. Evaluate the implication of this assumption for the scientific status of Freudian psychoanalysis.

2   Does the **Freudian model** of the mind satisfy any criteria of scientific status?

3   Explain and evaluate **Freud's claim** to have completed the Copernican revolution.

4   Explain the structure of **Freudianism**. What speaks in favor of and against the theory?

5   Are Copernicanism, Darwinism, and Freudianism best characterized as instances of **falsificationism** or **eliminative inductivism**?

6   What does adherence to **scientism** mean for Freud's psychoanalytic theories?

7   Why is there dispute about whether **Freud's theory** was scientific or not?

8   Couldn't we regard Freud's theory as just a **model**? What implication would this decision have for the status of psychoanalysis?

9   Critically discuss **Freud's claim** that "we are not even masters in our own house."

10   Critically discuss the notion of the **Unconscious** in the context of Freud's theory.

11   Is it reasonable to make a distinction between **"revolutions in science"** and **"revolutions in thought"**? Illustrate your answer.

12   Explain the major achievements of the **Copernican** or **Darwinian** or **Freudian revolution**, respectively.

13   Explain how the issues of **instrumentalism** and **realism** arise, respectively, in Copernicanism, Darwinism, and Freudianism.

14   Why does the issue of **realism** versus **instrumentalism** run through Copernicanism, Darwinism, and Freudianism alike?

15   Analyze the challenge which the Freudian model of the mind poses for **Enlightenment views** of human nature.

16 Explain the structure of **evolutionary psychology**.

17 Explain why **evolutionary psychology** emphasizes the universality of the human mind.

18 Discuss the similarities and dissimilarities between **sociobiology** and **evolutionary psychology**.

19 State the **main differences** between the *hermeneutic* and the *empirical* model in the social sciences. Evaluate the strengths and weaknesses of each of these models.

20 Discuss and illustrate the importance of **ideal types** as a tool in the social sciences. How do ideal types relate to both the *empirical* and *hermeneutic* models?

21 Use examples from astronomy, biology, and the social sciences to evaluate the appropriateness of **eliminative inductivism**.

22 Discuss the problem of **underdetermination** with respect to three revolutions in thought.

23 How does the problem of **relativism** arise in the context of the social sciences? Is it a pseudo-problem? Or does it have an effect on the social world?

24 Critically discuss the distinction between **instrumental reason** and **objective reason** in the *Critical model* of the social sciences.

25 Critically discuss Weber's views on **objectivity** and **value neutrality** in the social sciences.

26 Would it be true to say, according to **holism**, that "society is more than the sum of its individual members"?

27 Would it be true to say, according to **individualism**, that "society is nothing but the sum of its individual members"?

28 Is it possible to answer **causal questions** in the social sciences?

29 Discuss what **scientism** means, as defended by Comte and Durkheim. Consider whether it is an appropriate approach in the social sciences.

30 Explain Dilthey's method of *Verstehen* (Understanding).

31 What implications would **relativism** have in the practice of the social sciences?

32 Examine the major tenets of the **Critical model** in the social sciences and discuss whether it can make contributions over and above the standard models.

33 Explain the *DN* and *IS* models of scientific **explanation**. Can they be used in the social sciences?

34 Why should social scientists be particularly worried about the distinction between **facts** and **values** in social research?

35 Could there be **laws** in the social world?

Name Index

Subject Index